Lecture Notes in Computer Science   8973

Commenced Publication in 1973
Founding and Former Series Editors:
Gerhard Goos, Juris Hartmanis, and Jan van Leeuwen

T0234068

M. Sohel Rahman · Etsuji Tomita (Eds.)

WALCOM: Algorithms and Computation

9th International Workshop, WALCOM 2015
Dhaka, Bangladesh, February 26-28, 2015
Proceedings

 Springer

Volume Editors

M. Sohel Rahman
BUET, Department of CSE
ECE Building, West Palasi, Dhaka 1205, Bangladesh
E-mail: msrahman@cse.buet.ac.bd

Etsuji Tomita
The University of Electro-Communications
The Advanced Algorithms Research Laboratory
Chofugaoka 1-5-1, Chofu, Tokyo 182-8585, Japan
E-mail: tomita@ice.uec.ac.jp

ISSN 0302-9743   e-ISSN 1611-3349
ISBN 978-3-319-15611-8   e-ISBN 978-3-319-15612-5
DOI 10.1007/978-3-319-15612-5
Springer Cham Heidelberg New York Dordrecht London

Library of Congress Control Number: 2015930429

LNCS Sublibrary: SL 1 – Theoretical Computer Science and General Issues

Typesetting: Camera-ready by author, data conversion by Scientific Publishing Services, Chennai, India

Printed on acid-free paper

Springer is part of Springer Science+Business Media (www.springer.com)

Preface

This proceedings volume contains papers presented at WALCOM 2015, the 9th International Workshop on Algorithms and Computation, held during February 26–28, 2015, at the Training Institute, Atomic Energy Research Establishment (AERE), Ganakbari, Savar, Dhaka, Bangladesh. The workshop covered diverse areas of algorithms and computation, namely, approximation algorithms, data structures, computational geometry, combinatorial algorithms, distributed and online algorithms, graph drawing, graph algorithms, combinatorial problems and computational complexity. The workshop was organized jointly by the Bangladesh Atomic Energy Commission, Bangladesh Academy of Sciences (BAS), and the Department of Computer Science and Engineering, BUET.

WALCOM is an annual conference series on all aspects of algorithms and computation. Since its inception in 2007, it has been held as a yearly event. WALCOM has grown substantially in reputation and has been able to attract researchers and scientists around the globe. WALCOM 2015 was organized in cooperation with IEICE Technical Committee on Theoretical Foundations of Computing (COMP) and the Special Interest Group for ALgorithms (SIGAL) of the Information Processing Society of Japan (IPSJ). This year, 88 manuscripts were submitted to WALCOM. After a few withdrawals, 85 papers with authors from 25 different countries were reviewed. Among these submissions, 26 were accepted as full papers and three were accepted as short papers. Papers were selected based on a thorough reviewing (usually, at least three review reports per paper) followed by in-depth discussion sessions by the WALCOM Program Committee comprising 29 researchers of international repute from Australia, Bangladesh, Canada, France, Germany, Hong Kong, India, Italy, Japan, Korea, The Netherlands, Poland, Sweden, Switzerland, Taiwan, and USA. This year, for the first time in the history of WALCOM, Best Paper Awards were given. We are happy to highlight that "Edge-Colorings of Weighted Graphs" authored by Yuji Obata and Takao Nishizeki and "An Almost Optimal Algorithm for Voronoi Diagrams of Non-Disjoint Line Segments" authored by Sang Won Bae were selected for the Best Paper Awards by the Program Committee. We are also delighted to announce that following the tradition of the previous years, two special issues—one in Journal of Graph Algorithms and Applications and the other in Journal of Discrete Algorithms—are being organized comprising the extended versions of selected papers from WALCOM 2015.

In addition to the 29 contributed talks, the scientific program of the workshop included three invited talks by the 2006 Gödel Prize winner Prof. Manindra Agrawal of the Indian Institute of Technology Kanpur, India, Prof. Shin-ichi Minato of Hokkaido University, Japan, and Prof. Rajeev Raman, University of Leicester, UK. We are extremely grateful to our invited speakers for their excellent talks at the workshop. We thank all the authors who submitted their work for consideration to WALCOM 2015. We deeply appreciate the competent and timely handling of the submissions of all Program Committee members and external reviewers, despite their extremely busy schedule. In this connection we must also acknowledge the EasyChair conference management system for providing a beautiful platform for conference administration. We must thank Springer as well for publishing the proceedings of WALCOM 2015 in their prestigious LNCS series. We are indebted to the WALCOM Steering Committee and the Advisory Committee for their continuous guidance and support. Above all, we are extremely grateful to the Organizing Committee of WALCOM 2015 for making the event a grand success. Last but not the least, we express our heartiest gratitude to the kind and generous support of the sponsors.

February 2015

M. Sohel Rahman
Etsuji Tomita

Organization

WALCOM Steering Committee

Kyung-Yong Chwa	Korea Advanced Institute of Science and Technology
Costas S. Iliopoulos	King's College London, UK
M. Kaykobad	Bangladesh University of Engineering & Technology
Petra Mutzel	TU Dortmund, Germany
Shin-ichi Nakano	Gunma University, Japan
Subhas Chandra Nandy	Indian Statistical Institute, India
Takao Nishizeki	Tohoku University, Japan
C. Pandu Rangan	Indian Institute of Technology, Madras, India
Md. Saidur Rahman	Bangladesh University of Engineering & Technology

WALCOM 2015 Organizers

WALCOM 2015 Supporters

WALCOM 2015 Program Committee

Guillaume Blin	Université de Bordeaux, France
Hans L. Bodlaender	Utrecht University, The Netherlands
Francis Y.L. Chin	The University of Hong Kong, Hong Kong, SAR China
Rezaul A. Chowdhury	Stony Brook University, USA
Naveen Garg	Indian Institute of Technology Delhi, India
Mohammad T. Hajiaghayi	University of Maryland, USA
Seok-Hee Hong	University of Sydney, Australia
Kazuo Iwama	Kyoto University, Japan
Ming-Yang Kao	Northwestern University, USA
Ralf Klasing	CNRS, University of Bordeaux, France
Michael Langston	University of Tennessee, USA
Andrzej Lingas	Lund University, Sweden
Giuseppe Liotta	University of Perugia, Italy
Ian Munro	University of Waterloo, Canada
Petra Mutzel	TU Dortmund, Germany
Subhas Chandra Nandy	Indian Statistical Institute, India
Kunsoo Park	Seoul National University, Korea
Michał Pilipczuk	University of Warsaw, Poland
M. Sohel Rahman	Bangladesh University of Engineering & Technology (BUET), Bangladesh (Co-chair)
Md. Saidur Rahman	Bangladesh University of Engineering & Technology (BUET), Bangladesh
Wojciech Rytter	University of Warsaw, Poland
Kunihiko Sadakane	The University of Tokyo, Japan
Sandeep Sen	Indian Institute of Technology Delhi, India
Bill Smyth	McMaster University, Canada
Etsuji Tomita	The University of Electro-Communications, Japan (Co-chair)
Osamu Watanabe	Tokyo Institute of Technology, Japan
Sue Whitesides	University of Victoria, Canada
Peter Widmayer	ETH Zürich, Switzerland
Hsu-Chun Yen	National Taiwan University, Taiwan

WALCOM 2015 Advisory Committee

Mesbahuddin Ahmad	President, Bangladesh Academy of Sciences
M. Shamsher Ali	Fellow, Bangladesh Academy of Sciences
Khawaja Muhammed Sultanul Aziz	Secretary, Bangladesh Academy of Sciences
Naiyyum Choudhury	Fellow, Bangladesh Academy of Sciences

Khaleda Ekram	Vice-Chancellor, Bangladesh University of Engineering & Technology
Md. Monzurul Haque	Member (Engineering), Bangladesh Atomic Energy Commission
Mahmudul Hasan	Member (Physical Science), Bangladesh Atomic Energy Commission
Mohammad Mahfuzul Islam	Head, Bangladesh University of Engineering & Technology
Md. Monirul Islam	Chairman, Bangladesh Atomic Energy Commission
M. Kaykobad	Professor, Bangladesh University of Engineering & Technology
Harun-Or-Rashid	Director General, Bangladesh Atomic Energy Commission
M. Ali Zulquarnain	Member (Planning), Bangladesh Atomic Energy Commission

WALCOM 2015 Organizing Committee

Md. Shakil Ahmed	Bangladesh Atomic Energy Commission (Joint Secretary)
Md. Mostafa Akbar	Bangladesh University of Engineering & Technology
Addris Ali	Bangladesh Atomic Energy Commission
Mohammed Eunus Ali	Bangladesh University of Engineering & Technology
Md. Ashikur Rahman Azim	Bangladesh University of Engineering & Technology
Khawaja Muhammed Sultanul Aziz	Bangladesh Academy of Sciences
Mohammad Al-Mahmud	Bangladesh University of Engineering & Technology
Madhusudan Basak	Bangladesh University of Engineering & Technology
Md. Muradul Bashir	Bangladesh University of Engineering & Technology
Naiyyum Choudhury	Bangladesh Academy of Sciences
Anada Kumar Das	Bangladesh Atomic Energy Commission
Mohammod Abu Sayid Haque	Bangladesh Atomic Energy Commission
Md. Mahbubul Hoq	Bangladesh Atomic Energy Commission
Md. Dulal Hossain	Bangladesh Atomic Energy Commission
Md. Iqbal Hossain	Bangladesh University of Engineering & Technology

Shauli Sarmin Sumi   Bangladesh Atomic Energy Commission
Etsuji Tomita   The University of Electro-Communications,
                                 Japan
Md. Meshbah Uddin   Bangladesh Atomic Energy Commission
Md. Shuza Uddin   Bangladesh Atomic Energy Commission
Sk. Md. Yunus   Bangladesh Atomic Energy Commission
Md. Anzan-Uz-Zaman   Bangladesh Atomic Energy Commission

WALCOM 2015 Additional Reviewers

Agarwal, Pankaj
Aspnes, James
Bacher, Axel
Bannai, Hideo
Baswana, Surender
Bateni, Mohammadhossein
Bishnu, Arijit
Brinda, Karel
Böckenhauer, Hans-Joachim
Cai, Jin-Yi
Chitnis, Rajesh
Czyżowicz, Jerzy
de Rugy Altherre, Nicolas
Dehghani, Sina
Devismes, Stéphane
Didimo, Walter
Diwan, Ajit
Du, Hai
Ehsani Banafati, Soheil
Esfandiari, Hossein
Evans, William
Fischer, Johannes
Flocchini, Paola
Foucaud, Florent
Gambette, Philippe
Ganapathi, Pramod
Grilli, Luca
Grytczuk, Jarosław
Gudmundsson, Joachim
Hagan, Ron
Harsha, Prahaladh
Jansson, Jesper
Kakugawa, Hirotsugu
Karim, Md. Rezaul

Kortsarz, Guy
Kowaluk, Mirosław
Krasikov, Ilia
Kriege, Nils
Kurz, Denis
Kutzkov, Konstantin
Lu, Allan
Manlove, David
Manne, Fredrik
Mccauley, Samuel
Mitchell, Joseph
Mondal, Debajyoti
Monemizadeh, Morteza
Montecchiani, Fabrizio
Morris, Tim
Mömke, Tobias
Nakano, Shin-Ichi
Nishat, Rahnuma Islam
Nomikos, Christos
Nöllenburg, Martin
Pajak, Dominik
Papadopoulou, Evanthia
Persson, Mia
Phillips, Charles
Pilipczuk, Marcin
Purohit, Manish
Radoszewski, Jakub
Roselli, Vincenzo
Roy, Sasanka
Rutter, Ignaz
Sikora, Florian
Sledneu, Dzmitry
Smid, Michiel
Sommer, Christian

Soueidan, Hayssam
Su, Hsin-Hao
Suri, Subhash
Szreder, Bartosz
Tixeuil, Sébastien

Vialette, Stéphane
Wang, Kai
Wolff, Alexander
Zhang, Yong
Żyliński, Paweł

WALCOM 2015 Sponsors

Invited Talks
(Abstracts)

Polynomial Identity Testing

Manindra Agrawal*

1 Introduction

Polynomial Identity Testing (PIT in short) is the problem of checking if a polynomial of n variables with coefficients from r a field F is *identically zero*, i.e., if all its terms cancel each other out. The problem is very simple to solve efficiently if the polynomial is given in the usual sum-of-products form:

$$P(x_1, x_2, \ldots, x_n) = \sum_{0 \leq i_1, i_2, \ldots, i_n \leq d} \alpha_{i_1, i_2, \ldots, i_n} x_1^{i_1} x_2^{i_2} \cdots x_n^{i_n};$$

simply check whether all coefficients $\alpha_{i_1, i_2, \ldots, i_n} \in F$ are zero. It becomes non-trivial when the polynomial is given in a form different from sum-of-products. For example:

$$P(u, v, x, y) = (ux + vy)^2 + (vx - uy)^2 - (u^2 + v^2) \cdot (x^2 + y^2).$$

One can try expressing such polynomials as sum-of-products and then checking if they are zero, however, the size of the resulting polynomial can become exponential and so this method is not efficient. A general representation of polynomials is via *arithmetic circuits*: these define a sequence of addition and multiplication operations starting from variables and ending in the desired polynomial. For example, the second polynomial is expressed as arithmetic circuit below:

* N Rama Rao Professor, Indian Institute of Technology, Kanpur. Research supported by J C Bose Fellowship FLW/DST/CS/20060225

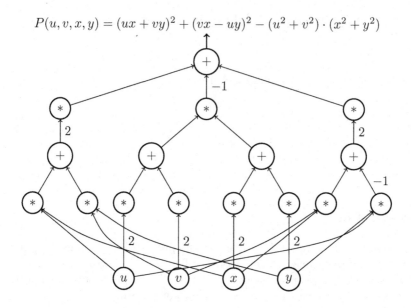

$$P(u,v,x,y) = (ux + vy)^2 + (vx - uy)^2 - (u^2 + v^2) \cdot (x^2 + y^2)$$

In the above figure, the operations are inside circles and take as input polynomials on arrows coming into the circles and output the resulting polynomial on arrows going out of the circles. If a constant c is present on an arrow carrying polynomial Q into an addition operation, the polynomial is replaced by cQ, and if the arrow is going into a multiplication operation, the polynomial is replaced by Q^c. The *size* of an arithmetic circuit is defined to be the number of operations in the circuit. So, for example, the size of the above circuit is 16. Another important parameter associated with an arithmetic circuit is its depth: *depth* of an arithmetic circuit is the length of the longest chain of arrows from an input variable to the output polynomial. In the above example circuit, the depth is 4.

PIT has had long and interesting history. A randomized polynomial time algorithm for solving it was given by Schwartz and Zippe[Sch80,Zip79]. Since then, several other randomized polynomial-time algorithms have been shown [CK97,LV98,AB03], but no deterministic polynomial time algorithm for the problem is known till date. In 2002, two important results were shown for PIT: Kabanets and Impagliazzo [KI04] showed that a determinsitic algorithm for PIT would imply a lower bound on arithmetic complexity of an explicit polynomial computable in NEXP(this was subsequently extended in [Agr05,AV08,GKKS13]); Agrawal, Kayal, and Saxena [AKS04] derandomized the randomized polynomial time algorithm for a special kind of PITs, resulting in a deterministic polynomial time algorithm for primality testing. Since then, the problem has come to occupy center-stage in complexity theory and a lot of subsequent development has taken place [KS07,SS13,ASSS12].

References

[AB03]   Agrawal, M., Biswas, S.: Primality and identity testing via chinese remaindering. J. ACM 50(4), 429–443 (2003)

[Agr05]   Agrawal, M.: Proving lower bounds via pesudo-random generators. In: Proceedings of the FST&TCS, pp. 96–105 (2005)

[AKS04]   Agrawal, M., Kayal, N., Saxena, N.: PRIMES is in P. Annals of Mathematics 160(2), 781–793 (2004)

[ASSS12]   Agrawal, M., Saha, C., Saptharishi, R., Saxena, N.: Jacobian hits circuits: Hitting sets, lower bounds for depth-d occur-k formulas and depth-3 transcendence degree-k circuits. In: Proceedings of Annual ACM Symposium on the Theory of Computing, pp. 599–614 (2012)

[AV08]   Agrawal, M., Vinay, V.: Arithmetic circuits: A chasm at depth four. In: Proceedings of Annual IEEE Symposium on Foundations of Computer Science, pp. 67–75 (2008)

[CK97]   Chen, Z.-Z., Kao, M.-Y.: Reducing randomness via irrational numbers. In: Proceedings of Annual ACM Symposium on the Theory of Computing, pp. 200–209 (1997)

[GKKS13]   Gupta, A., Kamath, P., Kayal, N., Saptharishi, R.: Arithmetic circuits: A chasm at depth three. In: Proceedings of Annual IEEE Symposium on Foundations of Computer Science, pp. 578–587 (2013)

[KI04]   Kabanets, V., Impagliazzo, R.: Derandomizing polyonmial identity tests means proving circuit lower bounds. Computational Complexity 13, 1–46 (2004)

[KS07]   Kayal, N., Saxena, N.: Polynomial identity testing for depth 3 circuits. Computational Complexity 16(2), 115–138 (2007)

[LV98]   Lewin, D., Vadhan, S.: Checking polynomial identities over any field: Towards a derandomization? In: Proceedings of Annual ACM Symposium on the Theory of Computing, pp. 428–437 (1998)

[Sch80]   Schwartz, J.T.: Fast probabilistic algorithms for verification of polynomial identities. J. ACM 27(4), 701–717 (1980)

[SS13]   Saxena, N., Seshadri, C.: From Sylvester-Gallai configurations to rank bounds: Improved black-box identity test for depth3 circuits. J. ACM 60(5), article 33 (2013)

[Zip79]   Zippel, R.E.: Probabilistic algorithms for sparse polynomials. In: Ng, K.W. (ed.) EUROSAM 1979 and ISSAC 1979. LNCS, vol. 72, pp. 216–226. Springer, Heidelberg (1979)

Power of Enumeration — BDD/ZDD-Based Methods for Indexing Combinatorial Patterns

Shin-ichi Minato

Graduate School of Information Science and Technology, Hokkaido University/
JST ERATO MINATO Discrete Structure Manipulation System Project,
Sapporo, 060-0814, Japan

Abstract

Discrete structures are foundational material for computer science and mathematics, which are related to set theory, symbolic logic, inductive proof, graph theory, combinatorics, probability theory, etc. Many problems are decomposed into discrete structures using simple primitive algebraic operations.

A Binary Decision Diagram (BDD) is a representation of a Boolean function, one of the most basic models of discrete structures. After the epoch-making paper [1] by Bryant in 1986, BDD-based methods have attracted a great deal of attention. The BDD was originally invented for the efficient Boolean function manipulation required in VLSI logic design, but Boolean functions are also used for modeling many kinds of combinatorial problems. A Zero-suppressed BDD (ZDD) [3] is a variant of the BDD, customized for representing and indexing combinatorial patterns. ZDDs have been successfully applied not only to VLSI design, but also for solving various combinatorial problems, such as constraint satisfaction, frequent pattern mining, and graph enumeration. Recently, ZDDs have become more widely known, since D.E. Knuth intensively discussed ZDD-based algorithms in the latest volume of his famous series of books [2].

Although a quarter of a century has passed since Bryant first put forth his idea, there are still many interesting and exciting research topics related to BDDs and ZDDs [4]. One of the most important topics would be that Knuth presented a surprisingly fast algorithm "Simpath" [2] to construct a ZDD which represents all the paths connecting two points in a given graph structure. This work is important because many kinds of practical problems are efficiently solved by some variations of this algorithm. We generically call such ZDD construction methods "frontier-based methods."

In this talk, we present recent research activity related to BDDs and ZDDs. We first briefly explain the basic techniques for BDD/ZDD manipulation, and we then show an overview of the frontier-based method for efficiently enumerating and indexing the solutions of combinatorial problems. We also present several topics on various applications of those state-of-the-art techniques.

References

1. Bryant, R.E.: Graph-based algorithms for Boolean function manipulation. IEEE Transactions on Computers C-35(8), 677–691 (1986)
2. Knuth, D.E.: The Art of Computer Programming: Bitwise Tricks & Techniques; Binary Decision Diagrams, vol. 4, fascicle 1. Addison-Wesley (2009)
3. Minato, S.-I.: Zero-suppressed BDDs for set manipulation in combinatorial problems. In: Proc. of 30th ACM/IEEE Design Automation Conference (DAC 1993), pp. 272–277 (1993)
4. Minato, S.-I.: Techniques of BDD/ZDD: Brief history and recent activity. IEICE Transactions on Information and Systems E96-D(7), 1419–1429 (2013)

Encoding Data Structures*

Rajeev Raman

University of Leicester, UK

Abstract. In recent years, there has been an explosion of interest in *succinct* data structures, which store the given data in compact or compressed formats and answer queries on the data rapidly while it is still in its compressed format. Our focus in this talk is to introduce *encoding* data structures. Encoding data structures consider the data together with the queries and aim to store only as much information about the data as is needed to store the queries. Once this is done, the original data can be deleted. In many cases, one can obtain space-efficient encoding data structures even when the original data is incompressible.

---

* A full version of the invited talk can be found on p. 1.

Table of Contents

Combinatorial Algorithms

Distributed and Online Algorithms

Graph Drawing and Algorithms

Combinatorial Problems and Complexity

Graph Enumeration and Algorithms

Encoding Data Structures

Rajeev Raman

University of Leicester, UK

Abstract. In recent years, there has been an explosion of interest in *succinct* data structures, which store the given data in compact or compressed formats and answer queries on the data rapidly while it is still in its compressed format. Our focus in this talk is to introduce *encoding* data structures. Encoding data structures consider the data together with the queries and aim to store only as much information about the data as is needed to store the queries. Once this is done, the original data can be deleted. In many cases, one can obtain space-efficient encoding data structures even when the original data is incompressible.

1 Introduction

The need for performing complex processing on ever-larger volumes of data has led to the re-evaluation of the space usage of data structures. Whereas in classical data structures, a linear space usage, namely using $O(n)$ words of space, is considered to be optimal, this is often too much for very large data. For example, a suffix tree on a string of n characters from a fixed alphabet requires $\Theta(n)$ words. A usual assumption is that a computer word must be of length $\Omega(\log n)$ bits, so that one can work with numbers such as the input size n, and be able to address enough memory to hold the input. Thus, a suffix tree requires $\Theta(n \log n)$ bits of memory, while the input requires only $O(n)$ bits. This asymptotic blow-up also manifests itself in practice: even a even a highly optimized implementation of a suffix tree requires $20n$ bytes in the worst case [15] to index a string of n bytes. This level of internal memory usage is unacceptable if we wish to index gigabytes or terabytes of string data.

In response to this issue, there has been a great deal of research into *succinct* and *compressed* data structures [2] building upon the early work of Jacobson [12] and contemporaries. In succinct data structures, we view the given instance x of the data on which we wish to build a data structure as coming from a set S of objects, and the aim is to represent x using space as close to the information-theoretic bound of $\lceil \log_2 |S| \rceil$ bits as possible. In compressed data structures, we postulate a probability distribution on S and aim to represent x using as close to the Shannon bound of $\lceil \log_2 1/\Pr(x) \rceil$ bits as possible.

However, there are cases where the succinct approach does not offer any asymptotic improvements. Consider the well-known *range maximum query* problem, which is, given a static array $A[1..n]$, to pre-process A to answer queries:

RMQ(l, r): return $\max_{l \le i \le r} A[i]$.

M.S. Rahman and E. Tomita (Eds.): WALCOM 2015, LNCS 8973, pp. 1–7, 2015.
© Springer International Publishing Switzerland 2015

Assume, for simplicity, that A contains a permutation of $\{1, \ldots, n\}$. Observe that since $\mathsf{RMQ}(i, i)$ queries can be used to reconstruct A, any data structure for answering RMQ on A must contain all the information contained in A, and hence use $\Omega(n \log n)$ bits.

In order to get around this, we modify the RMQ slightly:

$\mathbf{RMQ}(l, r)$: return $\arg\max_{l \leq i \leq r} A[i]$.

In other words, we only seek the index in the range $\{l, \ldots, r\}$ where the maximum value among $A[l], \ldots, A[r]$ lies. In many applications of the RMQ problem, knowing the index where the maximum value lies is sufficient. With this modified version, it is no longer possible to reconstruct A is by performing RMQs. For example, if $n = 3$ then the arrays $A = (1, 3, 2)$ and $A' = (2, 3, 1)$ will give exactly the same answer to any RMQ operation, and the space lower bound of $\Omega(n \log n)$ bits no longer applies.

2   Encoding Data Structures

We now define some terminology regarding encoding data structures.

Effective entropy. Given a set of objects S, and a set of queries Q, consider the equivalence class $\mathcal{C}$ on S induced by Q, where two objects from S are equivalent if they provide the same answer to all queries in Q. We define the quantity $\lceil \log_2 |\mathcal{C}| \rceil$ bits to be the *effective entropy* of S with respect to Q. In case it is possible to reconstruct the given object $x \in S$ by means of the queries Q, we have $|\mathcal{C}| = |S|$ and there is no advantage to be gained. However, if $|\mathcal{C}| \ll |S|$, as is sometimes the case, then there can be substantial savings. For example, it is known [7] that for the RMQ problem, $|\mathcal{C}| \leq 4^n \ll n! = |S|$, so the empirical entropy of the class of arrays A containing permutations, with respect to RMQ, is only about $2n$ bits, as opposed to the entropy of the arrays A which is $\sim n \log n$ bits. In what follows, we will abbreviate "the effective entropy of S with respect to Q" as "the effective entropy of Q."

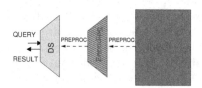

Fig. 1. Schematic illustation of the preprocessing steps in an encoding data structure

Encoding and Encoding Data Structure. Given an object $x \in S$, instead of storing x directly, we can store a representation of the equivalence class $y \in C$ that x belongs to. By the definition of C, queries can be answered correctly using y, rather than x; we call y an *encoding* of x. Note that since all queries can be answered using y, there is no need to store x, and it can be deleted, as depicted diagrammatically in Fig 1. The encoding y can then be converted into an *encoding data structure* that not only answers queries correctly but rapidly. Ideally the space usage of this data structure should be $(1 + o(1)) \log_2 |C|$ bits.

Expected effective entropy. One can also define the *expected* empirical entropy of a class of objects S with respect to Q: postulate a distribution on S, which induces a distribution on C. The effective empirical entropy is then defined as $\sum_{y \in C} \Pr(y) \cdot \log_2 1/\Pr(y)$. An encoding that aims to achieve the expected effective entropy then tries to represent an encoding y using as close to $\log_2 1/\Pr(y)$ bits as possible.

Minimal Encodings. As stated above, it is assumed that there is an effective characterization of the equivalence class C induced on S by Q, and that given an input $x \in S$, it is possible to constructively and fairly quickly find a $y \in C$ to represent x. We will henceforth call such encodings *minimal encodings*. A property of minimal encodings is that the encoding only contains the information about x that could be inferred via queries in Q, and an encoding data structure that is built upon y has the same property, provided the pre-processing does not refer to the input x. Characterizing the C that leads to a minimal encoding can be a non-trivial enumeration task, leading to objects studied by combinatorial mathematicians such as Baxter permutations [9] and Schröder trees [6].

Minimality is, however, a stringent requirement, often an element from a set E that is larger than C is used to represent the given input x. Provided that $\log_2 |E| = o(\log_2(|S|))$ we will still consider this to be an encoding.

3   Results on Encoding Data Structures

We now discuss some recent results on encoding data structures.

3.1   Range Statistics on 1D-Arrays

Range Maximum Queries. For the RMQ problem defined above, the non-encoding solution [8] is obtained via the Cartesian tree [17], a binary tree on n nodes. Fischer and Heun [7] observed that the Cartesian tree gives a 1-1 correspondence between binary trees on n nodes and equivalence classes for the RMQ problem, thus giving a minimal encoding for RMQ. Since there are $\frac{1}{2n+1}\binom{2n}{n}$ binary trees on $n + 1$ nodes, the effective entropy works out to be $2n - O(\log n)$ bits, and Fischer and Heun gave a $2n + o(n)$-bit data structure that answers RMQ in $O(1)$ time[1]. Davoodi et al. [4,3] gave alternative $2n + o(n)$-bit data structures.

---

[1] This result, as do all results in this abstract, use the word RAM model with word size $\Theta(\log n)$ bits.

The expected effective entropy (for uniform random permutations in A) of RMQ is approximately $1.736n$ bits [10]. Davoodi et al. [3] gave an encoding data structure that answers RMQ in $O(1)$ time, using $1.919n + o(n)$ bits on average.

Range Top-k and Range Selection. We are given an array $A[1..n]$ that contains a permutation of $\{1,\ldots,n\}$ and an integer k specified at pre-processing time, and need to answer the query:

top-k-pos(l,r): return positions of the k largest values in $A[l..r]$.

This is a generalization of the RMQ problem, which is the case $k = 1$. Grossi et al. [11] showed that any encoding must have size $\Omega(n \log k)$ bits and gave an encoding data structure that uses $O(n \log k)$ bits and answers queries in $O(k)$ time. For the case $k = 2$, Davoodi et al. [3] gave a minimal encoding, but were unable to obtain from this a closed-form expression for the empirical entropy of the top-2 problem. They showed that empirical entropy is at least $2.656n$ bits by a computational case analysis, and gave a data structure that took at most $3.272n + o(n)$ bits and answered top-2 queries in $O(1)$ time.

Recently, Gawrychowski and Nicholson [9] gave a different encoding for the top-k problem. Using their encoding, they were able to obtain tight upper and lower bounds of $\frac{1}{k+1}nH(\frac{1}{k+1})$ and $(1 - o(1))\frac{1}{k+1}nH(\frac{1}{k+1})$ bits on the effective entropy for all values of k. Here $H(x) = x \log_2(1/x) + (1-x) \log_2 1/(1-x)$ for any $0 \le x \le 1$. For $k = 2$, this gives the encoding complexity of the top-2 problem to be approximately $2.755n$ bits. However, it is not clear that their encoding is minimal, and they do not give an encoding data structure that answers top-k-pos queries rapidly.

The range selection problem is as follows. We are again given an array $A[1..n]$ that contains a permutation of $\{1,\ldots,n\}$ and an integer k specified at pre-processing time, and need to answer the query

select(i,l,r): return the position of the i-th largest value in $A[l..r]$, for any $i \le k$.

Clearly, since by repeated select operations, we can obtain the top-k in a given range, the effective entropy of range selection is no lower than that of the top-k problem. Navarro et al. [16] gave an encoding that takes $O(n \log k)$ bits of space, which is asymptotically optimal, and answers queries in optimal $O(1 + \log i / \log \log n)$ time.

Range Majority. We are given an array $A[1..n]$ that contains (wlog) values from $\{1,\ldots,n\}$, and a number $0 < \tau \le 1/2$, specified at pre-processing time. We wish to answer the following query:

majority$_\tau(l,r)$: If some value occurs at least $\tau(r-l+1)$ times in $A[l..r]$, return any index $i \in \{l,\ldots,r\}$ such that $A[i]$ contains this value. If no value occurs with this frequency, return null.

Navarro and Thankachan show that the encoding complexity is $\Omega(\tau \log(1/\tau)n)$ bits and give a data structure that takes $O((n/\tau) \log^* n)$ bits of space and answers queries in $O(\log n)$ time.

Range Maximum-segment Sum. We are given an array $A[1..n]$ that contains positive and negative numbers. We wish to answer the following query:

RMSS(l, r): Return l', r', $l \leq l' \leq r' \leq r$ such that $\sum_{i=l'}^{r'} A[i]$ is maximised.

Nicholson and Gawrychowski [9] showed that an encoding using $\Theta(n)$ bits can be used to answer such queries in $O(1)$ time.

3.2   2D Range Maximum Queries

The input to this problem is a two dimensional $m \times n$ array A, containing a permutation of $\{1, \ldots, N\}$ where $N = m \cdot n$. Assume that $m \leq n$. We wish to answer the following query:

RMQ(q): where $q = [i_1 \cdots i_2] \times [j_1 \cdots j_2]$ returns the position of the maximum element in the query range, i.e., $\mathsf{RMQ}(q) = \mathrm{argmax}_{(i,j) \in q} A[i, j]$.

Brodal et al. [1], following on the work of Demaine et al. [5] showed that the encoding complexity must be $\Omega(N \log m)$ bits. Brodal et al. [1] later gave an encoding of size $O(N \log m)$ bits, but this encoding does not yield a fast data structure. Golin et al. [10] showed that the expected effective entropy (assuming a random permutation in A) is $O(N)$ bits and gave a constant-time data structure with this space usage. Finally, for the case $m = 2$, Golin et al. gave a minimal encoding using $5n - O(\log n)$ bits. They also gave a data structure that takes $(5 + \epsilon)n + o(n)$ bits, for any $0 < \epsilon \leq 1$ and answers queries in $O(1/\epsilon)$ time.

3.3   Nearest Larger Values

Again, given an array $A[1..n]$ containing (not necessarily distinct) values from $\{1, \ldots, n\}$, we wish to answer the following query:

BNLV(i): return $j > i$ such that $A[j] > A[i]$ and $j - i$ is minimized, and $j' < i$ such that $A[j'] > A[i]$ and $i - j'$ is minimized.

Fischer [6] gave a minimal encoding for this problem that required at most $2.54n$ bits, and gave a corresponding data structure that answers queries in $O(1)$ time. The two-dimensional version of this problem, where A is an $n \times n$ matrix, was recently considered by Jayapaul et al. [13] and Jo et al. [14]. The latter authors gave an asymptotically optimal encoding data structure using $O(n^2)$ bits that answers queries in $O(1)$ time.

4   Conclusion

We have introduced the topic of *encoding* data structures. This topic is recently gaining interest, not only as a way to obtain more space-efficient data structures, but also due to the interesting combinatorial questions that arise. As can be seen even at this early stage, the tight space restrictions of encodings sometimes make it challenging to create efficient data structures with these space bounds. The topic is wide open – any data structuring question can be cast into the encoding framework, provided only that the queries considered do not allow the input to be reconstructed completely, no matter how many queries are asked of the data structure.

References

1. Brodal, G.S., Davoodi, P., Rao, S.S.: On space efficient two dimensional range minimum data structures. Algorithmica 63(4), 815–830 (2012), http://dx.doi.org/10.1007/s00453-011-9499-0
2. Brodnik, A., López-Ortiz, A., Raman, V., Viola, A. (eds.): Ianfest-66. LNCS, vol. 8066. Springer, Heidelberg (2013), http://dx.doi.org/10.1007/978-3-642-40273-9
3. Davoodi, P., Navarro, G., Raman, R., Satti, S.R.: Encoding range minima and range top-2 queries. Philosphical Transactions of the Royal Society A 372, 20130131 (2014), http://hdl.handle.net/2381/28856
4. Davoodi, P., Raman, R., Satti, S.R.: On succinct representations of binary trees. CoRR abs/1410.4963 (2014), http://arxiv.org/abs/1410.4963, preliminary version in COCOON 2012, LNCS 7434
5. Demaine, E.D., Landau, G.M., Weimann, O.: On cartesian trees and range minimum queries. In: Albers, S., Marchetti-Spaccamela, A., Matias, Y., Nikoletseas, S., Thomas, W. (eds.) ICALP 2009, Part I. LNCS, vol. 5555, pp. 341–353. Springer, Heidelberg (2009), http://dx.doi.org/10.1007/978-3-642-02927-1_29
6. Fischer, J.: Combined data structure for previous- and next-smaller-values. Theor. Comput. Sci. 412(22), 2451–2456 (2011), http://dx.doi.org/10.1016/j.tcs.2011.01.036
7. Fischer, J., Heun, V.: Space-efficient preprocessing schemes for range minimum queries on static arrays. SIAM J. Comput. 40(2), 465–492 (2011), http://dx.doi.org/10.1137/090779759
8. Gabow, H.N., Bentley, J.L., Tarjan, R.E.: Scaling and related techniques for geometry problems. In: DeMillo, R.A. (ed.) Proceedings of the 16th Annual ACM Symposium on Theory of Computing, April 30 - May 2, 1984, Washington, DC, USA. pp. 135–143. ACM (1984), http://doi.acm.org/10.1145/800057.808675
9. Gawrychowski, P., Nicholson, P.K.: Optimal encodings for range min-max and top-k. CoRR abs/1411.6581 (2014), http://arxiv.org/abs/1411.6581
10. Golin, M.J., Iacono, J., Krizanc, D., Raman, R., Rao, S.S.: Encoding 2-d range maximum queries. CoRR abs/1109.2885 (2011), http://arxiv.org/abs/1109.2885, preliminary version in Asano, T., Nakano, S.-I., Okamoto, Y., Watanabe, O. (eds.) ISAAC 2011. LNCS, vol. 7074, pp. 180–189. Springer, Heidelberg (2011)
11. Grossi, R., Iacono, J., Navarro, G., Raman, R., Rao, S.S.: Encodings for range selection and top-k queries. In: Bodlaender, H.L., Italiano, G.F. (eds.) ESA 2013. LNCS, vol. 8125, pp. 553–564. Springer, Heidelberg (2013), http://dx.doi.org/10.1007/978-3-642-40450-4_47
12. Jacobson, G.: Space-efficient static trees and graphs. In: 30th Annual Symposium on Foundations of Computer Science, Research Triangle Park, North Carolina, USA, October 30-November 1, pp. 549–554. IEEE Computer Society (1989), http://doi.ieeecomputersociety.org/10.1109/SFCS.1989.63533

13. Jayapaul, V., Jo, S., Raman, V., Satti, S.R.: Space efficient data structures for nearest larger neighbor. In: Proc. IWOCA 2014 (to appear, 2014)
14. Jo, S., Raman, R., Satti, S.R.: Optimal encodings and indexes for nearest larger value problems. In: Rahman, M.S., Tomita, E. (eds.) WALCOM 2015. LNCS, vol. 8973, Springer, Heidelberg (2015)
15. Kurtz, S.: Reducing the space requirement of suffix trees. Softw., Pract. Exper. 29(13), 1149–1171 (1999), http://dx.doi.org/10.1002/(SICI)1097-024X(199911)29:13<1149::AID-SPE274>3.0.CO;2-0
16. Navarro, G., Raman, R., Satti, S.R.: Asymptotically optimal encodings for range selection. In: Proc. 34th IARCS Annual Conference on Foundations of Software Technology and Theoretical Computer Science (FSTTCS 2014). Schloss Dagstuhl - Leibniz-Zentrum fuer Informatik (2014)
17. Vuillemin, J.: A unifying look at data structures. Commun. ACM 23(4), 229–239 (1980), http://doi.acm.org/10.1145/358841.358852

Fast Algorithms for Constrained Graph Density Problems[*]

Venkatesan Chakaravarthy[1], Neelima Gupta[2], Aditya Pancholi[2], and Sambuddha Roy[3]

[1] IBM Research, Bangalore
tcvenkat@in.ibm.com
[2] University of Delhi, Delhi
ngupta@cs.du.ac.in, aditya.cs.du@gmail.com
[3] Amazon, Bangalore
shombuddho@gmail.com

Abstract. We consider the question of finding communities in *large* social networks. In literature and practice, "communities" refer to a *well-connected* subgraph of the entire network. For instance, the notion of *graph density* has been considered as a reasonable measure of a community. Researchers have also looked at the *minimum degree* of a subgraph as a measure of the connectedness of the community.

Typically, a community is meaningful in the context of a social network if it is of somewhat significant size. Thus, earlier work has considered the densest graph problem subject to various *co-matroid constraints*. Most of these algorithms utilize an exact dense subgraph procedure as a subroutine; such a subroutine involves computing maximum flows or solving LPs. Consequently, they are rather inefficient when considered for massive graphs. For massive graphs, we are constrained to run in near-linear time, while producing subgraphs that provide reasonable approximations to the optimal solutions.

Our current work presents efficient greedy algorithms for the problem of graph density subject to an even more general class of constraints called *upward-monotone* constraints (these subsume co-matroid constraints). This generalizes and extends earlier work significantly. For instance, we are thereby able to present near-linear time 3-factor approximation algorithms for density subject to co-matroid constraints; we are also able to obtain 2-factor LP-based algorithms for density subject to 2 co-matroid constraints.

Our algorithms heavily utilize the *core decomposition* of a graph.

1 Introduction

Given an undirected graph $G = (V, E)$, the density $d(S)$ of a subgraph on vertex set S is defined as the quantity $\frac{|E(S)|}{|S|}$, where $E(S)$ is the set of edges in the subgraph induced by the vertex set S. The densest subgraph problem is to find the subgraph S of G that maximizes the density.

---

[*] Work done while at IBM Research, India.

M.S. Rahman and E. Tomita (Eds.): WALCOM 2015, LNCS 8973, pp. 8–19, 2015.

The concept of graph density is ubiquitous, more so in the context of social networks. In the context of social networks, the problem is to detect *communities*: collections of individuals who are relatively well connected as compared to other parts of the social network graph.

The notion of graph density has been fruitfully applied to finding communities in the social network graph (or even web graphs, gene annotation graphs [17], problems related to the formation of most effective teams [8], etc.). Also, note that graph density appears naturally in the study of threshold phenomena in random graphs, see [1].

Motivated by applications in social networks, the graph density problem and its variants have been well studied. Goldberg [9] (also see [14]) proved that the densest subgraph problem can be solved optimally in polynomial time: he showed this via a reduction to a series of max-flow computations. Later, others [5,12] have given new proofs for the above result, motivated by considerations to extend the result to some generalizations and variants.

Researchers also looked at the graph density problem subject to *size constraints*. In the k-densest subgraph problem, the objective is to find the densest subgraph induced over precisely k vertices. The problem is notoriously hard: while it was shown to be NP-hard by [7], Khot [11] shows that that there does not exist any PTAS for the k-densest subgraph problem under a reasonable complexity assumption. In the direction of upper bounds, [7] provide an algorithm with an approximation factor of $\mathcal{O}(n^{\theta})$ where $\theta < 1/3$. More recently, [3] give an algorithm for the k-densest subgraph with an approximation factor of $\mathcal{O}(n^{1/4})$.

Andersen and Chellapilla [2] considered two variants of the k-densest subgraph problem (they show both of the variants to be NP-hard). In one variant, the integer k provided along with the input specifies an *upper bound* on the size of the dense subgraph. They prove that this problem is almost as hard as the k-densest subgraph problem. The other variant they consider is where k specifies a *lower bound* on the size of the dense subgraph. Thus, here, the input includes an integer k and the goal is to find the densest subgraph S subject to the constraint $|S| \geqslant k$. This corresponds to finding *sufficiently large* dense subgraphs in social networks. Khuller and Saha [12] give two alternative algorithms for this problem: one of them is a greedy procedure that involves a dense subgraph subroutine, while the other is LP-based. Both the algorithms have 2-factor guarantees. However, these algorithms do not scale to *large* graphs such as web graphs because of the prohibitive runtime for the (exact) dense subgraph routine (or the LP solving routine). In this context, Andersen and Chellapilla [2] give an efficient approximation for this problem. They show that the problem may be approximated within a factor of 3 in linear time. Their algorithm is based on the *core decomposition* of a graph (see [13]).

Gajewar and Sarma [8] consider a further generalization, motivated by aspects of team formation (also see [16]). Call this the *team-formation* variant of graph density. Here, the input includes a partition of the vertex set into $U_1, U_2, \cdots, U_t$, and non-negative integers $r_1, r_2, \cdots, r_t$. The goal is to find the densest subgraph S subject to the constraint that for all $1 \leqslant i \leqslant t$, $|S \cap U_i| \geqslant r_i$.

They gave a 3-approximation algorithm by extending the greedy procedure of Khuller and Saha [12] (and thus involves an exact dense subgraph routine and has a prohibitive runtime).

Chakaravarthy et al. [4] vastly generalize the Gajewar & Sarma setting to that of *co-matroid constraints* (the precise definition of co-matroid constraints appears in Section 3). Co-matroid constraints capture both the cardinality constraints considered in [2,12] as well as the partition constraints considered in [8]. Chakaravarthy et al. provide a 2-factor approximation algorithm for the densest subgraph problem subject to arbitrary co-matroid constraints thereby improving the approximation factor of [8]. Their algorithm heavily utilizes an exact dense subgraph routine.

The work in [4] also considers another class of constraints called *dependency constraints* (this class of constraints generalize *subset constraints* as considered by [17]). They show that the problem of finding the densest subgraph subject to such constraints is in polynomial time - this involves finding solutions to LPs.

Summarizing the above discussion, we see that efficient linear time algorithms exist for the vanilla version of graph density and the variant where the size of the solution subgraph is lower bounded by k. For other variants, such as the team-formation variant, or the more general co-matroid constraint variant, the existing algorithms depend on an *exact graph density* routine. In turn, the exact graph density routines typically depend on certain *maximum flow* formulations on derived directed graphs (see, for instance [12]). To date, we do not know linear time (or even nearly linear time) algorithms for these specific maximum flow formulations; this is why the exact graph density routines (and consequently, the algorithms for the co-matroid constrained versions of graph density), to have prohibitive runtimes.

In this context, it ought to be mentioned that various recent papers (e.g. [19,10], see also [6]) exhibit an exciting line of work that provide *near-linear time* algorithms for *nearly* maximum flow; however, these results apply to *undirected graphs*.

To summarise, the principal motivation of the current work is the following: to devise efficient (i.e. *near linear*) algorithms for these general variants of graph density.

We may also ask a stronger question: can we replace the *exact* densest subgraph routine with an efficient *approximate* routine and still preserve the overall approximation guarantee?

Two Related Objectives. As an aside, let us remark on a related objective that has been looked at previously in the context of community detection. Given a subgraph H of G, note that *twice* the density of H is precisely the *average* degree of the subgraph H. Thus the densest subgraph problem may also be restated as the problem of finding the subgraph *maximizing* the *average* degree. Given this, a related objective may also be of finding the subgraph *maximizing* the *minimum* degree.

Thus, while the densest subgraph problem corresponds to the problem of finding a subgraph where the degrees of the vertices are large, *on average*, the problem of maximizing the minimum degree corresponds to finding a subgraph where the degrees of the vertices are large *in the worst-case*.

It turns out that the worst-case problem is significantly simpler than the average case question; in fact the *core decomposition* procedure [13] yields a subgraph with the maximum minimum degree - this was first proven by Matula & Beck [15]. Thus, the minimum degree objective may be solved in *linear time*, while an (exact) solution to the average degree objective involves solving a maximum flow problem (see [12]).

Since a subgraph with the maximum minimum degree is also a reasonable notion of *community*, various authors have studied this problem under different constraints. An *upward-monotone* constraint is such that if S is feasible, then any superset $S' \supseteq S$ is also feasible. Sozio & Gionis [20] show that the problem of maximizing the minimum degree subject to *any* upward-monotone constraint is solvable in linear time.

While we have generalizations to this latter result (to be presented in the full version of this paper), in this paper we primarily focus on the average degree objective.

2   Main Contributions

The principal message of our work is conceptual. Thus far in literature, most of the work on graph density has focused on specific constraints such as *co-matroid* constraints, *dependency* constraints, etc. (see, for instance, [4]). We explore the nature of constraints subject to which we may expect to have reasonable approximations for graph density. In the realm of *upward-monotone* constraints $\mathcal{C}$, the answer that we obtain has the following pleasing form. We are able to demonstrate a reduction from the problem of constrained graph density to that of a certain *Extension problem* for the constraints $\mathcal{C}$ (denoted as **Extend**$_\mathcal{C}$). The **Extend**$_\mathcal{C}(S)$ problem is to extend a partial set S to one that is *minimally feasible* for the constraints $\mathcal{C}$, the objective being to minimize the number of extra elements added; see Figure 3 for details.

Theorem 1. *Suppose for some class of upward-monotone constraints $\mathcal{C}$, the problem* **Extend**$_\mathcal{C}(S)$ *admits a f-factor approximation. Then there is a $(f+1)$-factor approximation algorithm for the densest subgraph problem subject to constraints $\mathcal{C}$.*

Similarly, suppose the problem **Extend**$_\mathcal{C}(S)$ *admits an **efficient** f-factor approximation. Then there is an **efficient** $(f+2)$-factor approximation algorithm for the densest subgraph problem subject to constraints $\mathcal{C}$.*

We believe that the aspect that the result holds even when we are considering *efficient* (i.e. near-linear time) algorithms is the novel technical contribution of this paper.

This allows us to approach the domain of *web-scale* graphs in the context of such *constrained* graph density problems. We prove Theorem 1 in Section 4.

We may instantiate the above general result with various constraints as in the corollaries below.

Corollary 1. *There is a linear time algorithm for the densest subgraph problem subject to co-matroid constraints that achieves a 3-factor approximation guarantee.*

Note that the problem $\mathbf{Extend}_C(S)$ in this case has an *optimal* (i.e. 1-factor) linear-time algorithm; we observe that the algorithm presented in [4] (see Lemma 9 therein) actually runs in linear time.

To exhibit the generality of our result, consider the following (perhaps contrived) problem. Given a graph, we are given the task of finding the densest subgraph that also forms a *vertex cover*. We can prove:

Corollary 2. *There is a 3-factor LP-based algorithm and a 4-factor linear time algorithm for this problem.*

In this case, the $\mathbf{Extend}_C(S)$ problem is a vertex cover instance and has a 2-factor approximation.

To the best of our knowledge, the question of achieving approximation algorithms for graph density subject to a *combination* of *two* co-matroid constraints has not been considered before.

Two Co-matroid Constraints

- Consider the densest subgraph problem subject to *two* co-matroid constraints. Hithertofore, no approximation algorithms were given for this. We prove that this densest subgraph variant admits an LP-based 2-factor approximation (here, the $\mathbf{Extend}_C$ routine corresponds to matroid intersection).
- For the specific case of the 2 co-matroid constraints corresponding to a bipartite matching, the densest subgraph problem admits a 3-factor *linear time* algorithm.

Knapsack Constraints

For a knapsack constraint (see Section 1.2 in [4]), the densest subgraph problem admits a 3-factor greedy linear time algorithm (the corresponding $\mathbf{Extend}_C$ routine admits an efficient 1-factor approximation).

3   Preliminaries

Given a graph G, we will denote the set of vertices of the graph as $V(G)$ and the set of edges as $E(G)$. If the graph G is clear from the context, then we may refer to the set of vertices (or edges) as V (or E). Given a graph G, the minimum degree in G is denoted by $\mathrm{MinDeg}(G)$.

Supermodular Functions

A set function $f : 2^U \to \mathbb{R}^+$ over a universe U is called *supermodular* if the following holds for any two sets $A, B \subseteq U$: $f(A) + f(B) \leqslant f(A \cup B) + f(A \cap B)$.

In this paper, we will use the following equivalent definition of supermodularity (the "increasing marginal returns" perspective). Given disjoint sets A, B

and C: $f(A + C) - f(A) \leqslant f(A + B + C) - f(A + B)$. The main fact that we will use in this paper is that: given a graph $G = (V, E)$, the set function $E(S)$ consisting of the edges within vertices of the set S is supermodular.

Core Decomposition

Our algorithms heavily use the concept of a *core*. Given a graph G and a degree parameter $d \in \mathbb{N}$, the core $C(G, d)$ of the graph G is the *maximal* subgraph H of G such that the minimum degree of a vertex $v \in V(H)$ in the induced subgraph H is at least d. It is easy to notice that there is a *unique* such maximal subgraph, so that the notion of a core is well defined. We will utilize the actual core decomposition as a subroutine in our algorithm for Theorem 1, and the core decomposition algorithm is presented there in Figure 2.

The core decomposition of a graph is obtained via the *minimum degree* (MD) ordering of the graph. The MD ordering of the vertices of the graph is one in which the vertices are ordered sequentially by successively removing the minimum degree vertex in the residual graph, see [15] for details.

It is well known that every core is *some suffix* of the MD ordering (see [15,13,5] for more facts about cores).

Upward-Monotone Constraints

An *upward-monotone* constraint is one such that if S is feasible, every superset of S is also feasible.

Co-matroid constraints (defined next) are a prominent subclass of upward-monotone constraints.

Co-matroid Constraints

A matroid is a pair $\mathcal{M} = (U, \mathcal{I})$ where $\mathcal{I} \subseteq 2^U$, and

1. (Hereditary Property) $\forall B \in \mathcal{I}, A \subset B \implies A \in \mathcal{I}$.
2. (Extension Property) $\forall A, B \in \mathcal{I} : |A| < |B| \implies \exists x \in B \setminus A : A + x \in \mathcal{I}$

Typically the sets in $\mathcal{I}$ are called *independent sets*, in keeping with the notions of linear algebra (see the excellent text by Schrijver [18] for details). A *co-matroid* constraint is defined as follows. Given a matroid $\mathcal{M} = (U, \mathcal{I})$, a set S is considered feasible iff the complement of S is *independent* in $\mathcal{I}$. Two commonly encountered matroids are cardinality matroids and partition matroids.

Structure of the Density Problem

It may be observed that most algorithms [4,12,2] for graph density (subject to upward-monotone constraints) work in *two* phases. Our main algorithm (see Figure 1) also follows a similar two-phase paradigm. The first phase of the algorithm may be viewed as aiming to construct the unconstrained densest subgraph, and the second phase performs *augmentations* to the candidate solutions from the first phase in order to obtain feasible solutions for the constrained problem. This raises the question: is it always true that there is a constrained optimum that *contains* the unconstrained optimum? The (affirmative) answer is given by the following lemma; this is new to the best of our knowledge.

Lemma 1. *Consider the upward-monotone constrained density problem for graph $G = (V, E)$ and let H denote an unconstrained optimum (i.e. a subgraph with the highest density). Then there exists a constrained optimum solution C such that $H \subseteq C$.*

This lemma may be viewed as justification for the two-phase approaches alluded to above. This may also be interpreted as a reason as to why density problems subject to upward-monotone constraints seem easier than when subject to upper bound constraints (as in the k-densest subgraph problem, which is known to be *equivalent* to the at-most-k densest subgraph problem, see [12]).

Proof. Consider a constrained optimum solution C. We will prove that one of two things has to happen: either $H \subseteq C$ in which case we have nothing to prove, or that $H \cup C$ is also a constrained optimum solution. We proceed to prove this last statement.

Clearly, $H \cup C$ is feasible for the upward-monotone constraints. We need to show that the subgraph induced by $H \cup C$ has density at least that of C. Let the unconstrained density be d_u and the constrained maximum density be d_c (where $d_u \geqslant d_c$). Note that

$$|E(H \cup C)| + |E(H \cap C)| \geqslant |E(H)| + |E(C)|$$
$$|E(H)| = d_u|H|$$
$$|E(C)| = d_c|C|$$
$$|E(H \cap C)| \leqslant d_u|H \cap C|$$

where the last inequality is because d_u is the maximum density. Thereby we get

$$|E(H \cup C)| \geqslant d_u(|H| - |H \cap C|) + d_c|C|$$
$$\geqslant d_c(|H| - |H \cap C|) + d_c|C|$$
$$= d_c|H \cup C|$$

Thus, $H \cup C$ has density at least as much as of C. Since C is the constrained optimum, so also is $H \cup C$. This proves a subclaim that there is a constrained optimum (namely $H \cup C$) that contains the unconstrained optimum. In order to prove that any constrained optimum contains H, we will have to work slightly more.

Note the equality cases of the inequalities in the lemma. Since C is the *optimum*, so also is the set $H \cup C$, and the inequalities above are all *tight*. But this implies that the supermodular inequality is tight, as also that $d_u = d_c$ or that $|H| = |H \cap C|$ (which means that H is already a subset of C). Thus, if $d_u \neq d_c$, then any constrained optimum *has to* contain the unconstrained optimum.

4   Proof of Theorem 1

The algorithm consists of two principal routines: **ConstructCandidates**(G) and **Extend$_{\mathcal{C}}$**(S). Interesting, the **ConstructCandidates** routine does not require the specific set of constraints $\mathcal{C}$. The **ConstructCandidates** routine outputs a list of candidate subsets $D_1, D_2, \cdots, D_r$. In fact, this is a **chain**: that

---

Density$(G, \mathcal{C})$
Input: a graph G, an upward-monotone constraint $\mathcal{C}$
Output: a (dense) subgraph H.

$D_1, \cdots, D_r \leftarrow$ **ConstructCandidates**(G)
for $i \in \{1, 2, \cdots, r\}$ **do**
    $D_i' \leftarrow$ **Extend**$_\mathcal{C}(D_i)$
end for
$H \leftarrow$ the subgraph among D_i' (for $i = 1, \cdots, r$) with the highest density
Output H

Fig. 1. Generic Algorithm

is, $D_1 \subseteq D_2 \subseteq \cdots D_r$. Although the generic algorithm as presented in Figure 1 does not explicitly utilize this fact, it turns out, that given this sequence one only need consider the *first feasible* solution in this chain along with the **Extend**$_\mathcal{C}$-ed solutions.

We also assume that there is a *near-linear* time algorithm for the **Extend**$_\mathcal{C}$ routine, that outputs a f-factor approximation.

We may implement the **ConstructCandidates** routine by repeatedly solving LP's; this is the approach adopted by earlier papers (see for instance [4]). Such an approach is possible for the current scenario with general upward-monotone constraints and would result in a $(f+1)$-factor approximation. However, such an algorithm would be costly, involving several LP-solving/max-flow steps. In the current work, we save over the costly LP-solving steps, incurring only a slightly worse $(f + 2)$-factor, achieving this in *near linear* time.

Our efficient algorithms use the *core decomposition* routine (presented in Figure 2). Given this, we prove only the latter half (the part dealing with *efficient* algorithms) of Theorem 1.

Proof. Let H^* be the optimal densest subgraph (of density d^*) subject to the constraints $\mathcal{C}$.

We will prove that some subgraph among D_i' (for $i = 1, \cdots, r$) has the correct density, i.e. has density $\geqslant d^*/(f + 2)$.

Note that, by construction, the sets in the sequence $D_1 \subseteq D_2 \subseteq \cdots \subseteq D_r$ satisfy the following relation: $\mathrm{MinDeg}(D_1) > \mathrm{MinDeg}(D_2) > \cdots > \mathrm{MinDeg}(D_r)$. We may also note that the maximum minimum degree over a subgraph of G is attained by D_1.

In fact, we will be able to prove a sharper statement. Given the ordered sequence $D_1 \subseteq D_2 \subseteq \cdots D_r$, let L be the least index $(1 \leqslant L \leqslant r)$ such that D_L is feasible. We will then prove that *some* subgraph among D_i' (for $i = 1, \cdots, (L-1)$) and D_L has the correct density (i.e. density $\geqslant d^*/(f + 2)$).

We will consider the subgraph D_ℓ such that the following holds (the *boundary condition*): The minimum degree in the subgraph D_i is at least $2d^*/(f + 2)$ for $i = 1 \cdots \ell$ and the minimum degree in the subgraph $D_{\ell+1}$ is $< 2d^*/(f + 2)$.

```
ConstructCandidates(G) ← Core(G)
Input: a graph G.
Output: subsets D_1, D_2, ··· , D_r of vertices of G.
i ← 1
K_i ← G
while K_i ≠ ∅ do
    H ← K_i
    while MinDeg(H)⩽MinDeg(K_i) do
        v ← v_min(H)
        H ← H \ {v}
    end while
    i ← i + 1
    K_i ← H
end while
Let the sets constructed be K_1 ⊇ ··· ⊇ K_r
Rename the sets as D_i = K_{r+1-i}, so that D_1 ⊆ ··· ⊆ D_r.
```

Fig. 2. Routine **Core**(G)

```
Extend_C(S)
Input: a set S, and a collection of upward-monotone constraints
C.
Output: a set T such that S ⊆ T and T is minimal feasible for
the constraints C.
```

Fig. 3. Routine **Extend**_C(S)

Why should such an ℓ even exist? Given that there is a subgraph in G of density d^*, the densest subgraph (without constraints) has density at least d^*. Let H denote the *unconstrained* densest subgraph of density at least d^*. It is easy to check that $\text{MinDeg}(H)\geqslant d^*$. Thus, G has a subgraph of minimum degree at least d^*; thus the maximum minimum degree of G (that is attained by the subgraph D_1) is at least $d^* > 2d^*/(f+2)$.

If $\ell = L$ (i.e. if D_ℓ is feasible), then note that $\text{MinDeg}(D_\ell)\geqslant 2d^*/(f+2)$; thus, the density of the subgraph D_ℓ is $\geqslant d^*/(f+2)$.

Thus, suppose that $\ell < L$, so that D_ℓ is *not* feasible. In this case, we will prove that the set D'_ℓ has the *correct* density i.e. $d(D'_\ell)\geqslant d^*/(f+2)$.

To this end, we will prove two lemmas about the set D_ℓ:

Lemma 2. *The following holds:*

$$|E(D_\ell)| - |E(D_\ell \cap H^*)| \geqslant \frac{d^*}{(f+2)}(|D_\ell| - |D_\ell \cap H^*|)$$

Proof. Let X denote the vertices in $D_\ell \setminus H^*$. Note that $|X| = |D_\ell| - |D_\ell \cap H^*|$. Also, since $X \subseteq D_\ell$, every vertex in X has degree at least $2d^*/(f+2)$, by definition of the subset D_ℓ.

The sum $\sum_{v \in X} \mathrm{Deg}_{D_\ell}(v)$ counts each edge $e \in E(D_\ell \setminus H^*)$ *twice* and every edge $e \in \delta(D_\ell \setminus H^*)$ once. Thus, we have that $\sum_{v \in X} \mathrm{Deg}_{D_\ell}(v) = 2|E(D_\ell \setminus H^*)| + |\delta(D_\ell \setminus H^*)|$. Hence, it also holds that

$$2(|E(D_\ell)| - |E(D_\ell \cap H^*)|) = 2|E(D_\ell \setminus H^*)| + 2|\delta(D_\ell \setminus H^*)| \geqslant \sum_{v \in X} \mathrm{Deg}_{D_\ell}(v)$$

$$\geqslant \frac{2d^*}{(f+2)}|X|.$$

This proves the statement of the lemma.

The next lemma attempts to lowerbound the value of $|E(D_\ell \cup H^*)|$.

Lemma 3. *The following holds:*

$$|E(D_\ell \cup H^*)| - |E(D_\ell)| < \frac{2d^*}{(f+2)}(|H^*| - |D_\ell \cap H^*|)$$

Proof. Consider the set $D_\ell \cup H^* \setminus D_\ell$. Assume that there are k vertices in this set; thus, $k = |H^*| - |D_\ell \cap H^*|$. We will prove that these vertices may be ordered as $\{h_1, h_2, \cdots, h_k\}$ such that for any i, the number of edges between h_{i+1} and $D_\ell \cup \{h_1, h_2, \cdots, h_i\}$ is *less than* $\frac{2d^*}{(f+2)}$. Note that this would prove the lemma; the quantity $|E(D_\ell \cup H^*)| - |E(D_\ell)|$ may be decomposed as $\sum_{i=1}^{k} |E(h_i, D_\ell \cup \{h_1, h_2, \cdots, h_{i-1}\})|$.

To this end, we will consider the *reverse of* the MD ordering of the vertices of $G = \{v_1, v_2, \cdots, v_n\}$. Thus v_n is the vertex of minimum degree in G. According to this notation let $D_\ell = \{v_1, v_2, \cdots, v_s\}$ for some s; also D_ℓ is the *core* corresponding to some degree $d \geqslant 2d^*/(f+2)$.

Also by definition of D_ℓ, for any set $\{v_1, v_2, \cdots, v_i\}$ (for $i > s$), it holds that the degree of v_i in the set is $< 2d^*/(f+2)$. Rephrased, this means that the number of edges between v_i and $\{v_1, v_2, \cdots, v_{i-1}\}$ is $< 2d^*/(f+2)$.

But now, the ordering of the h_i's is clear: it corresponds to the order in which the vertices h_i appear in the ordering $\{v_1, v_2, \cdots, v_n\}$. If a specific h_{i+1} is v_t in this ordering, then the number of edges between h_{i+1} and $D_\ell \cup \{h_1, h_2, \cdots, h_i\}$ is at most the number of edges between v_t and $\{v_1, v_2, \cdots, v_{t-1}\}$. However by the argument above, this is less than $2d^*/(f+2)$.

This completes the proof of this lemma.

Thus we may claim the following:

Lemma 4.

$$|E(D_\ell \cap H^*)| \geqslant \frac{2d^*}{(f+2)}|D_\ell \cap H^*| + \frac{fd^*}{(f+2)}|H^*|$$

Proof. Consider the following statement that follows from supermodularity of the $E(\cdot)$ function (using the *increasing marginal returns* perspective):

$$|E(H^*)| - |E(D_\ell \cap H^*)| \leqslant |E(D_\ell \cup H^*)| - |E(D_\ell)|$$

Thus, Lemma 3 implies that

$$|E(H^*)| - |E(D_\ell \cap H^*)| \leqslant \frac{2d^*}{(f+2)}(|H^*| - |D_\ell \cap H^*|)$$

Now, using that $E(H^*) = d^*|H^*|$ (since H^* is the densest constrained subgraph), and simplifying, we get that

$$|E(D_\ell \cap H^*)| \geqslant \frac{2d^*}{(f+2)}|D_\ell \cap H^*| + \frac{fd^*}{(f+2)}|H^*|$$

as required.

Now adding the statements of Lemma 4 and Lemma 2, we get that

$$|E(D_\ell)| \geqslant \frac{d^*}{(f+2)}(|D_\ell \cap H^*| + |D_\ell| + f|H^*|)$$

Given that D_ℓ is *infeasible*, the **Extend**$_C$ subroutine in the algorithm augments the set D_ℓ with some number of vertices in order to make it feasible (and the feasible solution is denoted as D'_ℓ).

We note that the number of extra vertices needed by the **Extend**$_C$ routine is upper bounded by $f|H^*|$ - this follows because, by assumption, we are given a f-factor approximation for the problem **Extend**$_C$ for the constraints C. Thus, this means that the number of vertices in the subgraph D'_ℓ is at most $|D_\ell| + f|H^*|$. The above inequality then implies that (after dropping the term corresponding to $\frac{d^*}{(f+2)}|D_\ell \cap H^*|$):

$$|E(D'_\ell)| \geqslant \frac{d^*}{(f+2)}(|D_\ell| + f|H^*|) \geqslant \frac{d^*}{(f+2)}|D'_\ell|$$

so that the augmented set D'_ℓ has density at least $d^*/(f+2)$.

This concludes the proof of Theorem 1.

Runtime Analysis
Deferred to the full version of the paper.

Remarks
It is interesting to compare the current result with the earlier work by Andersen & Chellapilla [2]. They showed a greedy 3-factor approximation algorithm for problem of finding a dense subgraph subject to a *cardinality* constraint $|S| \geqslant k$. Their algorithm in a nutshell is this: consider the core decomposition of the graph; each core is a candidate solution. Look only at the cores that are *feasible* (i.e. has size $\geqslant k$), and output the *densest* of these subgraphs as the final solution.

However, when we consider general upward-monotone constraints (in fact, even constraints as simple as $|S \cap A| \geqslant k$), we are unable to claim such a performance guarantee by just restricting ourselves to looking *only* at the feasible solutions from the core decomposition.

In the general scenario of *upward-monotone constraints*, it appears that it is essential that we consider the cores in the **reverse** order, i.e. consider the infeasible cores and augment them to render them feasible.

References

1. Alon, N., Spencer, J.H.: The Probabilistic Method. Wiley, New York (1992)
2. Andersen, R., Chellapilla, K.: Finding dense subgraphs with size bounds. In: Avrachenkov, K., Donato, D., Litvak, N. (eds.) WAW 2009. LNCS, vol. 5427, pp. 25–37. Springer, Heidelberg (2009)
3. Bhaskara, A., Charikar, M., Chlamtac, E., Feige, U., Vijayaraghavan, A.: Detecting high log-densities: an $o(n^{1/4})$ approximation for densest k-subgraph. In: STOC, pp. 201–210 (2010)
4. Chakaravarthy, V.T., Modani, N., Natarajan, S.R., Roy, S., Sabharwal, Y.: Density functions subject to a co-matroid constraint. In: FSTTCS, pp. 236–248 (2012)
5. Charikar, M.: Greedy approximation algorithms for finding dense components in a graph. In: Jansen, K., Khuller, S. (eds.) APPROX 2000. LNCS, vol. 1913, pp. 84–95. Springer, Heidelberg (2000)
6. Eisenstat, D., Klein, P.N.: Linear-time algorithms for max flow and multiple-source shortest paths in unit-weight planar graphs. In: Symposium on Theory of Computing Conference, STOC 2013, Palo Alto, CA, USA, June 1-4, pp. 735–744 (2013)
7. Feige, U., Kortsarz, G., Peleg, D.: The dense k-subgraph problem. Algorithmica 29, 2001 (1999)
8. Gajewar, A., Sarma, A.D.: Multi-skill collaborative teams based on densest subgraphs. In: SDM, pp. 165–176 (2012)
9. Goldberg, A.V.: Finding a maximum density subgraph. Technical report, UC Berkeley (1984)
10. Kelner, J.A., Lee, Y.T., Orecchia, L., Sidford, A.: An almost-linear-time algorithm for approximate max flow in undirected graphs, and its multicommodity generalizations. In: Proceedings of the Twenty-Fifth Annual ACM-SIAM Symposium on Discrete Algorithms, SODA 2014, Portland, Oregon, USA, January 5-7, pp. 217–226 (2014)
11. Khot, S.: Ruling Out PTAS for Graph Min-Bisection, Dense k-Subgraph, and Bipartite Clique. SIAM J. Comput. 36(4), 1025–1071 (2006)
12. Khuller, S., Saha, B.: On finding dense subgraphs. In: Albers, S., Marchetti-Spaccamela, A., Matias, Y., Nikoletseas, S., Thomas, W. (eds.) ICALP 2009, Part I. LNCS, vol. 5555, pp. 597–608. Springer, Heidelberg (2009)
13. Kortsarz, G., Peleg, D.: Generating sparse 2-spanners. J. Algorithms 17(2), 222–236 (1994)
14. Lawler, E.: Combinatorial optimization - networks and matroids. Holt, Rinehart and Winston, New York (1976)
15. Matula, D.W., Beck, L.L.: Smallest-last ordering and clustering and graph coloring algorithms. J. ACM 30(3), 417–427 (1983)
16. Rangapuram, S.S., Bühler, T., Hein, M.: Towards realistic team formation in social networks based on densest subgraphs. In: Proceedings of the 22nd International Conference on World Wide Web, WWW 2013, pp. 1077–1088 (2013)
17. Saha, B., Hoch, A., Khuller, S., Raschid, L., Zhang, X.-N.: Dense subgraphs with restrictions and applications to gene annotation graphs. In: Berger, B. (ed.) RECOMB 2010. LNCS, vol. 6044, pp. 456–472. Springer, Heidelberg (2010)
18. Schrijver, A.: Combinatorial Optimization - Polyhedra and Efficiency. Springer (2003)
19. Sherman, J.: Nearly maximum flows in nearly linear time. In: 54th Annual IEEE Symposium on Foundations of Computer Science, FOCS 2013, Berkeley, CA, USA, October 26-29, pp. 263–269 (2013)
20. Sozio, M., Gionis, A.: The community-search problem and how to plan a successful cocktail party. In: KDD, pp. 939–948 (2010)

The Directed Ring Loading with Penalty Cost

Li Guan, Jianping Li, Xuejie Zhang, and Weidong Li*

Yunnan University, Kunming 650091, PR China
{guanli,jianping,xjzhang,weidong}@ynu.edu.cn

Abstract. We study the directed ring loading problem with penalty cost, which is to select some of given multicast requests represented by hyperedges with different demands and embed them in a directed ring, such that the sum of the maximum congestion among all links on the ring and the total penalty cost of the unselected multicast requests is minimized. We prove that this problem is NP-hard even if the demand is divisible, and then design a 1.582-approximation algorithm for the demand divisible case and a 3-approximation algorithm for the demand indivisible case, respectively. As a consequence, for any $\varepsilon > 0$, we present a $(1.582+\varepsilon)$-approximation algorithm for the case where every multicast request contains exactly one sink.

Keywords: Approximation algorithms, Directed ring loading, Directed hypergraph embedding, Penalty cost.

1 Introduction

Arising in automated electronic circuit design, specifically in routing nets around a rectangle where the objective is to route within a minimum area rectangle [2], Ganley and Cohoon [3] proposed the problem of *hypergraph embedding in a cycle* (HEC, for short). The HEC problem is to embed the hyperedges in a hypergraph as paths in a cycle on the same number of vertices, such that the maximum *load* among all links in the cycle is minimized, where the load of a link is the number of paths that use it. HEC is a challenging problem and has many applications in various areas such as communication, computer networks, and parallel computation.

Ganley and Cohoon [3] proved that the HEC problem is NP-hard and gave a 3-approximation algorithm for it. Gonzalez [5] constructed an integer linear programming for the HEC problem and designed two 2-approximation algorithms. Gu and Wang [6] proposed an improved 1.8-approximation algorithm to solve the HEC problem by a reembedding technique. Finally, Li, Deng and Xu [9] designed a polynomial time approximation scheme (PTAS) for the HEC problem by a randomized rounding approach, which is generalized by Yang and Li [16] to a more general case where the links on the ring have different weights. If the hyperedges have different demands, Lee and Ho [8] proposed a 2-approximation algorithm, which is improved by the same authors [7]. When every hyperedge

---

* Corresponding author.

M.S. Rahman and E. Tomita (Eds.): WALCOM 2015, LNCS 8973, pp. 20–31, 2015.
© Springer International Publishing Switzerland 2015

contains exactly two nodes, the HEC problem with different demands is exactly the ring loading problem [13], which possesses a PTAS. Recently, Li, Li and Guan [12] studied a generalized version of the ring loading problem where the requests can be rejected, and presented some approximation algorithms.

Motivated by the multicast communication applications where a multicast request asks to send a message from a source to many sinks, Li and Wang [11] proposed the problem of directed hypergraph embedding on a directed ring (DHER), and presented a PTAS for it, which is generalized by Li, Li and Wang [10] to a more general case where the links on the ring have different weights. If the hyperedges have different demands, Wang et al. [14] designed a 2-approximation algorithm. When every directed hyperedge contains exactly one sink, the DHER problem is exactly the directed ring loading problem [1], which possesses a PTAS.

In this paper, we consider the directed ring loading problem with penalty cost (DRLPC, for short), which is defined as follows. Given a set of m multicast connection requests, each multicast request j has a demand d_j and a penalty p_j, and it must be routed in one of the several possible ways around the ring or be rejected with penalty p_j. The objective is to minimize the sum of the maximum load among all links on the ring and the total penalty cost of the requests rejected, where the load of a link is the sum of the demands of requests that use it.

This paper is organized as follows. In Section 2, we present some basic notations. In Section 3, we show that the DRLPC problem with divisible demand is NP-hard and present a 1.582-approximation algorithm by using a randomized rounding technique. In Section 4, we consider the DRLPC problem with indivisible demand, and design a 3-approximation algorithm for the general case and a $(1.582 + \varepsilon)$-approximation algorithm for the special case where each hyperedge contains exactly one sink. Finally, we summarize and describe some possible directions for future work.

2   Preliminaries

A *directed ring* is a directed network $R = (V, A)$, where $V = \{1, 2, \ldots, n\}$ is the set of n nodes, and $A = \{e_i^+ = (i, i+1), e_i^- = (i+1, i)| \ i = 1, 2, \ldots, n\}$ is the set of directed links. Throughout, we treat the node $n + i$ as the node i for $1 \leq i \leq n$. In communication applications, a request j is represented by a directed hyperedge $h_j = (u_j, S_j)$, where $u_j \in V$ is indicated as the *source* of the directed hyperedge h_j and $S_j = \{i_1^j, i_2^j, \ldots, i_{k_j}^j\} \subseteq V \setminus \{u_j\}$ is the set of *sinks* (here, $k_j = |S_j|$). The request j asks to send a message from u_j to every vertex in S_j. Let $D_H = \{h_1, h_2, \ldots, h_m\}$ be the set of m directed hyperedges (requests). For each hyperedge $h_j = (u_j, S_j) \in D_H$, let $d_j, p_j \in R^+$ be its demand and penalty cost, respectively.

For convenience, we use i_0^j to denote u_j. Assume that the $k_j + 1$ vertices $i_0^j, i_1^j, \ldots, i_{k_j}^j$ follow the clockwise order on the ring R. For each $k = 0, 1, \ldots, k_j - 1$, let P_k^j be the embedding of directed hyperedge h_j which is obtained by deleting the links on the ring between i_k^j and i_{k+1}^j in clockwise direction, and $P_{k_j}^j$ be the embedding which is obtained by deleting the links between $i_{k_j}^j$ and i_0^j in clockwise direction. For example, given a hyperedge $h_j = (1, \{2, 6\})$ with $V = \{1, 2, 3, 4, 5, 6\}$, we have $P_0^j = \{(1, 6), (6, 5), (5, 4), (4, 3), (3, 2)\}$, $P_1^j = \{(1, 2), (1, 6)\}$, and $P_2^j = \{(1, 2), (2, 3), (3, 4), (4, 5), (5, 6)\}$. Three embedding ways are described in Figure 1.

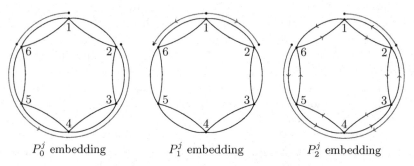

P_0^j embedding P_1^j embedding P_2^j embedding

Fig. 1. Three embedding ways

The problem of directed ring loading with penalty cost is to route the hyperedges accepted (may not all) such that the total flow for each hyperedge h_j is d_j. The objective is to minimize the sum of the maximum congestion of links and the total penalty cost of the rejected hyperedges, where the congestion of a link is the total flow of embedding paths that use it. We distinguish between *divisible* where each accepted hyperedge h_j can be embedded in several ways such that $\sum_{k=0}^{k_j} f(P_k^j) = d_j$ and *indivisible* where each accepted hyperedge h_j can only be embedded in one way P_k^j such that $f(P_k^j) = d_j$, where $f(P_k^j)$ is the amount of demands (or flows) sending along the path P_k^j.

Consider an example with $n = 6$, $m = 3$, $h_1 = \{1, \{2, 6\}\}$, $h_2 = \{2, \{1\}\}$, and $h_3 = \{6, \{2, 3, 5\}\}$. The demands are $d_1 = 7$, $d_2 = 3$, and $d_3 = 10$. The penalty costs are $p_1 = 6$, $p_2 = 2$, and $p_3 = 1$. When the demand is divisible, an optimal solution whose objective value is 4.5 with h_3 rejected is described in the Figure 2 (a). When the demand is indivisible, an optimal solution whose objective value is 8 with h_3 rejected is described in the Figure 2 (b).

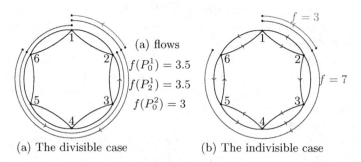

(a) flows

$f(P_0^1) = 3.5$
$f(P_2^1) = 3.5$
$f(P_0^2) = 3$

$f = 3$

$f = 7$

(a) The divisible case   (b) The indivisible case

Fig. 2. Optimal solutions

3   The DRLPC Problem with Divisible Demand

In this section, we consider the case where the demand is infinite divisible, which means each accepted request h_j can be routed in several ways satisfying that the sum of all flows is d_j. We prove that this problem is NP-hard, and formulate this problem as a mixed integer linear programming. Then, by rounding the optimal solution of its relaxation, we find a desired feasible solution whose objective value is close to the optimal value.

3.1 NP-hardness

Theorem 1. The DRLPC problem with divisible demand is NP-hard, even if $k_j = |S_j| = 1$ for every multicast request $h_j = (u_j, S_j)$, where $j = 1, 2, \ldots, n$.

Proof. We demonstrate a polynomial reduction from the PARTITIONproblem [4]. Given an instance I of the PARTITION problem with a set $S = \{a_1, a_2, \ldots, a_n\}$ of positive integers and a positive integer $a = \sum_{j=1}^{n} a_j / 2$, one should decide whether there is a subset $S' \subset S$ satisfying $\sum_{a_j \in S'} a_j = a$. Construct an instance $\tau(I)$ with $m = n + 4$ requests for the DRLPC problem with divisible demand as follows. Define a directed ring $R = (V, A)$ with $V = \{1, 2, 3\}$ and $A = \{(1,2), (2,3), (3,1), (2,1), (3,2), (1,3)\}$. For $j = 1, 2, \ldots, n$, the request $h_j = (1, \{2\})$ has a demand $d_j = 4a_j$ and a penalty cost $p_j = a_j$. The remaining requests $h_{n+1} = (2, \{3\})$, $h_{n+2} = (3, \{2\})$, $h_{n+3} = (3, \{1\})$ and $h_{n+4} = (1, \{3\})$ have the same demand $4a$ and penalty cost $5a + 1$. We claim that the instance I of the PARTITION problem has a feasible solution if and only if there is a feasible solution for the instance $\tau(I)$ with objective value no more than $5a$.

If the instance I has a feasible solution $S' \subset S$ satisfying $\sum_{a_j \in S'} a_j = a$, let $D(S') = \{j \mid a_j \in S'\}$. For each $j \in D(S')$, the request h_j is rejected. The total penalty cost of the rejected requests is a, as $p_j = a_j$. For $j \in D(S \setminus S')$, the request h_j is routed in the clockwise path $P_1^j = \{(1,2)\}$ with flow $4a_j$. The requests $h_{n+1} = (2, \{3\})$ and $h_{n+3} = (3, \{1\})$ are routed in the clockwise paths $P_1^{n+1} = \{(2,3)\}$ and $P_1^{n+3} = \{(3,1)\}$ with flow $4a$, respectively. The requests $h_{n+2} = (3, \{2\})$ and $h_{n+4} = (1, \{3\})$ are routed in the counterclockwise paths $P_0^{n+2} = \{(3,2)\}$ and $P_0^{n+4} = \{(1,3)\}$ with demand $4a$, respectively. The maximum load among all links on the ring is $4a$ (see Figure 3). Thus, we obtain a feasible solution for instance $\tau(I)$ with objective value $5a$.

If there is a feasible solution $\mathcal{F}$ for the instance $\tau(I)$ with objective value at most $5a$, the last four requests h_{n+j} $(j = 1, 2, 3, 4)$ can not be rejected. Clearly, the maximum load among all links used by the last four requests h_{n+j} $(j = 1, 2, 3, 4)$ in any feasible routing is at least $4a$. Let B denote the maximum load among all links in the feasible solution $\mathcal{F}$, then $B \geq 4a$. Let $\mathcal{R}$ be the set of the rejected requests in $\mathcal{F}$ and $D(\mathcal{R}) = \{j \mid h_j \in \mathcal{R}\}$. The total penalty cost of the rejected requests, denoted by X, satisfies

$$X \leq 5a - B \leq 5a - 4a = a. \tag{1}$$

Note that the sum of demands of the accepted requests h_j $(j \leq n)$ is $4(2a-X)$. Since all requests are parallel, by Proposition 2 in [15], we can assume that at most one accepted request h_j $(j \leq n)$ is routed in two ways. It implies h_{n+j} $(j = 1, 2, 3, 4)$ is routed in the shortest way as in Figure 3.

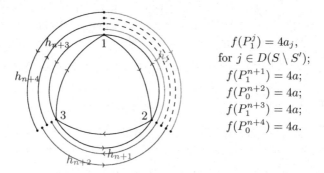

$$f(P_1^j) = 4a_j,$$
$$\text{for } j \in D(S \setminus S');$$
$$f(P_1^{n+1}) = 4a;$$
$$f(P_0^{n+2}) = 4a;$$
$$f(P_1^{n+3}) = 4a;$$
$$f(P_0^{n+4}) = 4a.$$

Fig. 3. Reduction

Since the accepted requests h_j $(j \leq n)$ must use at least one link in $\{e_1^+ = (1,2), e_3^- = (1,3)\}$, we have

$$B \geq 4a + \frac{4(2a - X) - 4a}{2} = 6a - 2X.$$

It implies that the objective value of $\mathcal{F}$ is at lest $X + 6a - 2X = 6a - X$. From the assumption that the objective value of $\mathcal{F}$ is at most $5a$, it follows that $6a - X \leq 5a$, and then $X \geq a$. Combining with (1), we have $X = a$. Let $S' = \{a_j \mid j \in D(\mathcal{R})\}$, and then $\sum_{a_j \in S'} a_j = X = a$. Thus S' is a feasible solution of the instance I. Since the PARTITION problem is NP-hard [4], so is the DRLPC problem. ∎

3.2   An Approximation Algorithm

For each multicast request h_j, we introduce $k_j + 1$ variables $x_k^j \in [0, 1]$ $(k = 0, 1, \ldots, k_j)$, where x_k^j means that the amount of flow $f(P_k^j)$ of path P_k^j used by h_j is $d_j x_k^j$, i.e., the source u_j sends $f(P_k^j) = d_j x_k^j$ messages to every sink using path P_k^j. For convenience, let $z_j = \sum_{k=0}^{k_j} x_k^j \in \{0, 1\}$. If the multicast request h_j is accepted, then $z_j = 1$. Otherwise, $z_j = 0$. For $i = 1, 2, \ldots, n$, let $\mathcal{P}_{ij}^+$ ($\mathcal{P}_{ij}^-$, resp.) be the set of embeddings P_k^j of h_j that use link e_i^+ (e_i^-, resp.), and $\mathcal{P}_i^+ = \cup_{j=1}^m \mathcal{P}_{ij}^+$ ($\mathcal{P}_i^- = \cup_{j=1}^m \mathcal{P}_{ij}^-$, resp.). Thus, the DRLPC problem with divisible demand can be formulated as the following mixed integer linear programming (MILP).

$$\min \quad B + \sum_{j=1}^m p_j(1 - z_j)$$

$$\sum_{P_k^j:P_k^j\in\mathcal{P}_i^+} d_j x_k^j \le B, i = 1,2,\ldots,n;$$

$$\sum_{P_k^j:P_k^j\in\mathcal{P}_i^-} d_j x_k^j \le B, i = 1,2,\ldots,n;$$

$$\sum_{k=0}^{k_j} x_k^j = z_j, j = 1,2,\ldots,m;$$

$$z_j \in \{0,1\}, j = 1,2,\ldots,m;$$

$$x_k^j \ge 0, j = 1,2,\ldots,m, k = 0,1,\ldots,k_j.$$

Replacing the constraints $z_j \in \{0,1\}$ by $0 \le z_j \le 1$, we obtain the relaxation of MILP, which is a linear programming, and can be solved in polynomial time. Let $(\tilde{x}, \tilde{z})$ be the optimal solution for the relaxation of MILP. We randomly choose a threshold α from the uniform distribution over $[1/e, 1]$. If $\tilde{z}_j \le \alpha$, set $\bar{z}_j = \bar{x}_0^j = \cdots = \bar{x}_{k_j}^j = 0$, and otherwise set $\bar{z}_j = 1$, and $\bar{x}_k^j = \tilde{x}_k^j/\tilde{z}_j$, for $k = 0,1,\ldots,k_j$. Thus, we obtain a feasible solution $(\bar{x}, \bar{z})$ for the MILP.

Lemma 1. The expected objective value of the solution $(\bar{x}, \bar{z})$ is at most $\frac{e}{e-1}OPT$, where OPT denotes the optimal value for the MILP.

Proof. Let $\tilde{B} = \max\{\sum_{P_k^j:P_k^j\in\mathcal{P}_i^+} d_j\tilde{x}_k^j, \sum_{P_k^j:P_k^j\in\mathcal{P}_i^-} d_j\tilde{x}_k^j | i = 1,2,\ldots,n\}$ be the maximum load among all links in the solution $(\tilde{x}, \tilde{z})$, and $\bar{B} = \max\{\sum_{P_k^j:P_k^j\in\mathcal{P}_i^+} d_j\bar{x}_k^j, \sum_{P_k^j:P_k^j\in\mathcal{P}_i^-} d_j\bar{x}_k^j | i = 1,2,\ldots,n\}$ the maximum load of the links in the solution $(\bar{x}, \bar{z})$. Clearly, for any $\alpha \in [1/e, 1]$, we have $\bar{x}_k^j \le \tilde{x}_k^j/\alpha$ for $k = 0,1,\ldots,k_j$, and then $\bar{B} \le \tilde{B}/\alpha$. Therefore, $E[\bar{B}] \le \frac{\int_{\frac{1}{e}}^{1}\frac{\tilde{B}}{\alpha}d\alpha}{1-\frac{1}{e}} = \frac{e}{e-1}\tilde{B}$.

If $\tilde{z}_j \le \frac{1}{e}$, we have $E[p_j(1-\bar{z}_j)] = p_j \le \frac{1-\tilde{z}_j}{1-\frac{1}{e}}p_j = \frac{e}{e-1}p_j(1-\tilde{z}_j)$; If $\tilde{z}_j > \frac{1}{e}$, we have $E[p_j(1-\bar{z}_j)] = p_j \cdot Pr[\tilde{z}_j \le \alpha] + 0 \cdot Pr[\tilde{z}_j > \alpha] = \frac{e}{e-1}p_j(1-\tilde{z}_j)$. Thus, $E[\bar{B} + \sum_{j=1}^{m} p_j(1-\bar{z}_j)] \le \frac{e}{e-1}\tilde{B} + \frac{e}{e-1}\sum_{j=1}^{m} p_j(1-\tilde{z}_j) \le \frac{e}{e-1}OPT.$ ∎

Note that there are at most m critical values $\tilde{z}_j$ $(j = 1,2,\ldots,m)$ for the threshold parameter α, which implies that the above algorithm can be derandomized by the standard method in polynomial time.

Theorem 2. There exists a $\frac{e}{e-1} \approx 1.582$-approximation algorithm for the DRLPC problem with divisible demand.

4   The DRLPC Problem with Indivisible Demand

In this section, we consider the DRLPC problem with indivisible demand, where each accepted multicast request h_j must be routed in only one of the $k_j + 1$ possible ways with flow d_j. If $p_j = \sum_{j=1}^{m} d_j + 1$ and $k_j = 1$ for each multicast request h_j, which implies that no requests will be rejected in the optimal solution, the DRLPC problem with indivisible demand is exactly the directed ring loading problem in [1]. Since the directed ring loading problem is NP-hard, then the DRLPC problem with indivisible demand is also NP-hard.

In this section, we design two approximation algorithms for the DRLPC problem with indivisible demand. For general integers k_j, we formulate this problem as a mixed integer linear programming as in Section 3. Then, by exploiting the properties of the hyperedge embedding, we round the optimal solution of its relaxation to find a desired feasible solution whose objective value is no more than $3OPT$. When $k_j = 1$ for all multicast requests h_j, combining the method in Section 3 and the rounding method in [1], we find a feasible solution whose objective value is no more than $1.582OPT$.

4.1   General Cases

As in Section 3, the DRLPC problem with indivisible demand can be formulated as the following integer linear programming (ILP), where we need to round more variables x_k^j.

$$\min \quad B + \sum_{j=1}^{m} p_j(1 - z_j)$$

$$\sum_{P_k^j : P_k^j \in \mathcal{P}_i^+} d_j x_k^j \le B, i = 1, 2, \ldots, n;$$

$$\sum_{P_k^j : P_k^j \in \mathcal{P}_i^-} d_j x_k^j \le B, i = 1, 2, \ldots, n;$$

$$\sum_{k=0}^{k_j} x_k^j = z_j, j = 1, 2, \ldots, m;$$

$$x_k^j, z_j \in \{0,1\}, j = 1, 2, \ldots, m, k = 0, 1, \ldots, k_j.$$

Replacing the constraints $x_k^j, z_j \in \{0,1\}$ by $0 \le z_j \le 1$ and $0 \le x_k^j \le 1$, we obtain the relaxation of ILP, which is a linear programming, and can be solved in polynomial time. Let $(\tilde{x}, \tilde{z})$ be the optimal solution for the relaxation of ILP. If $\tilde{z}_j \le 2/3$, set $\bar{z}_j = \bar{x}_0^j = \cdots = \bar{x}_{k_j}^j = 0$, and set $\bar{z}_j = 1$, otherwise. For each multicast request h_j with $\bar{z}_j = 1$, find the minimum $\tau \in \{0, 1, \ldots, k_j\}$ such that $\sum_{P_k^j : P_k^j \in \mathcal{P}_{i_{\tau}^j}} \tilde{x}_k^j \le 1/3$, and set $\bar{x}_\tau^j = 1$, $\bar{x}_k^j = 0$ for $k \in \{0, 1, \ldots, k_j\} \setminus \{\tau\}$.

For example, consider the multicast request $h_j = (1, \{2, 6\})$ satisfying $\tilde{x}_0^j = \tilde{x}_1^j = 1/4$, and $\tilde{x}_2^j = 1/3$, as shown in Figure 4. It is to verify that $\tilde{z}_j = 5/6 > 2/3$, $\tau = 1$, and $\bar{x}_1^j = 1$.

Theorem 3. The objective value of $(\bar{x}, \bar{z})$ is no more than $3OPT$, where OPT denotes the optimal value for the ILP.

Proof. For each multicast request h_j, if $\tilde{z}_j \le 2/3$, then $1 - \bar{z}_j = 1 \le 3(1 - \tilde{z}_j)$ and $\bar{x}_k^j = 0 \le 3\tilde{x}_k^j$ for $k = 0, 1, \ldots, k_j$. For convenience, let $Load(e_i^+, h_j) = \sum_{P_k^j : P_k^j \in \mathcal{P}_{ij}^+} d_j \tilde{x}_k^j$ $(Load(e_i^-, h_j) = \sum_{P_k^j : P_k^j \in \mathcal{P}_{ij}^-} d_j \tilde{x}_k^j$, resp.) denotes the load of link e_i^+ (e_i^-, resp.) resulting from the embedding of h_j in the solution $(\tilde{x}, \tilde{z})$.

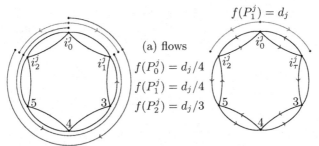

$$f(P_1^j) = d_j$$

(a) flows

$$f(P_0^j) = d_j/4$$
$$f(P_1^j) = d_j/4$$
$$f(P_2^j) = d_j/3$$

(a) The partial solution $\tilde{x}^j$ (b) The partial solution $\bar{x}^j$

Fig. 4. Rounding method

If $\tilde{z}_j > 2/3$, then $1 - \bar{z}_j = 0 \le 3(1 - \tilde{z}_j)$. Clearly, for each link $e_i^+ \notin P_\tau^j$, we have $Load(e_i^+, h_j) = 0 \le 3 \sum_{P_k^j : P_k^j \in \mathcal{P}_{ij}^+} d_j \tilde{x}_k^j$. It also holds for the links such that $e_i^- \notin P_\tau^j$.

If there is a link $e_i^+ \in P_\tau^j$, by the definition of P_τ^j, we have $i \in \{i_0^j, i_0^j + 1, \ldots, i_\tau^j - 1\}$. From the choice of τ, we have $\sum_{P_k^j : P_k^j \in \mathcal{P}_{i_{\tau-1}^j}^+} \tilde{x}_k^j > 1/3$. For every $i \in \{i_0^j, i_0^j + 1, \ldots, i_\tau^j - 1\}$, based on the observation that every embedding of h_j using $e_{i_{\tau-1}^j}^+$ must pass through the link e_i^+, we have

$$\sum_{P_k^j : P_k^j \in \mathcal{P}_{ij}^+} \tilde{x}_k^j \ge \sum_{P_k^j : P_k^j \in \mathcal{P}_{i_{\tau-1}^j}^+} \tilde{x}_k^j > 1/3.$$

Thus, for every link $e_i^+ \in P_\tau^j$,

$$Load(e_i^+, h_j) = d_j \le d_j \cdot 3 \sum_{P_k^j : P_k^j \in \mathcal{P}_{ij}^+} \tilde{x}_k^j = 3 \sum_{P_k^j : P_k^j \in \mathcal{P}_{ij}^+} d_j \tilde{x}_k^j.$$

If there is a link $e_i^- \in P_\tau^j$, by the definition of P_τ^j, we have $i \in \{i_{\tau+1}^j, i_{\tau+1}^j + 1, \ldots, i_0^j - 1\}$. From the choice of τ and the definition of embedding, we have

$$\sum_{P_k^j : P_k^j \in \mathcal{P}_{i_{\tau+1}^j - 1, j}^+} \tilde{x}_k^j = \sum_{P_k^j : P_k^j \in \mathcal{P}_{i_\tau^j j}^+} \tilde{x}_k^j \le 1/3.$$

Combining the fact that every embedding of h_j either pass through $e_{i_{\tau+1}^j - 1}^+$ or $e_{i_{\tau+1}^j}^-$, we have

$$\sum_{P_k^j : P_k^j \in \mathcal{P}_{i_{\tau+1}^j j}^-} \tilde{x}_k^j = \tilde{z}_j - \sum_{P_k^j : P_k^j \in \mathcal{P}_{i_{\tau+1}^j - 1, j}^+} \tilde{x}_k^j > 2/3 - \sum_{P_k^j : P_k^j \in \mathcal{P}_{i_{\tau+1}^j - 1j}^+} \tilde{x}_k^j > 1/3.$$

For every $i \in \{i^j_{\tau+1}, i^j_{\tau+1} + 1, \ldots, i^j_0 - 1\}$, based on the observation that every embedding of h_j using $e^-_{i^j_{\tau+1}}$ must pass through the link e^-_i, we have

$$\sum_{P^j_k : P^j_k \in \mathcal{P}^-_{ij}} \tilde{x}^j_k \geq \sum_{P^j_k : P^j_k \in \mathcal{P}^-_{i^j_{\tau+1}j}} \tilde{x}^j_k > 1/3.$$

Thus, for every link $e^-_i \in P^j_\tau$, we have

$$Load(e^-_i, h_j) = d_j \leq d_j \cdot 3 \sum_{P^j_k : P^j_k \in \mathcal{P}^-_{ij}} \tilde{x}^j_k = 3 \sum_{P^j_k : P^j_k \in \mathcal{P}^-_{ij}} d_j \tilde{x}^j_k.$$

Let $\bar{O}$ be the objective value of $(\bar{x}, \bar{z})$, we have

$$\bar{O} = \max\{\sum_{j=1}^m Load(e^+_i, h_j), \sum_{j=1}^m Load(e^-_i, h_j)\} + \sum_{j=1}^m p_j(1 - \bar{z}_j)$$

$$\leq 3 \max\{\sum_{j=1}^m \sum_{P^j_k : P^j_k \in \mathcal{P}^+_{ij}} d_j \tilde{x}^j_1, \sum_{j=1}^m \sum_{P^j_k : P^j_k \in \mathcal{P}^-_{ij}} d_j \tilde{x}^j_1 | i = 1, 2, \ldots, n\} + 3 \sum_{j=1}^m p_j(1 - \bar{z}_j)$$

$$\leq 3OPT.$$

$\blacksquare$

4.2   A Special Case when $k_j = 1$

In this subsection, we consider the case when $k_j = |S_j| = 1$ for every multicast request h_j, which implies that there are only two ways to embed h_j: clockwise and counterclockwise. From now on, for convenience, we use the request $r_j = (s_j, t_j)$ to denote the multicast request $h_j = (u_j, \{S_j\})$. We will combine the previous methods and the rounding technique in [1,13] to design a $(1.582 + \varepsilon)$-approximation algorithm for the DRLPC problem with $k_j = 1$, where $\varepsilon \in (0, 1)$ is a constant.

We say that the request $r_j = (s_j, t_j)$ is routed the long way if it is routed the longer of the two paths connecting s_j and t_j in the ring R. Using the method in the last subsection, we can find a feasible solution with objective value $\bar{O}$, which satisfies

$$\frac{\bar{O}}{3} \leq OPT \leq \bar{O}. \tag{2}$$

For any $\varepsilon > 0$, let $H = \{r_j \mid d_j > 2\varepsilon\bar{O}/9, j = 1, 2, \ldots, m\}$ be the set of *heavy* requests whose demands are more than $2\varepsilon\bar{O}/9$, and $L = \{r_j \mid d_j \leq 2\varepsilon\bar{O}/9, j = 1, 2, \ldots, m\}$ be the set of *light* requests. Since each long way in the ring uses at least $n/2$ links, we have

Lemma 2. In any optimal routing, there are at most $18/\varepsilon$ accepted requests in H routed in the long way.

Proof. Let q be the number of heavy requests that are routed long-way in the optimal solution. By pigeonhole principle, we have $OPT \geq \frac{q \cdot \frac{n}{2} \cdot \frac{2}{9} \varepsilon \bar{O}}{2n} = q \cdot \frac{\varepsilon \bar{O}}{18} \geq q \cdot \frac{\varepsilon OPT}{18}$, where the second inequality follows from the fact $\bar{O} \geq OPT$. Hence, $q \leq 18/\epsilon$. ∎

For a subset $S_{heavy} \subseteq H$, let $c_{heavy}(e_i^+)$ ($c_{heavy}(e_i^-)$, resp.) denote the congestion of the link e_i^+ (e_i^-, resp.) resulting from routing the requests in S_{heavy} in the long-way. For convenience, we assume the long way of every request $r_j \in \Omega = H \setminus S_{heavy}$ is in the clockwise direction, which implies that $x_1^j = 0$, for every $r_j \in \Omega$. For each subset $S_{heavy} \subseteq H$ with $|S_{heavy}| \leq 18/\varepsilon$, we construct an integer linear programming denoted as $ILP_{S_{heavy}}$ as follows:

$$\min \quad B + \sum_{r_j \in \Omega} p_j(1 - z_j) + \sum_{r_j \in L} p_j(1 - z_j)$$

$$c_{heavy}(e_i^+) + \sum_{j: e_i^+ \in P_1^j, r_j \in L} d_j x_1^j \leq B, i = 1, 2, \ldots, n;$$

$$c_{heavy}(e_i^-) + \sum_{j: e_i^- \in P_0^j, r_j \in \Omega} d_j x_0^j + \sum_{j: e_i^- \in P_0^j, r_j \in L} d_j x_0^j \leq B, i = 1, 2, \ldots, n;$$

$$x_0^j + x_1^j = z_j, r_j \in \Omega \cup L;$$

$$x_1^j = 0, r_j \in \Omega;$$

$$x_0^j, x_1^j, z_j \in \{0, 1\}, r_j \in \Omega \cup L.$$

Replacing the constraints x_0^j, x_1^j, $z_j \in \{0, 1\}$ by $0 \leq x_0^j$, x_1^j, $z_j \leq 1$, we obtain the relaxation of $ILP_{S_{heavy}}$, which is a linear programming, and can be solved in polynomial time. Let $(\tilde{x}, \tilde{z})_{S_{heavy}}$ with objective value $O\tilde{P}T_{S_{heavy}}$ be an optimal solution for the relaxation of the $ILP_{S_{heavy}}$. As in the last section, we randomly choose a threshold α from the uniform distribution over $[1/e, 1]$. For each request $r_j \in \Omega$, if $\tilde{z}_j \leq \alpha$, set $\bar{z}_j = \bar{x}_0^j = \bar{x}_1^j = 0$, and otherwise set $\bar{z}_j = \bar{x}_0^j = 1, \bar{x}_1^j = 0$. For each request $r_j \in L$, if $\tilde{z}_j \leq \alpha$, set $\bar{z}_j = \bar{x}_0^j = \bar{x}_1^j = 0$, and otherwise set $\bar{z}_j = 1, \bar{x}_0^j = \tilde{x}_0^j/\tilde{z}_j, \bar{x}_1^j = \tilde{x}_1^j/\tilde{z}_j$. Let $(\bar{x}, \bar{z})_{S_{heavy}}$ be the resulting solution whose expected objective value is not more than $1.582 O\tilde{P}T_{S_{heavy}}$, following from the proof of Lemma 1. By using the derandomization method, we can find a solution $(\bar{x}, \bar{z})_{S_{heavy}}$ with objective value no more than $1.582 O\tilde{P}T_{S_{heavy}}$ in deterministic polynomial time.

Lemma 3. For any subset $S_{heavy} \subseteq H$ with $|S_{heavy}| \leq 18/\varepsilon$, there is a polynomial rounding procedure of solution $(\bar{x}, \bar{z})_{S_{heavy}}$, that gives an integer solution $(\hat{x}, \hat{z})_{S_{heavy}}$ with objective value no more than $1.582 O\tilde{P}T_{S_{heavy}} + \varepsilon OPT$.

Proof. Note that in the solution $(\bar{x}, \bar{z})_{S_{heavy}}$, the request $r_j \in L$ satisfying $\bar{z}_j = 1$ may be *split*, which means $0 < \bar{x}_0^j, \bar{x}_1^j < 1$. Let $SL(\bar{x}) = \{r_j | 0 < \bar{x}_0^j < 1, r_j \in L\}$ be the set of split light requests, and $IL(\bar{x}) = S_{heavy} \cup \{r_j | \bar{z}_j = 1\} \setminus SL(\bar{x})$ be the set of accepted requests which are not split. Now, we only need to consider the requests in $SL(\bar{x})$, as for each request $r_j \notin SL(\bar{x})$, either r_j is rejected when $\bar{z}_j = 0$, or exactly one of $\bar{x}_0^j$ and $\bar{x}_1^j$ is 1 when $\bar{z}_j = 1$.

Let $\bar{c}_{IL(\bar{x})}(e_i^+)$ ($\bar{c}_{IL(\bar{x})}(e_i^-)$, resp.) be the congestion of e_i^+ (e_i^-, resp.) resulting from routing the requests in $IL(\bar{x})$ in only one way according to $\bar{x}^j$. For the split request set $SL(\bar{x})$, construct the following linear programming $LP_{SL(\bar{x})}$:

$$\min \quad B + \sum_{j=1}^{m} p_j(1 - \bar{z}_j)$$

$$\text{s.t.} \quad \bar{c}_{IL(\bar{x})}(e_i^+) + \sum_{j:e_i^+ \in P_1^j, r_j \in SL(\bar{x})} d_j x_1^j \leq B, i = 1, 2, \ldots, n;$$

$$\bar{c}_{IL(\bar{x})}(e_i^-) + \sum_{j:e_i^- \in P_0^j, r_j \in SL(\bar{x})} d_j x_0^j \leq B, i = 1, 2, \ldots, n;$$

$$x_0^j + x_1^j = 1, r_j \in SL(\bar{x});$$

$$x_1^j, x_0^j \in [0,1], r_j \in SL(\bar{x}).$$

As $\sum_{j=1}^{m} p_j(1 - \bar{z}_j)$ is a constant, it is easy to verify that $LP_{SL(\bar{x})}$ is equivalent to the linear program LP_S (Page 441 in [1]) and $\bar{x}^j$ ($r_j \in SL(\bar{x})$) is an optimal fractional solution to $LP_{SL(\bar{x})}$. Using the LP-rounding method in the proof of Lemma 12 [1], we can convert the fractional solution $\bar{x}^j$ ($r_j \in SL(\bar{x})$) into a feasible integer solution $\hat{x}^j$ ($r_j \in SL(\bar{x})$) with the maximum load among all links increasing at most an additional load of $3/2$ times the maximum demand of the requests in $SL(\bar{x})$. In other words, the objective value of the new solution $(\hat{x}, \bar{z})_{S_{heavy}}$ is no more than $1.582 O\tilde{P}T_{S_{heavy}} + 3/2 \cdot 2\varepsilon\bar{O}/9 \leq 1.582 O\tilde{P}T_{S_{heavy}} + \varepsilon OPT$, where the inequality follows from $\bar{O} \leq 3OPT$. ∎

Theorem 4. When $k_j = 1$ for every r_j, there exists a $(1.582 + \varepsilon)$-approximation algorithm for the DRLPC problem with indivisible demand, where $\varepsilon \in (0,1)$.

Proof. By Lemma 2 and Lemma 3, considering all possible subsets $S_{heavy} \subseteq H$ with $|S_{heavy}| \leq \frac{18}{\varepsilon}$ and choosing the best solution, we can find a feasible solution with objective value no more than $(1.582 + \varepsilon)OPT$. It is easy to verify that the running time is polynomial in m and n, for any fixed constant $\varepsilon > 0$. ∎

5 Conclusions

We have given a 1.582-approximation algorithm for the DRLPC problem with divisible demand where each multicast request can be routed in several ways. It is desired to design a better approximation algorithm. We guess that the DRLPC problem with divisible demand possesses a PTAS.

We have given a 3-approximation algorithm for the DRLPC problem with indivisible demand where each multicast request can only be routed in one way. It is desired to design an efficient algorithm with approximation ratio less than 2. When each hyperedge contains exactly one sink, it is desired to design a PTAS for the DRLPC problem with indivisible demand.

Acknowledgements. The work is supported in part by the Tianyuan Fund for Mathematics of the National Natural Science Foundation of China [No. 11126315], the National Natural Science Foundation of China [Nos. 11301466, 11461081, 61170222], and the Natural Science Foundation of Yunnan Province of China [No. 2014FB114].

References

1. Becchetti, L., Ianni, M.D., Spaccamela, A.M.: Approximation algorithms for routing and call scheduling in all-optical chains and rings. Theoretical Computer Science 287(2), 429–448 (2002)
2. Frank, A., Nishizeki, T., Saito, N., Suzuki, H., Tardos, E.: Algorithms for routing around a rectangle. Discrete Applied Mathematics 40(3), 363–378 (1992)
3. Ganley, J.L., Cohoon, J.P.: Minimum-congestion hypergraph embedding in a cycle. IEEE Transactions on Computers 46(5), 600–602 (1997)
4. Garey, M.R., Johnson, D.S.: Computer and Intractability: A Guide to The Theory of NP-Completeness. W. H. Freeman and Company, San Francisco (1979)
5. Gonzalez, T.: Improved approximation algorithm for embedding hyperedges in a cycle. Information Processing Letters 67(5), 267–271 (1998)
6. Gu, Q., Wang, Y.: Efficient algorithms for minimum congestion hypergraph embedding in a cycle. IEEE Transactions on Parallel and Distributed Systems 17(3), 205–214 (2006)
7. Ho, H., Lee, S.: Improved approximation algorithms for weighted hypergraph embedding in a cycle. SIAM Journal on Optimization 18(4), 1490–1500 (2008)
8. Lee, S., Ho, H.: On minimizing the maximum congestion for weighted hypergraph embedding in a cycle. Information Processing Letters 87(5), 271–275 (2003)
9. Li, G., Deng, X., Xu, Y.: A polynomial time approximation scheme for embedding hypergraph in a cycle. ACM Transactions on Algorithms 5(2), Article No 20 (2009)
10. Li, J., Li, W., Wang, L.: A polynomial time approximation scheme for embedding a directed hypergraph on a weighted ring. Journal of Combinatorial Optimization 24(3), 319–328 (2012)
11. Li, K., Wang, L.: A polynomial time approximation scheme for embedding a directed hypergraph on a ring. Information Processing Letters 97(5), 203–207 (2006)
12. Li, W., Li, J., Guan, L.: Approximation algorithms for the ring loading problem with penalty cost. Information Processing Letters 114(1-2), 56–59 (2014)
13. Schrijver, A., Seymour, P., Winkler, P.: The ring loading problem. SIAM Review 41, 777–791 (1999)
14. Wang, Q., Liu, X., Zheng, X., Zhao, X.: A 2-approximation algorithm for weighted directed hypergraph embedding in a cycle. In: The 4th International Conference on Natural Computation, pp. 377–381 (2008)
15. Wilfong, G., Winkler, P.: Ring routing and wavelength translation. In: Proceedings of the Ninth Annual ACM-SIAM Symposium on Discrete Algorithms, pp. 333–341 (1998)
16. Yang, C., Li, G.: A polynomial time approximation scheme for embedding hypergraph in a weighted cycle. Theoretical Computer Science 412(48), 6786–6793 (2011)

Edge-Colorings of Weighted Graphs

(Extended Abstract)

Yuji Obata and Takao Nishizeki

Kwansei Gakuin University, 2-1 Gakuen, Sanda 669-1337, Japan
{bnb86950,nishi}@kwansei.ac.jp

Abstract. Let G be a graph with a positive integer weight $\omega(v)$ for each vertex v. One wishes to assign each edge e of G a positive integer $f(e)$ as a color so that $\omega(v) \leq |f(e) - f(e')|$ for any vertex v and any two edges e and e' incident to v. Such an assignment f is called an ω-edge-coloring of G, and the maximum integer assigned to edges is called the span of f. The ω-chromatic index of G is the minimum span over all ω-edge-colorings of G. In the paper, we present various upper and lower bounds on the ω-chromatic index, and obtain three efficient algorithms to find an ω-edge-coloring of a given graph. One of them finds an ω-edge-coloring with span smaller than twice the ω-chromatic index.

1 Introduction

An ordinary *edge-coloring* of a graph G assigns different colors to any two adjacent edges. The paper extends the concept to an edge-coloring of a weighted graph.

Let $G = (V, E)$ be a graph with a positive integer weight $\omega(v) \in \mathbb{N}$ for each vertex $v \in V$, where $\mathbb{N}$ is the set of all positive integers. Indeed G may be a multigraph. Figure 1 illustrates such a graph G, in which each vertex v is drawn as a circle and the weight $\omega(v)$ is written in it. One wishes to assign each edge $e \in E$ a positive integer $f(e)$ as a color so that $\omega(v) \leq |f(e) - f(e')|$ for any vertex $v \in V$ and any two edges e and e' incident to v. Such a function $f : E \to \mathbb{N}$ is called an *edge-coloring of a graph G with a weight function ω* or simply an *ω-edge-coloring* of G. An ω-edge-coloring f of a graph G is illustrated in Fig. 1, where $f(e)$ is attached to each edge e.

The *span* $\mathrm{span}(f)$ of an ω-edge-coloring f of a graph G is the maximum integer assigned to edges by f, that is, $\mathrm{span}(f) = \max_{e \in E} f(e)$. An ω-edge-coloring f of G is called *optimal* if $\mathrm{span}(f)$ is minimum among all ω-edge-colorings of G. The ω-edge-coloring in Fig. 1 is optimal, and its span is 7. The span of an optimal ω-edge-coloring of a graph G is called the *ω-chromatic index $\chi'_\omega(G)$* of G. The *ω-edge-coloring problem* is to find an optimal ω-edge-coloring of a given graph.

An ω-edge-coloring often appears in a task scheduling problem [12]. Each vertex v of a graph G represents a processor, while each edge $e = (u, v)$ of G represents a task, which can be executed within a unit time with the cooperation of the two processors represented by vertices u and v. Each processor v needs an

M.S. Rahman and E. Tomita (Eds.): WALCOM 2015, LNCS 8973, pp. 32–43, 2015.

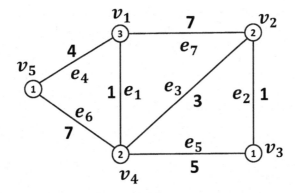

Fig. 1. An optimal ω-edge-coloring f of a graph G

idle time $\omega(v)$ between any two tasks executed by v. Then an optimal ω-edge-coloring of G corresponds to a scheduling with the minimum makespan.

If $\omega(v) = 1$ for every vertex v of a graph G, then an ω-edge-coloring of G is merely an ordinary edge-coloring of G and the ω-chromatic index $\chi'_\omega(G)$ of G is equal to the ordinary chromatic index $\chi'(G)$ of G. Since an ordinary edge-coloring problem is NP-complete [4], the ω-edge-coloring problem is strongly NP-complete and does not look to be solved in polynomial time or in pseudo polynomial time. So it is desired to obtain an efficient approximation algorithm for the ω-edge-coloring problem.

In this paper we present various upper and lower bounds on the ω-chromatic index, and obtain three efficient approximation algorithms for the ω-edge-coloring problem. The first algorithm **Delta** finds an ω-edge-coloring f of a given graph G such that $\mathrm{span}(f) \leq \Delta'_{1\omega}(G) + 1$, where $\Delta'_{1\omega}(G)$ is the maximum "unidirectional ω-edge-degree" of G. The second algorithm **Degenerate** finds an ω-edge-coloring f such that $\mathrm{span}(f) \leq k+1$ for any "k-edge-degenerated graph." **Delta** and **Degenerate** have approximation ratios smaller than two and four, respectively. We also show that an optimal ω-edge-coloring can be easily obtained for a graph G with the maximum degree $\Delta(G)$ at most two. The third algorithm **Factor** first decomposes a given graph G into several subgraphs $G_1, G_2, ..., G_r$, each having the maximum degree at most two, then finds optimal ω-edge-colorings of $G_1, G_2, ..., G_r$, and finally combines them to an ω-edge-coloring of G. The approximation ratio of **Factor** is near to $3/2$ for many graphs.

2   Preliminaries

In this section, we define several terms, present simple lower and upper bounds on the ω-chromatic index, and show that an optimal ω-edge-coloring of a graph G can be easily obtained if $\Delta(G) \leq 2$.

We denote by $G = (V, E)$ a graph with vertex set V and edge set E. G is a so-called *multigraph*, which has no selfloops but may have multiple edges.

We denote by (u, v) an edge joining vertices u and v. Let $n = |V|$ and $m = |E|$ throughout the paper. One may assume that G has no isolated vertex and hence $m \geq n/2$. Let $\omega : V \to \mathbb{N}$ be a *weight function* of G.

We denote by $E(v)$ the set of all edges incident to a vertex v in a graph $G = (V, E)$. The *degree* of a vertex v is $|E(v)|$ and is denoted by $d(v, G)$ or simply $d(v)$. The maximum degree of vertices in G is called the *maximum degree* of G, and denoted by $\Delta(G)$ or simply Δ. Every ω-edge-coloring f of G satisfies

$$1 + (d(v) - 1)\omega(v) \leq \max_{e \in E(v)} f(e)$$

for every vertex v. We thus define the ω-*degree* $d_\omega(v)$ of a vertex v as follows:

$$d_\omega(v) = 1 + (d(v) - 1)\omega(v). \tag{1}$$

The maximum ω-degree of vertices in G is called the *maximum* ω-*degree* $\Delta_\omega(G)$ of G. It should be noted that $\Delta_\omega(G) = \Delta(G)$ if $\omega(v) = 1$ for every vertex v. Clearly $\Delta_\omega(G) \leq \mathrm{span}(f)$ for every ω-edge-coloring f of G. Therefore, the following lower bound holds for the ω-chromatic index $\chi'_\omega(G)$:

$$\Delta_\omega(G) \leq \chi'_\omega(G). \tag{2}$$

The graph G in Fig. 1 satisfies $\Delta_\omega(G) = d_\omega(v_1) = 7$, the ω-edge-coloring f in Fig. 1 has span 7, and hence f is optimal. In Section 3 we will present an upper bound: $\chi'_\omega(G) \leq 2\Delta_\omega(G) - 1$ for every graph G.

Since the weight of a vertex of degree 1 is meaningless, we define the *largest weight* $\omega_l(G)$ of a graph G as follows:

$$\omega_l(G) = \max\{\omega(v) \mid v \in V, d(v) \geq 2\}$$

where $\omega_l(G)$ is defined to be zero if $\Delta(G) = 1$. Since $1 + \omega_l(G) \leq \Delta_\omega(G)$, Eq. (2) implies the following lower bound:

$$1 + \omega_l(G) \leq \chi'_\omega(G) \tag{3}$$

We often denote $\omega_l(G)$ simply by ω_l.

Suppose that a graph G is ordinarily edge-colored by colors 1, 2, ..., c, where $c \geq \chi'(G)$. Replace colors 1, 2, ..., c by 1, $1 + \omega_l$, ..., $1 + (c - 1)\omega_l$, respectively. Then the resulting coloring is an ω-edge-coloring of G. Thus we have an upper bound:

$$\chi'_\omega(G) \leq 1 + (\chi'(G) - 1)\omega_l \tag{4}$$

for every graph G.

V. G. Vizing showed that $\chi'(G) \leq \Delta(G) + 1$ for every *simple* graph G, which has no multiple edges [14,15]; and it is known that such an edge-coloring of G can be found in time $O(mn)$, $O(m\Delta \log n)$ or $O(m\sqrt{n \log n})$ [3]. Therefore, by Eq. (4) we have

$$\chi'_\omega(G) \leq 1 + \Delta \omega_l$$

for every simple graph G, and such an ω-edge-coloring can be found in time $O(mn)$, $O(m\Delta \log n)$ or $O(m\sqrt{n \log n})$.

D. König showed that $\chi'(G) = \Delta(G)$ for every bipartite graph G [14,15], and it is known that such an edge-coloring can be found in time $O(m \log \Delta)$ [1]. Therefore, by Eq. (4) we have

$$\chi'_\omega(G) \leq 1 + (\Delta - 1)\omega_l$$

for every bipartite graph G, and such an ω-edge-coloring can be found in time $O(m \log \Delta)$. Since $\Delta_\omega(G) \leq \chi'_\omega(G)$ by Eq. (2), such an ω-edge-coloring of a bipartite graph G is optimal if

$$\Delta_\omega(G) = 1 + (\Delta - 1)\omega_l. \tag{5}$$

A graph G satisfies Eq. (5) if and only if G has a vertex v such that $d(v) = \Delta$ and $\omega(v) = \omega_l$, and does for example if either G is a regular graph or $\omega(v) = 1$ for every vertex v.

We then present another lower bound $1 + \omega_s(G)$ on $\chi'_\omega(G)$. An *odd cycle* C in G has an odd number of vertices. We define $\omega_s(C)$ as follows:

$$\omega_s(C) = \min\{\omega(u) + \omega(v) \mid \text{vertices } u \text{ and } v \text{ consecutively appear in } C\}.$$

We define $\omega_s(G)$ as follows:

$$\omega_s(G) = \max\{\omega_s(C) \mid C \text{ is an odd cycle in } G\}$$

where $\omega_s(G)$ is defined to be zero if G has no odd cycle. We often denote $\omega_s(G)$ simply by ω_s. One can easily prove the following lemma for a lower bound on $\chi'_\omega(G)$.

Lemma 1. *For every graph G*

$$1 + \max\{\omega_l, \omega_s\} \leq \chi'_\omega(G).$$

We then show that if $\Delta(G) \leq 2$ then $\chi'_\omega(G)$ is equal to the rather trivial lower bound in Lemma 1 and an optimal ω-edge-coloring of G can be easily obtained. One may assume that G is connected. Then G is a path or cycle. If G is a path or an even cycle, then a coloring of G in which edges are colored alternately by 1 and $1+\omega_l$ is an optimal ω-edge coloring and hence $\chi'_\omega(G) = 1+\omega_l$. One may thus assume that G is an odd cycle. Let the vertices $v_1, v_2, ..., v_n$ appear in G in this order, where n (≥ 3) is an odd number. One may further assume that $\omega_s(G) = \omega(v_2) + \omega(v_3)$. Color the consecutive three edges $e_1 = (v_1, v_2), e_2 = (v_2, v_3)$ and $e_3 = (v_3, v_4)$ by 1, $1 + \omega(v_2)$ and $1 + \max\{\omega_l, \omega_s\}$, respectively, and color the remaining $n - 3$ edges alternately by 1 and ω_l. Then the resulting coloring f of G is obviously an ω-edge-coloring of G, and $\text{span}(f) = 1 + \max\{\omega_l, \omega_s\}$. Since $\chi'_\omega(G) \geq 1 + \max\{\omega_l, \omega_s\}$ by Lemma 2.1, f is optimal and $\chi'_\omega(G) = 1 + \max\{\omega_l, \omega_s\}$.

We thus have the following theorem.

Theorem 1. *If G is a graph with $\Delta(G) \leq 2$, then $\chi'_\omega(G) = 1 + \max\{\omega_l, \omega_s\}$ and an optimal ω-edge-coloring of G can be found in linear time.*

For two integers α and β, we denote by $[\alpha, \beta]$ the set of all integers z with $\alpha \leq z \leq \beta$. Let f be an ω-edge-coloring of a graph G. Let $e = (u, v)$ be an edge in G, let e' be an edge adjacent to e, and let x be a vertex to which both e and e' are incident. Then x is u or v. Neither the consecutive $\omega(x)$ integers greater than or equal to $f(e')$ nor those smaller than or equal to $f(e')$ can be assigned to e. Therefore

$$f(e) \notin B(e, e', x)$$

where

$$B(e, e', x) = [f(e') - \omega(x) + 1, f(e') + \omega(x) - 1].$$

Clearly $|B(e, e', x)| = 2\omega(x) - 1$. G has $d(u) - 1$ edges adjacent to e at end u and $d(v) - 1$ edges adjacent to e at end v. Therefore, there are at most $(d(u) - 1)(2\omega(u) - 1) + (d(v) - 1)(2\omega(v) - 1)$ integers that cannot be assigned to e. This number is called a *bi-directional ω-edge-degree* $d_{2\omega}(e, G)$ of e, and hence

$$d_{2\omega}(e, G) = (d(u) - 1)(2\omega(u) - 1) + (d(v) - 1)(2\omega(v) - 1). \tag{6}$$

The maximum bi-directional ω-edge-degree of edges in G is called the *maximum bi-directional edge-degree* $\Delta'_{2\omega}(G)$ of G. Then one can easily prove by induction on the number m of edges that the following upper bound on $\chi'_\omega(G)$ holds for every graph G:

$$\chi'_\omega(G) \leq \Delta'_{2\omega}(G) + 1.$$

For the graph G in Fig. 1, $\Delta'_{2\omega}(G) = d_{2\omega}(e_1) = 19$.

Let f be an ω-edge-coloring of a graph G. Let $e = (u, v)$, let e' be adjacent to e, and let x be a vertex to which both e and e' are incident. Suppose that $f(e') < f(e)$. Then

$$f(e) \notin B_1(e, e', x)$$

where

$$B_1(e, e', x) = [f(e'), f(e') + \omega(x) - 1]$$

and

$$|B_1(e, e', x)| = \omega(x).$$

Therefore we have

$$f(e) \notin \left(\bigcup_{e'} B_1(e, e', u)\right) \bigcup \left(\bigcup_{e'} B_1(e, e', v)\right) \tag{7}$$

where e' runs over every edge such that e' is adjacent to e and $f(e') < f(e)$. In this sense we define the *uni-directional* ω-*edge-degree* $d_{1\omega}(e)$ *of an edge* $e = (u, v)$ as follows:

$$d_{1\omega}(e) = (d(u) - 1)\omega(u) + (d(v) - 1)\omega(v). \qquad (8)$$

The maximum uni-directional ω-degree of edges in G is called the *maximum uni-directional* ω-*edge-degree* $\Delta'_{1\omega}(G)$ of G:

$$\Delta'_{1\omega}(G) = \max_{e \in E} d_{1\omega}(e).$$

Clearly $\Delta'_{1\omega}(G) \leq \Delta'_{2\omega}(G)$ for every graph G. For the graph G in Fig. 1, $\Delta'_{1\omega}(G) = d_{1\omega}(e_1) = 12$. We will show in Section 3 that the following upper bound holds for every graph G:

$$\chi'_\omega(G) \leq \Delta'_{1\omega}(G) + 1.$$

3   Algorithm Delta

In this section we present an algorithm **Delta** to find an ω-edge-coloring f of a given graph G such that $\mathrm{span}(f) \leq \Delta'_{1\omega}(G) + 1$, and show that the approximation ratio of **Delta** is smaller than two.

For an ω-edge-coloring f of a graph $G = (V, E)$, one may assume that

$$f(e_1) \leq f(e_2) \leq \dots \leq f(e_m) \qquad (9)$$

for some numbering $e_1, e_2, \dots, e_m$ of the edges in E. Let $2 \leq i \leq m$, and let $e_i = (u, v)$. We define $E_i(u)$ as follows:

$$E_i(u) = \{e_j \in E(u) \mid 1 \leq j < i\}.$$

We similarly define $E_i(v)$. Then Eq. (7) implies that

$$f(e_i) \geq \max\{ \max_{e_j \in E_i(u)} (f(e_j) + \omega(u)), \max_{e_j \in E_i(v)} (f(e_j) + \omega(v))\}. \qquad (10)$$

Algorithm **Delta** finds a numbering $e_1, e_2, \dots, e_m$ satisfying Eq. (9) and determines $f(e_1), f(e_2), \dots, f(e_m)$ in this order so that $f(e_1) = 1$ and Eq. (10) holds in equality, that is,

$$f(e_i) = \max\{ \max_{e_j \in E_i(u)} (f(e_j) + \omega(u)), \max_{e_j \in E_i(v)} (f(e_j) + \omega(v))\}.$$

Delta is similar to the Dijkstra's shortest path algorithm [2], and its details are as follows, where P is the set of edges e for which $f(e)$ have been decided.

Algorithm. Delta(G, f)

---

for every edge $e \in E$, let $f(e) := 1$; (initialization)
$P := \emptyset$;
for $i1$ **until** m **do**
    {
    let $e_i = (u, v)$ be an edge $e \in E \backslash P$ with minimum $f(e)$;
    $P := P \cup \{e_i\}$; ($f(e_i)$ is decided)
    for every edge $e \in E(u) \backslash P$, let $f(e) := \max\{f(e), f(e_i) + \omega(u)\}$; (update $f(e)$)
    for every edge $e \in E(v) \backslash P$, let $f(e) := \max\{f(e), f(e_i) + \omega(v)\}$; (update $f(e)$)
    }
end for

Clearly **Delta** correctly finds an ω-edge-coloring f of G. For the graph G in Fig. 1, **Delta** finds the coloring f in Fig. 1 such that $\mathrm{span}(f) = 7 = \Delta_\omega(G)$, and hence f happens to be optimal. **Delta** decides $f(e_1)$, $f(e_2)$, ..., $f(e_7)$ in this order for the edge-numbering $e_1, e_2, ..., e_7$ depicted in Fig. 1.

We then prove that the coloring f obtained by **Delta** satisfies

$$\mathrm{span}(f) \leq \Delta'_{1\omega}(G) + 1.$$

Obviously $f(e_1) = 1$ and $\mathrm{span}(f) = f(e_m)$. Let $e_m = (u, v)$, and let j be any integer in $[1, \mathrm{span}(f) - 1]$. Since j is not assigned to e_m by f, either $f(e_i) \leq j \leq f(e_i) + \omega(u) - 1$ for some edge $e_i \in E_m(u)$ or $f(e_i) \leq j \leq f(e_i) + \omega(v) - 1$ for some edge $e_i \in E_m(v)$. Therefore,

$$[1, \mathrm{span}(f) - 1] \subseteq \left(\bigcup_{e_i \in E_m(u)} B_1(e_m, e_i, u) \right) \bigcup \left(\bigcup_{e_i \in E_m(v)} B_1(e_m, e_i, v) \right)$$

and hence

$$\mathrm{span}(f) - 1 \leq (d(u) - 1)\omega(u) + (d(v) - 1)\omega(v) = d_{1\omega}(e_m).$$

We have thus proved

$$\mathrm{span}(f) \leq d_{1\omega}(e_m) + 1 \leq \Delta'_{1\omega}(G) + 1. \tag{11}$$

From Eqs. (1), (2), (8) and (11) we have

$$\begin{aligned}
\mathrm{span}(f) &\leq d_{1\omega}(e_m) + 1 \\
&\leq d_\omega(u) + d_\omega(v) - 1 \\
&\leq 2\Delta_\omega(G) - 1 \\
&\leq 2\chi'_\omega(G) - 1.
\end{aligned}$$

Thus **Delta** has an approximation ratio smaller than two.

Using a binary heap [2], one can implement **Delta** so that it takes time $O(m\Delta \log m)$, similarly as the Dijkstra's shortest path algorithm.

We thus have the following theorem.

Theorem 2. *For every graph* G

$$\chi'_\omega(G) \leq \Delta'_{1\omega}(G) + 1 \leq 2\Delta_\omega(G) - 1.$$

Algorithm **Delta** *finds in time* $O(m\Delta \log m)$ *an* ω-*edge-coloring of* G *such that* $\text{span}(f) \leq \Delta'_{1\omega}(G) + 1$, *and its approximation ratio is smaller than two.*

4   Edge-Degenerated Graphs

It is known that a "k-degenerated graph" has a vertex-coloring with $k+1$ colors [5]. In this section, we define a "k-edge-degenerated graph," and present an algorithm **Degenerate** to find an ω-edge-coloring f of a k-edge-degenerated graph such that $\text{span}(f) \leq k+1$.

A graph G is called *k-edge-degenerated* for a non-negative integer k if G has an edge-numbering $e_1, e_2, ..., e_m$ such that $d_{2\omega}(e_i, G_i) \leq k$ for every index i, $1 \leq i \leq m$, where G_i is a subgraph of G induced by edges $e_1, e_2, ..., e_i$.

Since G_1 consists of a single edge e_1, we have $d_{2\omega}(e_1, G_1) = 0 \leq k$ and hence $\text{span}(f) = 1 \leq k+1$ for an ω-edge-coloring f of G_1 such that $f(e_1) = 1$. This coloring f of G_1 can be extended to an ω-edge-coloring f of G_2 such that $\text{span}(f) \leq k+1$. Repeating such an extention, **Degenerate** obtains an ω-edge-coloring f of $G = G_m$ such that $\text{span}(f) \leq k+1$.

We shall prove that an ω-edge-coloring f of G_i, $i \geq 1$, with $\text{span}(f) \leq k+1$ can be extended to an ω-edge-coloring f of G_{i+1} with $\text{span}(f) \leq k+1$. Let $e_{i+1} = (u, v)$, then an integer $j \in [1, k+1]$ can be chosen as $f(e_{i+1})$ for the extention if and only if

$$j \notin \left(\bigcup_{e_l} B(e_{i+1}, e_l, u)\right) \bigcup \left(\bigcup_{e_l} B(e_{i+1}, e_l, v)\right) \tag{12}$$

where the unions are taken over all edges e_l of G_{i+1} that are adjacent to e_{i+1}, and hence $1 \leq l \leq i$. The cardinality of the set in the right hand side of Eq. (12) is bounded above by

$$d_{2\omega}(e_{i+1}, G_{i+1}) = (d(u, G_{i+1}) - 1)(2\omega(u) - 1) + (d(v, G_{i+1}) - 1)(2\omega(v) - 1),$$

and $d_{2\omega}(e_{i+1}, G_{i+1}) \leq k$ since G is k-edge-degenerated. Therefore, there always exists an integer $j \in [1, k+1]$ which can be chosen as $f(e_{i+1})$, and hence f can be extended to an ω-edge-coloring of G_{i+1} with $\text{span}(f) \leq k+1$.

Algorithm **Degenerate** successively finds ω-edge-colorings of G_1, G_2, ..., $G_m(= G)$ in this order. Indeed it employs a simple greedy technique; when extending an ω-edge-coloring of G_i to that of G_{i+1}, $1 \leq i \leq m-1$, **Degenerate** always chooses, as $f(e_{i+1})$, the *smallest* positive integer j satisfying Eq. (12). For every edge e_l adjacent to e_{i+1} in G_{i+1}, let

$$B(e_{i+1}, e_l, x) = [\alpha(e_l, x), \beta(e_l, x)]$$

where x is u or v, $\alpha(e_l, x) = f(e_l) - \omega(x) + 1$ and $\beta(e_l, x) = f(e_l) + \omega(x) - 1$. Sorting the set $\{\alpha(e_l, x) \mid x \text{ is } u \text{ or } v, e_l \text{ is adjacent to } e_{i+1} \text{ in } G_{i+1}\}$ of $d(u, G_{i+1}) +$

$d(v, G_{i+1})-2$ integers, one can find the smallest integer j above in time $O((d(u)+d(v))\log(d(u)+d(v)))$. Thus **Degenerate** takes time $O(m\Delta\log\Delta)$.

The ω-*edge-degeneracy* $k_\omega(G)$ of a graph G is defined to be the minimum integer k such that G is k-edge-degenerated. Then, similarly as the case of the "vertex-degeneracy" [5], one can compute $k_\omega(G)$ as follows. Let $G_m = G$, and let e_m be an edge e in G_m with minimum $d_{2\omega}(e, G_m)$. Let G_{m-1} be the graph obtained from G_m by deleting e_m, and let e_{m-1} be an edge e in G_{m-1} with minimum $d_{2\omega}(e, G_{m-1})$. Repeating the operation, one can obtain an edge-numbering $e_1, e_2, ..., e_m$ of G, and $k_\omega(G) = \max_{1\le i\le m} d_{2\omega}(e_i, G_i)$.

Using a binary heap, one can compute $k_\omega(G)$ in time $O(m\Delta\log m)$. Using a Fibonacci heap [2], one can improve the time complexity to $O(m\Delta + m\log m)$.

Clearly $k_\omega(G) \le \Delta'_{2\omega}(G)$. Let $\Delta'_{2\omega}(G) = d_{2\omega}(e)$ for an edge $e = (u, v)$, then by Eqs. (1), (2) and (6) we have

$$
\begin{aligned}
\Delta'_{2\omega}(G) + 1 &= (d(u) - 1)(2\omega(u) - 1) + (d(v) - 1)(2\omega(v) - 1) + 1 \\
&= 2(d_\omega(u) + d_\omega(v)) - d(u) - d(v) - 1 \\
&< 4\Delta_\omega(G) \\
&\le 4\chi_\omega(G).
\end{aligned}
$$

We thus have the following theorem.

Theorem 3. *Algorithm* **Degenerate** *finds in time* $O(m\Delta\log\Delta)$ *an* ω-*edge-coloring* f *of a* k-*edge-degenerated graph* G *such that* $\mathrm{span}(f) \le k + 1$. *When* $k = k_\omega(G)$, *the approximation ratio of* **Degenerate** *is smaller than four.*

5   Algorithm Factor

C. E. Shannon showed that every graph G can be edge-colored with at most $3\Delta(G)/2$ colors [13], and it is known that such a coloring can be found in time $O(m(n + \Delta))$ [9]. Therefore, by Eq. (4) we have

$$
\chi'_\omega(G) \le 1 + (3\Delta/2 - 1)\omega_l
$$

for every graph G, and an ω-edge-coloring f of G with $\mathrm{span}(f) \le 1+(3\Delta/2-1)\omega_l$ can be found in time $O(m(n+\Delta))$. In this section we present an algorithm **Factor** of time complexity $O(m\log\Delta)$.

One may assume that a graph $G = (V, E)$ is connected. Our third algorithm **Factor** finds an ω-edge-coloring f of G as follows.
(Step 1)
Partition E into $r(= \lceil\Delta/2\rceil)$ subsets E_i, $1 \le i \le r$, so that the subgraph G_i of G induced by E_i satisfies $\Delta(G_i) \le 2$, and hence G_i consists of vertex-disjoint paths and cycles. (Such a partition is called a *factorization* of G to subgraphs G_i with $\Delta(G_i) \le 2$.)
(Step 2)
Using the algorithm in Section 2, obtain an optimal ω-edge-coloring f_i of G_i for each index i, $1 \le i \le r$.

(Step 3)
Obtain an ω-edge-coloring f of G by combining f_i, $1 \leq i \leq r$.

We now describe the details of these three steps.

[Step 1]
G contains an even number of vertices of odd degree. Join them pairwise by dummy edges, and let G' be the resulting Eulerian graph. (G' may have multiple edges even if G has no multiple edges.) Then the maximum degree $\Delta(G')$ of G' is an even number. More precisely, $\Delta(G') = 2r$ for an integer

$$r = \lceil \Delta(G)/2 \rceil. \tag{13}$$

Let C be an Eulerian circuit of G', which passes through every edge of G' exactly once. We then construct a bipartite graph $B = (V_B, E_B)$ according to the direction of edges in C. The left vertices of B are the vertices of G, and the right vertices are their copies. All edges of B are copies of the edges of G. B has an edge joining a left vertex u and a right vertex v if and only if the Eulerian circuit C passes through an edge (u, v) of G from u to v. (A similar construction of B has appeared in [6].) For every vertex $v \in V$, at most r edges emanate from v in C and at most r edges enter to v. Therefore, $\Delta(B) \leq r$ and hence B has an ordinary edge-coloring with r colors. Let $E_{B_1}, E_{B_2}, ..., E_{B_r}$ be the color classes of the edge-coloring of B. Let $E_1, E_2, ..., E_r$ be the subsets of E which correspond to $E_{B_1}, E_{B_2}, ..., E_{B_r}$, respectively. Then the subgraph G_i, $1 \leq i \leq f$, of G induced by E_i satisfies $\Delta(G_i) \leq 2$ since E_{B_i} is a matching in B.

[Step 2]
By Theorem 1 one can find an optimal ω-edge-coloring $f_i : E_i \to \mathbb{N}$ of G_i in linear time, and f_i satisfies

$$\operatorname{span}(f_i) = 1 + \max\{\omega_l(G_i), \omega_s(G_i)\} \tag{14}$$

for every index i, $1 \leq i \leq r$.

[Step 3]
When combining f_i, $1 \leq i \leq r$, to f, we shift up $f_i(e)$ uniformly for every edge $e \in E_i$. More precisely, let

$$f_i(e) := f_i(e) + \operatorname{span}(f_1) + (\omega_l(G) - 1) + \operatorname{span}(f_2) + (\omega_l(G) - 1)$$
$$+ ... + \operatorname{span}(f_{i-1}) + (\omega_l(G) - 1)$$

for each index i, $2 \leq i \leq r$. Then, simply superimposing $f_1, f_2, ..., f_r$, one can obtain an ω-edge-coloring f of G; $f(e) = f_i(e)$ if $e \in E_i$.
  We then evaluate $\operatorname{span}(f)$ for the coloring f obtained by **Factor**. Clearly

$$\operatorname{span}(f) = \sum_{i=1}^{r} \operatorname{span}(f_i) + (r-1)(\omega_l(G) - 1). \tag{15}$$

Since $\omega_s(G) \le 2\omega_l(G)$ and $\omega_l(G_i) \le \omega_l(G)$ and $\omega_s(G_i) \le \omega_s(G)$ for every index i, $1 \le i \le r$, by Eqs. (13), (14) and (15) we have

$$
\begin{aligned}
\mathrm{span}(f) &\le r(1 + \max\{\omega_l(G), \omega_s(G)\}) + (r-1)(\omega_l(G) - 1) \\
&= 1 + r(\omega_l(G) + \max\{\omega_l(G), \omega_s(G)\}) - \omega_l(G) \\
&\le 1 + (3r - 1)\omega_l(G) \\
&= 1 + (3\lceil \Delta(G)/2 \rceil - 1)\omega_l(G).
\end{aligned}
\tag{16}
$$

Assume now that G satisfies Eq. (5). Then, by Eqs. (5) and (16) we have

$$
\mathrm{span}(f) \le \begin{cases} 3\Delta_\omega/2 + (\omega_l(G) - 1)/2 & \text{if } \Delta \text{ is even;} \\ 3\Delta_\omega/2 + 2\omega_l(G) - 1/2 & \text{otherwise.} \end{cases}
\tag{17}
$$

Since $\Delta_\omega \le \chi'_\omega$ by Eq. (2), the approximation ratio of **Factor** is near to $3/2$. Especially when Δ is even, one may assume that $\Delta \ge 4$, and hence by Eqs. (5) and (17) we have

$$
\mathrm{span}(f) \le (5\Delta_\omega - 2)/3 < 5\chi'_\omega/3
$$

and hence the approximation ratio is smaller than $5/3$.

The most time-consuming part of **Factor** is Step 1, in which one must find an ordinary edge-coloring of a bipartite graph $B = (V_B, E_B)$ with $\Delta(B)$ colors. The coloring can be found in time $O(|E_B| \log \Delta(B))$ [1]. Since $|E_B| = m$ and $\Delta(B) \le r = \lceil \Delta(G)/2 \rceil$, **Factor** takes time $O(m \log \Delta)$.

We thus have the following theorem.

Theorem 4. *For every graph G, algorithm* **Factor** *finds in time $O(m \log \Delta)$ an ω-edge-coloring f of G such that $\mathrm{span}(f) \le 1 + (3\lceil \Delta/2 \rceil - 1)\omega_l$. If $\Delta_\omega(G) = 1 + (\Delta - 1)\omega_l$, then*

$$
\mathrm{span}(f) \le \begin{cases} 3\Delta_\omega/2 + (\omega_l - 1)/2 \text{ if } \Delta \text{ is even;} \\ 3\Delta_\omega/2 + 2\omega_l - 1/2 \text{ otherwise.} \end{cases}
$$

If $\Delta_\omega(G) = 1 + (\Delta - 1)\omega_l$ and Δ is even, then the approximation ratio is smaller than $5/3$.

References

1. Cole, R., Ost, K., Schirra, S.: Edge-coloring bipartite multigraphs in $O(E \log D)$ time. Combinatorica 21(1), 5–12 (2001)
2. Corman, T.H., Leiserson, C.E., Rivest, R.L., Stein, C.: Introduction to Algorithms. MIT Press and McGraw Hill, Cambridge (2001)
3. Gabow, H.N., Nishizeki, T., Kariv, O., Leven, D., Terada, O.: Algorithms for edge-coloring graphs, Tech. Rept. TRECIS 41-85, Tohoku Univ. (1985)
4. Holyer, I.J.: The NP-completeness of edge coloring. SIAM J. on Computing 10, 718–721 (1981)
5. Jensen, T.R., Toft, B.: Graph Coloring Problems. John Wiley & Sons, New York (1995)

6. Karloff, H., Shmoys, D.B.: Efficient parallel algorithms for edge-coloring problems. J. of Algorithms 8(1), 39–52 (1987)
7. McDiamid, C.: On the span in channel assignment problems: bounds, computing and counting. Discrete Math 266, 387–397 (2003)
8. McDiamid, C., Reed, B.: Channel assignment on graphs of bounded treewidth. Discrete Math 273, 183–192 (2003)
9. Nakano, S., Nishizeki, T.: Edge-coloring problems for graphs. Interdisciplinary Information Sciences 1(1), 19–32 (1994)
10. Nishikawa, K., Nishizeki, T., Zhou, X.: Bandwidth consecutive multicolorings of graphs. Theoretical Computer Science 532, 64–72 (2014)
11. Obata, Y., Nishizeki, T.: Approximation Algorithms for Bandwidth Consecutive Multicolorings. In: Chen, J., Hopcroft, J.E., Wang, J. (eds.) FAW 2014. LNCS, vol. 8497, pp. 194–204. Springer, Heidelberg (2014)
12. Pinedo, M.L.: Scheduling: Theory. Springer Science, New York (2008)
13. Shannon, C.E.: A theorem on coloring the lines of a network. J. Math. Physics 28, 148–151 (1949)
14. Stiebitz, M., Scheide, D., Toft, B., Favrholdt, L.M.: Graph Edge Coloring. Wiley, Hoboken (2012)
15. West, D.B.: Introduction to Graph Theory. Prentice-Hall, Englewood Cliffs (1996)

Unit Covering in Color-Spanning Set Model

Ehsan Emamjomeh-Zadeh[1], Mohammad Ghodsi[2], Hamid Homapour[2],
and Masoud Seddighin[2]

[1] Department of Computer Science, University of Southern California,
Los Angeles, USA
`emamjome@usc.edu`
[2] Department of Computer Engineering, Sharif University of Technology,
Tehran, Iran
`ghodsi@sharif.edu`
`{homapour,mseddighin}@ce.sharif.edu`

Abstract. In this paper, we consider two new variants of the unit covering problem in color-spanning set model: Given a set of n points in d-dimensional plane colored with m colors, the *MinCSBC* problem is to select m points of different colors minimizing the minimum number of unit balls needed to cover them. Similarly, the *MaxCSBC* problem is to choose one point of each color to maximize the minimum number of needed unit balls. We show that MinCSBC is NP-hard and hard to approximate within any constant factor even in one dimension. For $d = 1$, however, we propose an $\ln(m)$-approximation algorithm and present a constant-factor approximation algorithm for fixed f, where f is the maximum frequency of the colors. For the MaxCSBC problem, we first prove its NP-hardness. Then we present an approximation algorithm with a factor of $1/2$ in one-dimensional case.

Keywords: Unit Covering, Color-Spanning Set, Computational Geometry, Computational Complexity, Approximation Algorithm.

1 Introduction

Given a set of n points, the **unit covering (UC)** problem is to cover them with minimum number of unit balls. This problem is NP-hard in Euclidean plane [3], while for constant-dimensional cases, it admits polynomial-time approximation schemes (PTAS) [5]. The UC problem has been studied extensively due to wide applications in many fields such as data management in terrains and wireless networks [2,1,4,11].

Recently, many researchers address geometric problems in the situation where the input data is imprecise [8]. One common approach for modeling imprecise points is to use a set of finite points for possible locations that a single imprecise point may appear. In computational geometry, this problem is named *color-spanning set model*. In this model, we are given n points colored with m colors. Points with the same color refer to possible locations of an imprecise

M.S. Rahman and E. Tomita (Eds.): WALCOM 2015, LNCS 8973, pp. 44–52, 2015.

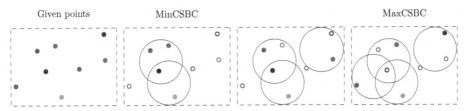

Fig. 1. Three different color selections of given points and their corresponding unit covering (i.e., to cover them using minimum number of unit balls)

point. Imprecise inputs lead to imprecision of output. One of the widely studied problems in this model is to compute bounds on output [8].

In this paper, we discuss the unit covering problem in color-spanning set model. This model can be applied to the case when for each term, at least one of its alternatives should be covered. As an example, consider n different networks and suppose that we want to connect these networks to the Internet with minimum number of access points. Each access point can cover nodes in the certain distance, and a network is connected to the Internet if and only if at least one of its nodes is close enough to an access point, i.e., it is "covered" by the ball corresponding to the access point.

Given a set $P = \{p_1, p_2, ..., p_n\}$ of n points in d dimensions colored with $m \leq n$ colors in $C = \{c_1, c_2, ..., c_m\}$, a **color selection** of P is a subset of m points, one from each color.

We define the following two problems:

Problem 1 (MinCSBC). Find a color selection S of P that minimizes the number of balls in unit covering of S. This problem is called the Minimum Color-spanning Set Ball Covering.

Problem 2 (MaxCSBC). Find a color selection S of P that maximizes the number of balls in unit covering of S. This problem is called the Maximum Color-spanning Set Ball Covering.

In Figure 1, three different color selections for a set of points and their corresponding unit covering depicted.

2   Preliminaries and Notation

Suppose that P is a set of points given as the input for either MinCSBC or MaxCSBC and $C = \{c_1, \ldots, c_m\}$ is the set of colors of elements in P. For each $c_i \in C$, define the *frequency* of c_i as the number of points colored with c_i. We refer to the maximum frequency as f_P (and omit the subscript P when it is clear from the context), so that no more than f_P points are of the same color. Since there exists only one color selection for $f = 1$, we assume that $f \geq 2$.

Except explicitly specified, we restrict our discussion to one-dimensional case. In this case, a unit ball turns into a unit interval.

$$G = \{x_1, x_2, x_3, x_4\} \quad F = \{S_1, S_2, S_3, S_4\}$$
$$S_1 = \{x_1, x_3\} \qquad S_2 = \{x_2, x_3\}$$
$$S_3 = \{x_2, x_3, x_4\} \quad S_4 = \{x_1, x_4\}$$

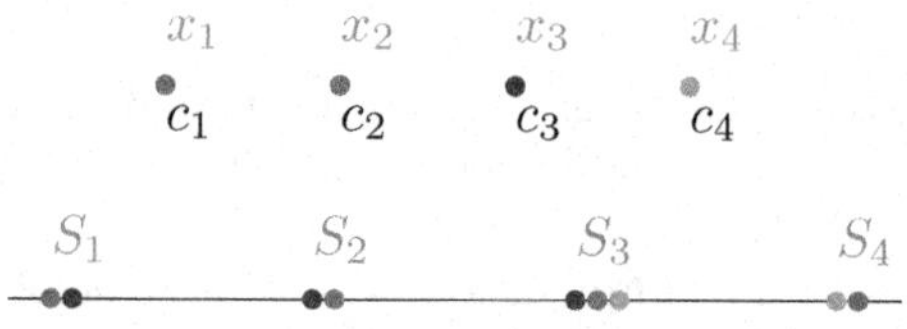

Fig. 2. An instance of MinCSBC

Given a color selection S of P, let $U(S)$ denote the set of intervals in the unit covering of S. Recall that unit covering uses the minimum number of intervals to cover the points. We have the following simple observation:

Observation 1. *There is an optimal covering $U(S)$ such that the left endpoint of each interval corresponds to a point in S and all intervals in $U(S)$ are disjoint.*

Proof. Let $U(S) = \{I_1, I_2, ..., I_k\}$ be the set of intervals sorted by their left endpoints. Starting from I_1, for each interval I_i, shift I_i until its left endpoint lies on the first point that is not covered by intervals $I_1, ..., I_{i-1}$. Clearly, the resulting set of shifted intervals satisfies the required property. ◻

In the rest of the paper, $U(S)$ refers to an optimal covering with the property in Observation 1. We define OPT_{min} and OPT_{max} as the color selection regarding MinCSBC and MaxCSBC, respectively (for explicitly mentioned P or whenever it is clear from the context).

3   MinCSBC

3.1   Hardness of MinCSBC

Theorem 1. *MinCSBC is NP-hard.*

Proof. We show that the problem is NP-hard even in one dimension using a reduction from the Set Cover. Consider an instance of Set Cover with ground set $G = \{x_1, x_2, ..., x_m\}$, covering family $F = \{S_1, S_2, ..., S_k\}$ and $OPT_{sc} \subseteq F$ as the optimal cover. For each $x_j \in G$, consider color c_j in MinCSBC instance, and for each subset S_i, specify a unit segment on x-axis $Cell_i$ in a way that the distance between the endpoints of different segments is more than 1. Next, for each element $x_j \in S_i$, put a point with color c_j in $Cell_i$ as illustrated in Figure 2.

Suppose that P is the set of created points in the MinCSBC instance. Since the distance between each two cells is more than 1, each interval in $U(OPT_{min})$

covers points in only one cell. Moreover, if two intervals intersect the same cell, it is possible to replace them with one interval which includes the whole cell contradicting the minimality of OPT_{min}. We return the sets whose corresponding cells in MinCSBC instance intersect the intervals in $U(OPT_{min})$. Let R denote the set of returned subsets. Since at least one point of each color is covered by intervals in $U(OPT_{min})$, R is a feasible solution to the Set Cover instance. Consequently, we have $|U(OPT_{min})| \geq |OPT_{sc}|$.

On the other hand, consider the cells corresponding to subsets in OPT_{sc} and find a color selection using points in these cells. Since OPT_{sc} covers all elements in G, such a color selection exists. Obviously, this color selection can be covered by $|OPT_{sc}|$ unit intervals, so $|U(OPT_{min})| \leq |OPT_{sc}|$.

As a consequence, $|U(OPT_{min})| = |OPT_{sc}|$ which, keeping in mind NP-hardness of the Set Cover problem, implies MinCSBC to be NP-hard as well. □

Note that the Set Cover problem is NP-hard even when the frequency of each $x_j \in G$ is at most 2, i.e., x_j appears in at most two subsets in F. Therefore, using the same reduction for this restricted version of the Set Cover problem, we can claim that one-dimensional MinCSBC is NP-hard even when $f = 2$.

Furthermore, it can be concluded from the above reduction that any constant-factor approximation algorithm for MinCSBC yields an approximation for the Set Cover problem with the same factor. Taking into account that there is no approximation algorithm with a constant factor for the Set Cover problem unless P=NP, we obtain the following corollary.

Corollary 1. *MinCSBC admits no polynomial-time approximation algorithm with a constant factor unless $P = NP$.*

3.2 Approximation Algorithms for MinCSBC

Theorem 2. *There is an $\ln(m)$-approximation algorithm for MinCSBC in one dimension.*

Proof. Let $I = \{I_1, I_2, ..., I_n\}$ be the set of intervals, where I_i is the unit interval whose left endpoint lies on point p_i. By Observation 1, for any color selection, there exists an optimal covering using intervals in I. Therefore, the MinCSBC problem is basically to find $\mathcal{I} \subseteq I$ of the minimum size such that $\mathcal{I}$ covers at least one point of each color, and then choose a color selection from the covered points.

In order to represent this problem with the Set Cover problem:

- let G be the set of all colors;
- for each $I_i \in I$, define a subset of G containing the colors covered by I_i.

There is a well-known greedy approximation algorithm for the Set Cover problem with factor $\ln(\Delta)$, where Δ is the maximum size of the subsets in the covering family [6]. Since in above reduction, the size of each covering subset is at most m, applying this $\ln(\Delta)$-approximation algorithm results in an $\ln(m)$-approximation algorithm for MinCSBC.

□

Theorem 3. *There is a 2f-approximation algorithm for MinCSBC in one dimension.*

Proof. First, find a set I of unit intervals, with $|I| \leq n$, satisfying the following two conditions.

- All of the n points are covered.
- No two intervals in I intersect, i.e., all the intervals are disjoint.

Then, consider the following problem which is similar to MinCSBC but with an additional restriction.

The Modified MinCSBC Problem: Find a subset of I with minimum size that covers at least one point of each color.

Lemma 1. $|S_I| \leq 2|U(OPT_{min})|$, *where S_I is the optimal solution to the modified MinCSBC problem with respect to I.*

Proof. Since intervals lie on x-axis, for each $u \in U(OPT_{min})$, there exists $I' \subseteq I$ with $|I'| \leq 2$ such that I' covers all points covered by u. Consequently, by replacing each interval in $U(OPT_{min})$ with at most two intervals in I, one can obtain a family $S' \subseteq I$ of intervals covering all points that are covered by $U(OPT_{min})$. Since $|S_I| \leq |S'|$ and $|S'| \leq 2|U(OPT_{min})|$, we obtain that $|S_I| \leq 2|U(OPT_{min})|$. □

As a consequence of Lemma 1, any f-approximation algorithm for modified MinCSBC yields an approximation algorithm for MinCSBC with factor $2f$.

Note that, an instance of the modified MinCSBC problem can be considered as an instance of the Set Cover problem. So, the f-approximation algorithm for Set Cover [10] leads to an $2f$-approximation algorithm for MinCSBC. □

4   MaxCSBC

4.1   Hardness of MaxCSBC

Theorem 4. *MaxCSBC is NP-hard.*

Proof. We show that MaxCSBC is NP-hard even in one dimension for $f > 2$ by reduction from a restricted version of normal CNF 3SAT in which each variable occurs at most twice in positive form and once in negative form. This problem which we name *3-Occurrence SAT* is known to be NP-complete [1].

Given an instance of 3-Occurrence SAT, for each variable x_i, specify two disjoint segments $Cell_i$ and $\overline{Cell_i}$, each of length 3. These two segments correspond

---

[1] It is worth mentioning that in 3-Occurrence SAT problem, if each clause has to have exactly 3 distinct variables, the formula is always satisfiable and thus, the problem is not hard anymore. However, we allow clauses to have less than 3 variables, see [9] for more details.

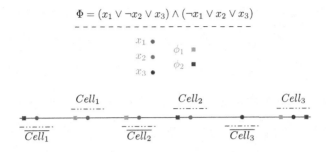

Fig. 3. Reduction to an instance of MaxCSBC

to x_i and $\overline{x_i}$, respectively. The cells have to be placed in a way that no two cells intersect. Next, assign distinct colors to each variable and each clause. Denote the color assigned to variable x_i by c_i and the color which is corresponding to clause ϕ_j by c'_j.

For each variable x_i, place two points colored with c_i at the middle of $Cell_i$ and $\overline{Cell_i}$. Since these two points are the only points which are colored with c_i, any color selection must include at least one of them. Selecting the middle point of a cell is interpreted as setting the corresponding literal to 0. In other words, if the middle point of $Cell_i$ is selected, then $x_i = 0$, while selecting the middle point of $\overline{Cell_i}$ means that $\overline{x_i} = 0$ or, equivalently, $x_i = 1$.

Next, for each clause ϕ_j, place three points colored with c'_j in the cells corresponding to its literals at the distance of $\frac{3}{4}$ from the middle. Note that at most two clause-points are placed in the same cell (by the definition of 3-Occurrence SAT). If two points corresponding to different clauses are placed in the same cell, they have to be placed at the different sides of the middle point. See the example depicted in Figure 3.

Lemma 2. *The instance of 3-Occurrence SAT is satisfiable if and only if there exists a color selection for the corresponding MaxCSBC instance in which the distance between any pair of points is greater than 1.*

Proof. Suppose that a color selection S exists with the distance between each two points in S more than 1. For each color c_i, if the middle point of $Cell_i$ is in S, set $x_i = 0$, otherwise (i.e., if the middle point of $\overline{Cell_i}$ is in S) set $\overline{x_i} = 0$. Note that for each clause ϕ_j, there is one point p_j of color c'_j in S. Since the distance between p_j and the middle point of the cell that p_j lies in, is less than 1, this middle point cannot be in S, and so there exists a literal in ϕ_j whose value is 1.

On the other hand, we prove that any satisfying assignment for the 3-Occurrence SAT instance can result in a color selection in which the distance between any pair of points is greater than 1. For each variable x_i, the middle point of either $Cell_i$ or $\overline{Cell_i}$ should be chosen in order to have a color selection. If $x_i = 0$, select the middle point of $Cell_i$. Otherwise, select the middle point of $\overline{Cell_i}$. Since for each clause ϕ_j, there is at least one literal in ϕ_j satisfying it, a cell containing a point with color c'_j exists whose middle point is not selected, so it is possible to select a point of color c'_j. □

Observation 2. *In an instance of MaxCSBC, $|U(OPT_{max})| = m$ if and only if there exists a color selection in which the distance between any pair of points is greater than 1.*

Taking into account Observation 2, we can claim that the instance of 3-Occurrence SAT is satisfiable if and only if $|U(OPT_{max})| = m$. Notice that $|U(OPT_{max})|$ is never strictly larger than m. □

4.2 Approximation Algorithm for MaxCSBC

Now, we present an $O(n \log n)$-time approximation algorithm with factor $\frac{1}{2}$ for MaxCSBC in one dimension.

Algorithm 1. Approximation Algorithm for MaxCSBC

Input: A set P of n points colored with m colors
Output: A color selection of P
1: $\mathcal{M} = \emptyset$, $T = \emptyset$, $T' = \emptyset$
2: **while** $|\mathcal{M}| < n$ **do**
3:      $p = $ the leftmost point in $P \setminus \mathcal{M}$
4:      $\mathcal{M} = \mathcal{M} \cup \{p\}$
5:      $T = T \cup \{p\}$
6:      **for each** point $q \in P \setminus \mathcal{M}$ with the same color as p **do**
7:          $\mathcal{M} = \mathcal{M} \cup \{q\}$
8:      **end for**
9:      **for each** point $q \in P \setminus \mathcal{M}$ where $dist(p, q) \leq 1$ **do**
10:          $\mathcal{M} = \mathcal{M} \cup \{q\}$
11:      **end for**
12: **end while**
13: **for each** color c with no candidate in T **do**
14:     insert an arbitrary point of color c in T'
15: **end for**
16: **return** $T \cup T'$

Theorem 5. *Algorithm 1 is a $\frac{1}{2}$-approximation algorithm.*

Proof. Clearly $T \cup T'$ needs at least $|T|$ unit intervals to be covered since the distance between any two points in T is greater than 1. By Observation 1, all the intervals in $U(OPT_{max})$ are disjoint and the left endpoint of any interval in $U(OPT_{max})$ is one of the input points. Let $\mathcal{T}$ be a set of these points. We claim that $|T| \geq \frac{|\mathcal{T}|}{2}$.

To this end, we show that by adding p to T, at most two points of $\mathcal{T}$ that have been left *unmarked*[2] yet, can be inserted to $\mathcal{M}$. Note that when we add point p to T,

---

[2] A point is *marked* if it is in set $\mathcal{M}$ and *unmarked* otherwise.

- only one of the points in $\mathcal{T}$ can be of the same color with p because all points in $\mathcal{T}$ have different colors;
- there is at most one unmarked point in $\mathcal{T}$ within the distance at most 1 to p as any two points in $\mathcal{T}$ are within the distance greater than 1. Recall that p is the left-most unmarked point, so all points at the left-hand side of p have been already marked.

Therefore, by adding p to T, at most two unmarked points of $\mathcal{T}$ might be inserted to $\mathcal{M}$. At the end of the algorithm, all points in $\mathcal{T}$ are marked, so $|T| \geq \frac{|\mathcal{T}|}{2}$. Thus the output of Algorithm 1 is within a factor $\frac{1}{2}$ of the optimal solution.   □

5   Conclusion

In this paper, we investigated on the problem of unit covering in the color-spanning set model.

For MinCSBC, we showed the NP-hardness and also hardness of approximation within any constant factor. In addition, we presented an $\ln(m)$-approximation algorithm for this problem and also an approximation algorithm for one-dimensional case with factor $2f$. While one-dimensional MinCSBC is NP-hard even when $f = 2$, the latter algorithm results in a constant-factor approximation algorithm for fixed f.

For MaxCSBC, we proved the NP-hardness and proposed an approximation algorithm with constant factor 2 when $d = 1$.

Here are some open questions.

1. Is there any algorithm with approximation factor better than $2f$ for MinCSBC? For special case when $f = 2$, the proposed algorithm leads to a 4-approximation algorithm. In this case ($f = 2$), a reduction from the Vertex Cover problem shows that assuming the Unique Game Conjecture, the problem does not admit any approximation algorithm with a factor better than 2 [7]. There is still a gap between these lower bounds and our factor, however.
2. Is there any approximation algorithm for MinCSBC and MaxCSBC in higher dimensions?
3. Having considered our reduction from 3-Occurrence SAT, we showed that MaxCSBC is NP-hard for $f > 2$ even in one-dimensional case, but the complexity of the problem for $f = 2$ is still unknown.

References

1. Claude, F., Das, G.K., Dorrigiv, R., Durocher, S., Fraser, R., López-Ortiz, A., Nickerson, B.G., Salinger, A.: An improved line-separable algorithm for discrete unit disk cover. Discrete Mathematics, Algorithms and Applications 2(01), 77–87 (2010)

2. Das, G.K., Fraser, R., Lòpez-Ortiz, A., Nickerson, B.G.: On the discrete unit disk cover problem. In: Katoh, N., Kumar, A. (eds.) WALCOM 2011. LNCS, vol. 6552, pp. 146–157. Springer, Heidelberg (2011)
3. Fowler, R.J., Paterson, M.S., Tanimoto, S.L.: Optimal packing and covering in the plane are np-complete. Information Processing Letters 12(3), 133–137 (1981)
4. Funke, S., Kesselman, A., Kuhn, F., Lotker, Z., Segal, M.: Improved approximation algorithms for connected sensor cover. Wireless Networks 13(2), 153–164 (2007)
5. Hochbaum, D.S., Maass, W.: Approximation schemes for covering and packing problems in image processing and vlsi. Journal of the ACM (JACM) 32(1), 130–136 (1985)
6. Johnson, D.S.: Approximation algorithms for combinatorial problems. In: Proceedings of the Fifth Annual ACM Symposium on Theory of Computing, pp. 38–49. ACM (1973)
7. Khot, S., Regev, O.: Vertex cover might be hard to approximate to within 2- ε. Journal of Computer and System Sciences 74(3), 335–349 (2008)
8. Löffler, M.: Data imprecision in computational geometry (2009)
9. Tovey, C.A.: A simplified np-complete satisfiability problem. Discrete Applied Mathematics 8(1), 85–89 (1984)
10. Vazirani, V.V.: Approximation algorithms, pp. 118–119. Springer (2001)
11. Yang, D., Misra, S., Fang, X., Xue, G., Zhang, J.: Two-tiered constrained relay node placement in wireless sensor networks: efficient approximations. In: 2010 7th Annual IEEE Communications Society Conference on Sensor Mesh and Ad Hoc Communications and Networks (SECON), pp. 1–9. IEEE (2010)

Compact Encodings and Indexes for the Nearest Larger Neighbor Problem

Seungbum Jo[1], Rajeev Raman[2], and Srinivasa Rao Satti[1]

[1] Seoul National University, Seoul, South Korea
sbcho@tcs.snu.ac.kr, ssrao@cse.snu.ac.kr
[2] University of Leicester, Leicester, UK
r.raman@leicester.ac.uk

Abstract. Given a d-dimensional array, for any integer $d > 0$, the *nearest larger value* (NLV) query returns the position of the element which is closest, in L_1 distance, to the query position, and is larger than the element at the query position. We consider the problem of preprocessing a given array, to construct a data structure that can answer NLV queries efficiently. In the 2-D case, given an $n \times n$ array A, we give an asymptotically optimal $O(n^2)$-bit encoding that answers NLV queries in $O(1)$ time. When A is a binary array, we describe a simpler $O(n^2)$-bit encoding that also supports NLV queries in $O(1)$ time. Using this, we obtain an index of size $O(n^2/c)$ bits that supports NLV queries in $O(c)$ time, for any parameter c, where $1 \le c \le n$, matching the lower bound. For the 1-D case we consider the *nearest larger right value* (NLRV) problem where the nearest larger value to the right is sought. For an array of length n, we obtain an index that takes $O((n/c) \log c)$ bits, and supports NLRV queries in $O(c)$ time, for any any parameter c, where $1 \le c \le n$, improving the earlier results of Fischer et al. and Jayapaul et al.

1 Introduction and Motivation

We consider cases of the following general problem. We are given a d-dimensional array A consisting of (possibly not all distinct) items from an ordered universe. After preprocessing A we are given a series of queries, each of which specifies an element of A, and our objective is to return the element of A nearest to the query element that is strictly larger than the query element. One may also restrict that the answer to the query comes from some particular sub-array (e.g. a quadrant) of A. Specifically, we consider the two queries below:

NLRV: Given a 1-D array A and an index i, returns the first larger element to i's right, i.e., returns $\min\{j > i | A[j] > A[i]\}$ (and is undefined if this set is empty). The query NLLV is defined analogously to i's left.
NLV: Given a d-dimensional array A and an index $p = (i_1, i_2 \ldots, i_d)$, returns an index $q = (i'_1, i'_2 \ldots, i'_d)$ such that $A[q] > A[p]$ and the distance between p and q, $dist(p,q) = |i_1 - i'_1| + |i_2 - i'_2| + \cdots + |i_d - i'_d|$, is minimized (note that we use the L_1 metric). In case of many equidistant larger values, ties can be broken arbitrarily. If there is no larger value, then it is undefined.

M.S. Rahman and E. Tomita (Eds.): WALCOM 2015, LNCS 8973, pp. 53–64, 2015.
© Springer International Publishing Switzerland 2015

Encoding and indexing models. We consider these problems in two different models that have been studied in the succinct data structures literature, namely the *indexing* and *encoding* models. In both these models, the data structure is created after preprocessing A. In the indexing model, the queries can be answered by probing the data structure as well as the input data, whereas in the encoding model, the query algorithm cannot access the input data.

Previous Work and Motivation. The off-line version of this problem: given A, to compute nearest larger values for *all* entries of A (the ANLV problem), has been studied previously[1]. In the 1-D case, Berkman et al. [3] noted that the best highly-parallel solutions to a number of tasks including answering range minimum queries, triangulating monotone polygons and matching parentheses, are obtained by reducing to the ANLV problem, and efficient parallel solutions to the ANLV problem were also given by the same authors. A number of plausible applications, and algorithms, for the ANLV problem in 2 and higher dimensions were given by Asano et al. [1], and time-space tradeoffs for the 1-D case were given by Asano and Kirkpatrick [2].

Fischer et al. [10] considered the problem of supporting NLRV and NLLV in the 1-D case, and showed how a data structure supporting these two queries is essential to a space-efficient compressed suffix tree. They also considered the problem of supporting NLRV and NLLV in the indexing model, and gave a space-time tradeoff (the precise result is given later). Fischer [9] gave a structure in the encoding model that uses $2.54n + o(n)$ bits and supports NLRV and NLLV queries in $O(1)$ time.

Jayapaul et al. [12] considered the problem of encoding and indexing NLV in the 2-D case. Below, we describe the directly relevant results from their work.

Our results. We obtain new results for encoding and/or indexing 1-D and 2-D nearest larger value queries. In all the 2-D results we assume L_1 distances.

- We show that 2-D NLV can be encoded in the asymptotically optimal $O(n^2)$ bits in the *general* case. Jayapaul et al. showed this only for the case where all elements of A are distinct. Distinctness is a strong assumption in these kinds of problems. For example, in the 1-D case with distinct values, NLRV and NLLV can both be trivially encoded by the Cartesian tree (giving a $2n - O(\log n)$ bit encoding). By contrast, if we do not assume distinctness, the optimal space is about $2.54n$ bits, and the data structure achieving this bound is also more complex [9]. Also, Asano et al. [1] remark that the ANLV problem for any dimension is "simplified considerably" if one assumes distinctness. In fact, for the general case, Jayapaul et al. were only able to give an encoding with size $\Theta(n^2 \log \log n)$ bits and $O(1)$ query time.

---

[1] The terminology varies considerably. Berkman et al. studied the all nearest *smaller* values (ANSV) problem, which is symmetric to the ANLV problem. The previous/next smaller value (PSV/NSV) problems of Fischer et al. are symmetric to the NLLV/NLRV problems. Asano et al. and Jayapaul et al. call the NLV problem the *nearest larger neighbour* (NLN) problem: we consider the term "neighbour" to be mildly misleading, as the answer may not be a neighbour of the query element in A.

We also remark that in 1-D case, the NLRV and NLLV problems are closely connected to the *range minimum query (RMQ)* problem (see e.g. [9]), another problem of wide interest. In the 1-D case, there is no asymptotic difference between the encoding complexity of RMQ and NLRV/NLLV. The 2-D RMQ problem has received a great deal of attention lately [7,6,5,4]. It is known that any 2-D RMQ encoding takes $\Omega(n^2 \log n)$ bits [7,6]; thus, this result shows that NLV is different from RMQ in the 2-D encoding scenario.

- For the special case where A comprises 0-1 values, we provide an optimal trade-off. Specifically, given an $n \times n$ array A and any $1 \leq c \leq n$, we describe an index of size $O(n^2/c)$ bits that can answer NLV queries in $O(c)$ time.
- For indexing 1-D NLRV and NLLV, we give an index that takes $O((n/c) \log c)$ bits and answers these queries in $O(c)$ time. This improves the two previous trade-offs for this problem:
  - Fischer et al. [10] showed that, for any $1 \leq c, \ell \leq n$, one can use:

$$O\left(\frac{n}{c} \log c + \ell \frac{n \log \log n}{\log n} + \frac{n \log n}{c^\ell}\right)$$

    bits of space and answer queries in $O(c\ell)$ time. As given, they are unable to go below $O(n \log \log n / \log n)$ space, and use more space than we do whenever $c = \omega(\log n)$. To attain $O((n/c) \log c)$ space for $c = O(\log n)$, observe that for $c = (\log n)^{\Omega(1)}$, one can choose $\ell = O(1)$ and obtain $O(c)$ time. For smaller values of c, the middle term in the space usage will never dominate for reasonable values of ℓ (clearly, we must always choose $c \geq 2$ and $\ell = O(\log \log n)$ in this range) and it suffices (and is optimal) to choose $\ell = O(\log_c \log n) = O(\log \log n - \log \log c)$. Thus, for any $c = O(\log n)$ their running time for space $O((n/c) \log c)$ is $O(c(\log \log n - \log \log c))$, and our solution is better for small enough c.
  - Jayapaul et al. [12] gave a solution that uses $O((n/c) \log c + (n \log n)/c^2)$ bits and $O(c)$ time; this space usage equals ours for $c = \Omega(\log n / \log \log n)$ but is worse otherwise.

Our solution is a minor modification of the approach of Fischer et al.: we replace a data structure they use by a *slower* one to obtain the result.

We assume a standard word RAM model with word size $\Theta(\lg n)$, and we count space in terms of the number of bits used.

2 Indexing NLRV on 1-dimensional Arrays

In this section, we give a new time-space trade-off for the indexing model for supporting NLRV queries in 1-D arrays. The approach follows closely the proof of Fischer et al. [10], which in turn adapts ideas from Jacobson's representation of balanced parentheses sequences [11], and is given in full for completeness.

We begin with some definitions. Given a string X over an alphabet Σ, define the following operations:

- $\mathsf{rank}_\alpha(X, i)$ returns the number of occurrences of α in the first i positions of X, for any $\alpha \in \Sigma$.
- $\mathsf{select}_\alpha(X, i)$ returns the position of the ith α in X, for any $\alpha \in \Sigma$.

Lemma 1 ([13]). *Given a string X over $\{0,1\}$ with length n, containing m 1s, it can be represented in $O(m \log(n/m))$ bits such that rank_1 and select_1 can be supported in $O(n/m)$ time.*

Lemma 2 ([10]). *There exists a data structure in the encoding model that solves NLRV queries using $2n + o(n)$ bits in $O(1)$ time.*

Theorem 1. *Given a 1-D array A of size n, there exists a data structure which supports NLRV queries in the indexing model in $O(c)$ time using $O((n/c) \lg c)$ bits for any parameter $2 \le c \le n$.*

Proof. Divide A into n/c blocks of size c. For any value $1 \le i \le n$, if i and $\mathsf{NLRV}(i)$ are in the same block, say that i is a *near* value, otherwise say that i is a *far* value. Consider a block B and suppose that one or more of its far values have an NLRV in a block B'. Then the leftmost far value in B whose NLRV is in B' is called a *pioneer*, and its NLRV is called its *match*. It is known that there are $O(n/c)$ pioneers in A [11].

We maintain a bit-vector V in which the i-th bit is a 1 if $A[i]$ is a pioneer or a match of one, and 0 otherwise. This bit-vector has length n and weight $O(n/c)$ so by Lemma 1, we can store it in $O((n/c) \lg c)$ bits and perform rank/select queries on it in $O(c)$ time. Next, we take the sub-sequence S_P consisting of all pioneers and their matches. This subsequence is of length $O(n/c)$. We represent this sequence using Lemma 2 using $O(n/c)$ bits, to support NLRV queries in $O(1)$ time. We claim that for any pioneer in the list, its NLRV in the sequence of pioneers/matches is the same as its NLRV in the original sequence. Suppose that this claim is not true. This means there is a pioneer i_p such that $\mathsf{NLRV}(i_p)$ is the value between i_p and the match of i_p. It cannot be the case that i_p and $\mathsf{NLRV}(i_p)$ are in the same block, since i_p is a far value. If i_p and $\mathsf{NLRV}(i_p)$ are in different blocks, then $\mathsf{NLRV}(i_p)$ is the match of i_p. So the claim is true.

To answer the query $\mathsf{NLRV}(i)$, we first check to see if the answer is in the same block as i taking $O(c)$ time. If so, we are done. Else, (assuming wlog that $A[i]$ is not a pioneer value) we find the first pioneer p_i before position i by doing rank/select on V. As $A[i] < A[p_i]$, $\mathsf{NLRV}(i)$ is less than or equal to the match of p_i. Since i is the far value in this case, $\mathsf{NLRV}(i)$ and $\mathsf{NLRV}(p_i)$ are in the same block. We find $\mathsf{NLRV}(p_i)$ using the NLRV encoding of S_P and find the corresponding position i_{ap} in A using rank/select on V. Finally we scan left from i_{ap} to find $\mathsf{NLRV}(i)$. The overall time taken to answer the query is $O(c)$. ☐

3   NLV on 2-D Binary Arrays

In this section, we first give an optimal encoding for NLV, and using this obtain an almost optimal trade-off for an NLV index for a 2-D binary array. We use the following lemma:

Lemma 3 ([8]). *Given a string X of length n over an alphabet Σ, $|\Sigma| = O(1)$, there is an encoding of X using $O(n)$ bits, that supports rank_α and select_α in $O(1)$ time, for any $\alpha \in \Sigma$.*

Theorem 2. *There is an encoding for an $n \times n$ binary array A which takes $O(n^2)$ bits and supports NLV queries in $O(1)$ time.*

Proof. We compute the NLV by computing the nearest larger value in all four *quadrants* induced by a vertical and a horizontal line that pass through the query position, and then returning the closest of these four positions to the query. Thus, it is enough to describe a structure that supports NLV in the upper-right quadrant. Given a query position $p = (i, j)$, let $q = (i', j')$ be its NLV in the upper-right quadrant. There are four possibilities: (1) $A[p] = 1$ (q is not defined); (2) $i = i'$; (3) $j = j'$; or (4) $i < i'$ and $j < j'$. The encoding will simply store, for each position, which case it belongs to. Then, in Case (2), we can find its answer by following the positions $(i, j + k)$, for $k = 1, 2, \ldots$ (i.e., elements in the same row) till we reach a position that belongs to Case (1). Also, one can easily show that all the intermediate elements also belong to Case (2). Analogously, in Case (3), we follow the positions in the same column until we reach a position that belongs to Case (1). Finally, in Case (4), we first follow the positions $(i+k, j+k)$, for $k = 1, 2, \ldots$ till we reach the first position $(i + \ell, j + \ell)$ that does not belong to Case (4), and then find the answer using the algorithm for Case (2) or (3), or return the position $(i + \ell, j + \ell)$ if it belongs to Case (1). To support the queries faster, we build rank/select structures (Lemma 3) for the encoding of each row, each column and each diagonal. The total space usage is clearly $O(n^2)$ bits. Now, queries can be supported in constant time by using rank/select to jump to the appropriate positions as described in the above procedures. □

Now we describe an index for a given 2-D binary array. We begin by introducing some notation that will be used later. Suppose we divide an $n \times n$ array A into blocks of size $c \times c$, for $0 < c \leq n$, and divide each block into c sub-blocks of size $\sqrt{c} \times \sqrt{c}$. We define an (i, j)-block as the sub-array $A[(i - 1)c + 1 \ldots ic][(j - 1)c \ldots jc]$ and an (i, j, k, l)-sub-block as the sub-array $A[(i - 1)c + (k - 1)\sqrt{c} \ldots (i - 1)c + k\sqrt{c}][(j - 1)c + (l - 1)\sqrt{c} \ldots (j - 1)c + l\sqrt{c}]$. For each (i, j)-block, we define eight regions, consisting of sets of blocks (some of which can be empty) as follows: the region
$N(i, j)$ consists of all (i, l)-blocks with $l > j$;
$S(i, j)$ consists of all (i, l)-blocks with $l < j$;
$E(i, j)$ consists of all (k, j)-blocks with $k > i$;
$W(i, j)$ consists of all (k, j)-blocks with $k < i$;
$NE(i, j)$ consists of all (k, l)-blocks with $k > i$ and $l > j$;
$SE(i, j)$ consists of all (k, l)-blocks with $k > i$ and $l < j$;
$NW(i, j)$ consists of all (k, l)-blocks with $k < i$ and $l > j$; and
$SW(i, j)$ consists of all (k, l)-blocks with $k < i$ and $l < j$.
Similarly, for each (i, j, k, l)-sub-block, we also define the regions $N_{i,j}(k, l)$, $S_{i,j}(k, l)$, $E_{i,j}(k, l)$, $W_{i,j}(k, l)$, $NE_{i,j}(k, l)$, $NW_{i,j}(k, l)$, $SE_{i,j}(k, l)$ and $SW_{i,j}(k, l)$.

We construct an $n/c \times n/c$ array $A'[1 \ldots n/c][1 \ldots n/c]$ such that $A'[i][j] = 1$ if there exists at least a single $\mathbf{1}$ in the (i,j)-block, and $\mathbf{0}$ otherwise. We also construct another $n/\sqrt{c} \times n/\sqrt{c}$ array $A''[1 \ldots n/\sqrt{c}][1 \ldots n/\sqrt{c}]$ such that $A''[i][j] = 1$ if there exists at least a single $\mathbf{1}$ in the $(\lfloor i/\sqrt{c} \rfloor, \lfloor j/\sqrt{c} \rfloor, i \bmod \sqrt{c}, j \bmod \sqrt{c})$-sub-block, and $\mathbf{0}$ otherwise.

Theorem 3. *Given an $n \times n$ binary array $A[1 \ldots n][1 \ldots n]$ one can construct an index of size $O(n^2/c)$ bits to support NLV queries in $O(c)$ time for $0 < c \leq n$.*

Proof. We divide the array A into blocks and sub-blocks as mentioned earlier. Suppose the query q is in the (i,j,k,l)-sub-block. If $A''[ic+k, jc+l] = 1$, scanning $O(1)$ sub-blocks is enough to find the NLV of q, and this takes $O(c)$ time.

Now, consider the case when $A''[ic+k, jc+l] = 0$ but $A'[i,j] = 1$. In this case, it is clear that we can identify $O(c)$ sub-blocks in which the answer may lie – namely all the sub-blocks in its block, and in the eight neighbouring blocks. We find the potential answer in each of the eight directions (E, W, N, S, NE, NW, SE, and SW), and then compare their positions to find the actual answer. To find the the answer in E direction, we scan the bits in A'' that are to the right of the current sub-block, till we find a $\mathbf{1}$. We then scan this sub-block, and the sub-block to its right to find the potential answer in this direction. Similarly, we can find the potential answers in the W, S, and N directions. Next, we find the nearest $\mathbf{1}$ to the query in the $NE_{i,j}(k,l)$ region. This element is the nearest $\mathbf{1}$ from the bottom-left conner of $(i,j,k+1,l+1)$-sub-block. The nearest $\mathbf{1}$ from the bottom-left conner of (a,b,c,d)-sub-block in the $NE_{a,b}(c,d)$ region is same as either the nearest $\mathbf{1}$ in the same block, or is the nearest $\mathbf{1}$ from the bottom-left corners of one of these three blocks: (1) $(a,b,c+1,d)$-sub-block, (2) $(a,b,c,d+1)$-sub-block, or (3) $(a,b,c+1,d+1)$-sub-block. Therefore we encode each sub-blocks using 2 bits indicating the case it belongs to ((1), (2) or (3)), which takes a total of $O(n^2/c)$ bits. Now, to find the answer in the NE direction, we scan $O(c)$ sub-blocks to find the sub-block which contains the nearest $\mathbf{1}$ from q in $NE(i,j,k,l)$. Once we find the corresponding sub-block, finding the nearest $\mathbf{1}$ from the bottom-left conner in the sub-block takes $O(c)$ time. We can find the nearest $\mathbf{1}$ in the $NW_{ij}(k,l)$, $SE_{ij}(k,l)$ and $SW_{ij}(k,l)$ regions in the same way. Then NLV of q is the closest one among these eight candidates.

Finally, consider the case when $A'[i,j] = 0$. By storing the encoding of Theorem 2 for the array A' using $O(n^2/c^2)$ bits, we can find the nearest block to the query position which contains a $\mathbf{1}$, in $O(1)$ time. Let this block be the (i',j')-block, let ℓ be the L_1 distance from (i,j) to (i',j') in A'. The value ℓc is an estimate (within an additive factor of $2c$) for the L_1 distance from q to its NLV. Assume, wlog, that (i',j') is in the $NE(i,j)$ region. We first describe how to find the nearest $\mathbf{1}$ in $NE(i,j)$ region. Define $d(i,j)$ as the set of blocks in the top-left to the bottom-right diagonal that contains the (i,j)-block and define the array $D_{(i,j)}$ of size at most n/c such that $D_{(i,j)}[m]$ is the distance from the bottom-left element to the nearest $\mathbf{1}$ in the m-th block in $d(i,j)$. Now we construct a linear-bit RMQ (range minimum query) data structure for each $D_{(i,j)}$ (using a total of $O(n^2/c)$ bits), so that RMQ queries can be supported in $O(1)$ time. Now, we find the two potential blocks in $NE(i,j)$ region that may have

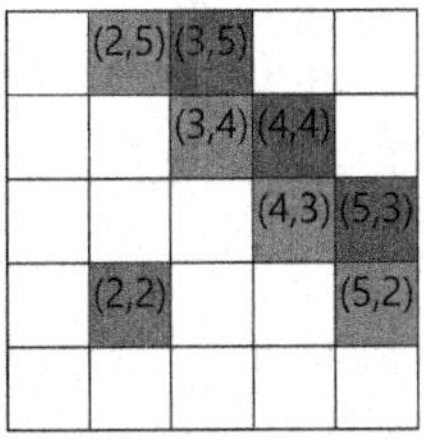

Fig. 1. Suppose the nearest block that contains **1** from (2,2)-block is (4,3)-block, Then $d(4,3)$ are the blocks colored by green and we can find the nearest **1** in NE(2,2) using RMQ for $D_{4,3}[2,3]$ and $D_{4,4}[1\ldots3]$

the nearest **1** from q by performing RMQs on $D_{(i',j')}$ and $D_{(i',j'+1)}$ among all the blocks that are contained in the $NE(i,j)$ region (it is easy to see that they form a consecutive range). We then choose the closer one between these two from the q. (Figure 1 shows the example). Note that if (i',j') is in a different region from $NE(i,j)$, then we may not find any potential answer in $NE(i,j)$, as all the 'relevant' blocks in $D_{(i',j')}$ and $D_{(i',j'+1)}$ may be empty. We can find the nearest **1** in $NW(i,j)$, $SE(i,j)$ and $SW(i,j)$ in a similar way.

Next, we describe how to find the nearest **1** in the $N(i,j)$ region (analogous for $S(i,j)$, $E(i,j)$ and $W(i,j)$ regions). For each position in the bottom row of an (a,b)-block with $A'[a,b] = \mathbf{1}$, we store two bits indicating whether its answer within the block is in (1) the same column (H), or (2) some column to the left (L), or (3) some column to the right (R). (The query algorithm simply "follows" the L or R "pointers" till it reaches a H, and then scans the column upwards till it finds a **1** in that column. Note that L and R cannot be in two adjacent columns.) This takes $O(c \times n^2/c^2) = O(n^2/c)$ bits over all the blocks. This encoding enables us to find the closest **1** within the block from any column in the bottom row of that block in $O(c)$ time. Since ℓ is the L_1 distance between (i,j) and (i',j') in A', we know that all the blocks $A[i,j-r]$, for $1 \le r < \ell$ are empty (otherwise, we have a closer non-empty block than (i',j')). Let k be the column corresponding to the query position q. We claim that the closest **1** to q in the $N(i,j)$ region is closest **1** to the bottom row and column k of the either $(i,j+\ell)$-block or $(i,j+\ell+1)$-block. These can be computed in $O(c)$ time using the above encoding, and then compared to find the required answer. Finally we can find NLV of q by comparing these eight candidate answers. □

4   Encoding of **NLV** on 2-D Arrays

In this section, we give an encoding which supports NLV queries in 2-D array with $O(n^2)$ bits. We consider the 1-D array case first. Jayapaul et al. [12] showed how to encode an array A with n distinct items using $O(n)$ bits to answer NLV queries. We give an alternate proof of this, based on ideas from [2,12]:

Lemma 4. *There exists an encoding of an array $A[1 \dots n]$ that uses $O(n)$ bits while supporting NLV queries, provided all elements are distinct.*

Proof. We write down the sequence $d(1), d(2), \dots, d(n)$ explicitly, where $d(i) = |i - \mathsf{NLV}(i)|$, for $1 \le i \le n$, together with a sequence of n bits that indicate if $i < \mathsf{NLV}(i)$ or $i > \mathsf{NLV}(i)$. Because for $k > 0$, the elements in A are distinct, there are $\ge n/2^k$ elements for which $d(i) \le 2^k$, and $d(i)$ for these elements can be encoded in $O(k)$ bits. In all, $\sum_{k=1}^{\lg n}(n/2^k \cdot O(k)) = O(n)$ bits are used. $\qquad \square$

If the elements in A are not distinct, the above argument does not hold. So instead of encoding the NLV of a position i explicitly as in Lemma 4, we encode the distance between i and the nearest value which is $\ge A[i]$ in the same direction as $\mathsf{NLV}(i)$. Formally, we define $d_l(i) = i - (\max_{j<i, A[j] \ge A[i]} j)$ and $d_r(i) = (\min_{j>i, A[j] \ge A[i]} j) - i$ and $d(i) = d_l(i)$ if $\mathsf{NLV}(i) < i$ and $d(i) = d_r(i)$ otherwise. For each i, we encode $d(i)$ and store a bit stating whether $d(i) = d_r(i)$ or $d(i) = d_l(i)$, and view this as a "pointer" to $j = i + d_r(i)$ or $j = i - d_l(i)$ respectively. Finally, we also store a bit indicating whether or not $A[i] = A[j]$. With this encoding, $\mathsf{NLV}(i)$ can be easily found by following the $d(\cdot)$ "pointers" from i until we reach a position that is greater than $A[i]$.

The following lemma says that this encoding still uses $O(n)$ bits:

Lemma 5. *For any array $A[1 \dots n]$, $\sum_{i=1}^{n} \lg d(i) = O(n)$.*

Proof. Consider the array $A'[1 \dots n]$ of size n, where $A'[i] = A[i] + \epsilon i$ if $\mathsf{NLV}(i) > i$ and $A'[i] = A[i] - \epsilon i$ if $\mathsf{NLV}(i) < i$ for some $\epsilon > 0$. If we set ϵ small enough then if $A[i] > A[j]$ for some i, j then $A'[i] > A'[j]$ as well, but all elements in A' are distinct. So if we define $d'(i)$ and NLV' on A' analogously to $d(i)$ and NLV on A, $D' = \sum_{i=1}^{n} \lg d'(i) = O(n)$ by Lemma 4. We now show that $D = \sum_{i=1}^{n} \lg d(i) \le 2D'$.

To prove this claim, let $0 \le i_0 < \cdots < i_r \le n$ with $r > 0$ be a maximal sequence of indices such that $A[i_0] > A[i_1]$, $A[i_{r-1}] < A[i_r]$, $A[i_1] = A[i_2] = \cdots = A[i_{r-1}]$, and $A[j] < A[i_1]$ for all $i_0 < j < i_r$ and $j \notin \{i_k | 1 \le k \le r-1\}$. For $0 \le k \le r$, let i_k be the index such that $\mathsf{NLV}(i_l) = i_0$ for all $0 < l \le k$ and $\mathsf{NLV}(i_l) = i_r$ for all $k < l \le r-1$. Then by the definition of A', for all $k < l \le r-1$, $\mathsf{NLV}'(i_l) = i_{l+1}$ so $d(i_l) = d'(i_l)$. For the elements to the left of i_k, we can consider the case that there exist $0 < m \le k$ such that $\mathsf{NLV}'(i_l) = i_{l-1}$ for all $0 < l \le m-1$ and $\mathsf{NLV}'(i_l) = i_{l+1}$ for $m \le l \le k$. Then:

$$D - D' = \sum_{i=1}^{n} \lg d(i) - \sum_{i=1}^{n} \lg d'(i)$$

$$= \left(\sum_{i=1}^{m-1} \lg d(i) + \sum_{i=k+1}^{r-1} \lg d(i) + \sum_{j=m}^{k} \lg(i_j - i_{j-1})\right)$$

$$- \left(\sum_{i=1}^{m-1} \lg d'(i) + \sum_{i=k+1}^{r-1} \lg d'(i) + \sum_{j=m}^{k} \lg(i_{k+1} - i_j)\right)$$

$$= \sum_{j=m}^{k} \lg(i_j - i_{j-1}) - \sum_{j=m}^{k} \lg(i_{k+1} - i_j)$$

$$\le \lg(i_m - i_{m-1}) - \lg(i_{k+1} - i_k) \ (\because i_j - j_{j-1} \le i_{k+1} - i_{j-1} \text{ for all } 0 \le j \le k)$$

$$\le \lg(i_m - i_{m-1}) \le \lg(i_m - i_0) \le \lg(i_r - i_m) \ (\because \mathsf{NLV}(i_m) = i_0)$$

$$\leq \lg(i_{k+1} - i_m) + \sum_{j=k+1}^{r-1} \lg(i_{j+1} - i_j) \text{ (by the concavity of lg function)}$$

$$\leq \sum_{i=1}^{n} \lg d'(i) = D' \qquad \qquad \square$$

We now extend this encoding to encode NLVs for a 2-D array $A[1 \ldots n][1 \ldots n]$ of size n^2. In our encoding, each (i, j) "points to" another location (i', j'), such that $A[i', j'] \geq A[i, j]$, as follows: $|i - i'|$ is encoded using $O(1 + \lg |i' - i|)$ (the *row cost* of the pointer) and $|j - j'|$ is encoded using $O(1 + \lg |j' - j|)$ bits (the *column cost* of the pointer), the direction from (i, j) to (i', j') is given using two bits, and finally one extra bit indicates whether or not $A[i', j'] > A[i, j]$. Now we explain how to specify the pointers. Pick an element $A[i, j]$ and wlog assume that $\text{NLV}(i, j) = (i^*, j^*)$ with $i^* \geq i, j^* \geq j$. We choose pointers as follows:
Case (1) Let $i' > i$ be the smallest value such that $i' \leq i^*$ and $A[i, j] = A[i', j]$. If i' exists, we store a pointer to (i', j) and set the extra bit to 0.
Case (2) If not, let $j' > j$ be the smallest value such that $j' \leq j^*$ and $A[i, j] = A[i, j']$. If j' exists, we store a pointer to (i, j') and set the extra bit to 0.
Case (3) Otherwise we store a pointer to (i^*, j^*) and set the extra bit to 1.

We call this encoding scheme $encoding_{2D}$. To obtain $\text{NLV}(i, j)$, we follow pointers starting from (i, j) until we follow one with the extra bit set to 1, and return the position pointed to by this pointer. The correctness of this procedure can be proved by induction on k; we omit the details due to lack of space.

Theorem 4. *There exists an encoding of 2-D array $A[1 \ldots n][1 \ldots n]$ that uses $O(n^2)$ bits while supporting NLV queries.*

Proof. We describe an encoding, called $encoding_{grid}$ as follows. We first encode each column and row of A using Lemma 5, using $O(n^2)$ bits. These pointers are called *grid* pointers. However, the maximal values in each row and column do not have pointers by Lemma 5, as their NLV is not defined. So, in addition, for each row r which has (locally) maximum values in columns $i_1 < \ldots < i_k$, we store *extra* pointers in both directions from (r, i_j) to (r, i_{j+1}) for $j = 0, \ldots, k$, taking $i_0 = 0$ and $i_{k+1} = n + 1$. The space taken by these extra pointers is $O(\lg i_1 + \sum_{j=2}^{k-1} \lg (i_j - i_{j-1}) + \lg (n + 1 - i_k)) = O(n)$ bits for row r. We do this for all rows and columns, at a cost of $O(n^2)$ bits overall.

Although $encoding_{grid}$ does not encode NLV, we use it to upper bound the space used by $encoding_{2D}$. Let a *grid pointer* and a *2D pointer* refer to a pointer in $encoding_{grid}$ and $encoding_{2D}$ respectively. For any 2D pointer, the cost of encoding it can be upper-bounded by the cost of encoding (one or more) grid pointers. Each grid pointer will be used $O(1)$ times this way. Below, we show how to upper bound all Case 2) 2D pointers and the row cost of all Case 3) 2D pointers by grid pointers in rows, using each grid pointer at most thrice. The costs of Case (1) 2D pointers and the column cost of Case (3) 2D pointers can similarly be bounded by the costs of grid pointers in the columns. This will prove the theorem.

We consider a fixed location (i, j), and assume wlog that $\text{NLV}(i, j) = (i^*, j^*)$ with $i^* \geq i$ and $j^* > j$ (if $j^* = j$ then the pointer from (i, j) will have row distance 0 and there is nothing to bound). There are four cases to consider.

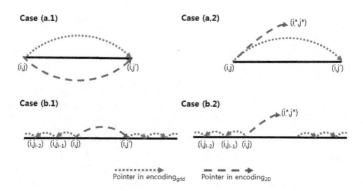

Fig. 2. Pointers in $encoding_{2D}$ and $encoding_{grid}$

Case (a) Let $j' > j$ be the minimum index such that $A[i, j'] \geq A[i, j]$. Suppose that j' exists and there and there is a grid pointer from (i, j) to (i, j') or vice versa. There are two sub-cases:

(a.1) The 2D pointer from (i, j) points to (i, j'). We use the cost of the grid pointer to upper bound the cost of this 2D pointer. Observe that if there is a 2D pointer from (i, j) to (i, j'), there cannot be a 2D pointer from (i, j') to (i, j), so the grid pointer is used for upper-bounding only once in this case (Case (a,1) in Figure 2).

(a.2) The 2D pointer from (i, j) points to (i^*, j^*). Observe that $j' \geq j^*$, since otherwise either (i, j') is a larger value that is closer than (i^*, j^*), a contradiction, or we would have a Case (2) 2D pointer from (i, j) to (i, j'). The grid pointer between (i, j) and (i, j') will only be used twice for upper-bounding in this case (Case (a,2) in Figure 2).

Case (b) There is either no value $A[i, j'] \geq A[i, j]$ for $j' > j$, or if there is, then there are no grid pointers either from (i, j) to (i, j') or vice versa. As before, we consider two sub-cases.

(b.1) First suppose that the 2D pointer from (i, j) points to (i, j'), where $j' > j$ is the smallest index such that $A[i, j'] \geq A[i, j]$, and there is no grid pointers between (i, j) and (i, j') in either direction.

If $A[i, j]$ is a maximal value in row i, the cost of the pointer is upper-bounded by the extra pointer between (i, j) and (i, j'). If not, the absence of grid pointers between (i, j) and (i, j') implies that the NLV of (i, j) in the i-th row is (i, j_0) for some $j_0 < j$. Note that $|j_0 - j| \geq dist((i, j), (i^*, j^*))$, otherwise NLV$(i, j)$ would be (i, j_0). As $dist((i, j), (i^*, j^*)) = |j - j'| + |j' - j^*| + |i^* - i|$, $|j_0 - j| \geq |j' - j|$. The path p between (i, j) and (i, j_0) in $encoding_{grid}$ may comprise a number of grid edges. We can bound the cost of the 2D edge from (i, j) to (i, j') by the total cost of the grid edges on the path p (since the log function is concave, the sum of the costs of the path p is no less than the cost of a single edge from (i, j) to (i, j_0)). Let p comprise the elements $j = j_l, j_{l-1}, \ldots, j_1, j_0$ (omitting the row number for brevity).

Note that for any $0 < k < l$, no 2D pointer from (i, j_k) can end up in Case
(b), so this path can only be used twice to upper-bound the cost of a 2D edge:
once from (i, j) and once (possibly) from (i, j_0) (Case (b,1) in Figure 2).
(b.2) The 2D pointer from (i, j) points to (i^*, j^*), and either there is no value
$\geq A[i, j]$ in locations $A[i, j']$ for $j' > j$, or if there is, and j' is the minimum
such value, then there is no grid pointer between (i, j) and (i, j'). If $A[i, j]$
is a maximal value in row i, then if j' exists, then it must be that $j' > j^*$,
and the row cost of the 2D pointer is bounded by the extra pointer between
(i, j) and (i, j'). On the other hand, if j' does not exist, then the row cost of
the 2D pointer is bounded by the extra pointer from (i, j) to $(i, n + 1)$.

If $A[i, j]$ is not maximal, then arguing as above, we see that the NLV of
(i, j) in the i-th row is (i, j_0) for some $j_0 < j$, that $|j_0 - j| \geq |j - j^*|$, and so
we can upper-bound the row cost of this 2D pointer by the total cost of all
the grid pointers between j and j_0, and each of these grid pointers is used
at most twice (once each for the pointers out of (i, j) and (i, j_0) in Case (b)
to upper bound a 2D pointer (Case (b,2) in Figure 2). □

Now we describe the $O(n^2)$-space data structure that supports NLV query in
constant time on 2-D array $A[1 \ldots n][1 \ldots n]$. To support this, first divide A into
blocks of size $b \times b$ and divide each block into sub-blocks of size $s \times s$. By following
lemma, we can bound the number of distinct NLVs for all the maximal elements
in the block.

Lemma 6 ([12]). *Given a block of size $k \times k$, the maximal elements in the block
have at most $O(k)$ distinct NLVs.*

We divide the area outside each blocks (sub-blocks) into 8 regions (N, S, W,
E, NW, NE, SE, SW) as defined in Section 3. Now we prove the main theorem.

Theorem 5. *There exists an encoding of 2-D array $A[1 \ldots n][1 \ldots n]$ that uses
$O(n^2)$ bits while supporting NLV queries in $O(1)$ time.*

Proof. For each element $A[i, j]$, we assign a color from the set $\{C_1, C_2, \ldots, C_9\}$,
as follows. If $NLV(i, j)$ is in one of the 8 regions, we give one of the colors from
C_1 to C_8; and if the answer is within the block containing (i, j), then we give the
color C_9. Also for block B, let B_{max} be the set of NLV of the maximal elements
in B. Then for each boundary element e_B in block B, we store a pointer to an
element in B_{max} which is closest to e_B. These structures take $O(b^2 + b \lg n)$ bits,
for each block by Lemma 6. For each sub-block, we maintain similar structure as
above using $O(s^2 + s \lg b)$ bits. (Note that a maximal element in a sub-block which
is not a maximal element in its block can have its answer outside the block, but
its distance to NLV is bounded by $O(b)$. So, we can store the answer explicitly.)
Also for each element, we assign the color c_1 to c_9 by the position of its NLV
analogous to C_1 to C_9. To support NLV quries for non-maximal elements in a
sub-block, we encode each sub-block together with its 8 neighboring sub-blocks,
using the encoding of Theorem 4, using $O(s^2)$ bits. In addition, we construct a
precomputed table that is indexed by the above $O(s^2)$-bit encoding of a sub-
block and a position within it, and returns the NLV for that position.

We now describe the query algorithm. Consider the query $q = A[i,j]$ and let B_q (b_q) be the block (sub-block) that contains q. We first check whether q is a maximal element in b_q. If q is not a maximal element in b_q, we use the precomputed table to find the answer, in $O(1)$ time. Otherwise, if q is not a maximal element in B_q, then NLV(q) can be answered in $O(1)$ time by finding the nearest boundary in the direction corresponding to q's assigned color. If q is a maximal element in B_q, we can find its NLV in $O(1)$ time by a similar procedure using colors C_1 to C_9. So total query time is $O(1)$, and total space is $O(n^2/b^2 \times (b^2 + b \lg n) + n^2/s^2 \times (s^2 + s \lg b) + 2^{O(s^2)})$ bits. If we set $b = \lg n$ and $s = c\sqrt{\lg n}$, we can encode A with supporting NLV queries in $O(n^2)$ bits.   □

References

1. Asano, T., Bereg, S., Kirkpatrick, D.: Finding nearest larger neighbors. In: Albers, S., Alt, H., Näher, S. (eds.) Efficient Algorithms. LNCS, vol. 5760, pp. 249–260. Springer, Heidelberg (2009)
2. Asano, T., Kirkpatrick, D.: Time-space tradeoffs for all-nearest-larger-neighbors problems. In: Dehne, F., Solis-Oba, R., Sack, J.-R. (eds.) WADS 2013. LNCS, vol. 8037, pp. 61–72. Springer, Heidelberg (2013)
3. Berkman, O., Schieber, B., Vishkin, U.: Optimal doubly logarithmic parallel algorithms based on finding all nearest smaller values. J. Algorithms 14(3), 344–370 (1993)
4. Brodal, G.S., Brodnik, A., Davoodi, P.: The encoding complexity of two dimensional range minimum data structures. In: Bodlaender, H.L., Italiano, G.F. (eds.) ESA 2013. LNCS, vol. 8125, pp. 229–240. Springer, Heidelberg (2013)
5. Brodal, G.S., Davoodi, P., Lewenstein, M., Raman, R., Srinivasa Rao, S.: Two dimensional range minimum queries and fibonacci lattices. In: Epstein, L., Ferragina, P. (eds.) ESA 2012. LNCS, vol. 7501, pp. 217–228. Springer, Heidelberg (2012)
6. Brodal, G.S., Davoodi, P., Rao, S.S.: On space efficient two dimensional range minimum data structures. Algorithmica 63(4), 815–830 (2012)
7. Demaine, E.D., Landau, G.M., Weimann, O.: On cartesian trees and range minimum queries. Algorithmica 68(3), 610–625 (2014)
8. Ferragina, P., Manzini, G., Mäkinen, V., Navarro, G.: Compressed representations of sequences and full-text indexes. ACM Transactions on Algorithms 3(2) (2007)
9. Fischer, J.: Combined data structure for previous- and next-smaller-values. Theor. Comput. Sci. 412(22), 2451–2456 (2011)
10. Fischer, J., Mäkinen, V., Navarro, G.: Faster entropy-bounded compressed suffix trees. Theor. Comput. Sci. 410(51), 5354–5364 (2009)
11. Jacobson, G.: Space-efficient static trees and graphs. In: FOCS, pp. 549–554. IEEE Computer Society (1989)
12. Jayapaul, V., Jo, S., Raman, V., Satti, S.R.: Space efficient data structures for nearest larger neighbor. In: Proc. IWOCA 2014 (to appear, 2014)
13. Okanohara, D., Sadakane, K.: Practical entropy-compressed rank/select dictionary. In: ALENEX (2007)

A Practical Succinct Data Structure
for Tree-Like Graphs

Johannes Fischer[1] and Daniel Peters[2]

[1] TU Dortmund, Germany
johannes.fischer@cs.tu-dortmund.de
[2] Physikalisch-Technische Bundesanstalt (PTB), Germany
daniel.peters@ptb.de

Abstract. We present a new succinct data structure for graphs that are "tree-like," in the sense that the number of "additional" edges (w.r.t. a spanning tree) is not too high. Our algorithmic idea is to represent a BFS-spanning tree of the graph with a succinct data structure for trees, and enhance it with additional information that accounts for the non-tree edges. In practical tests, our data structure performs well for graphs containing up to 10% of non-tree edges, reducing the space of a pointer-based representation by a factor of ≈ 20, while increasing the worst-case running times for the operations by roughly the same factor.

1 Introduction

Succinct data structures have been one of the key contributions to the algorithms community in the past two decades. Their goal is to represent objects from a universe of size u in information-theoretical optimal $\lg u$ bits of space.[1] Apart from the bare representation of the object, fast operations should also be supported, ideally in time no worse than with a "conventional" data structure for the object. For this, one usually allows extra space $o(\lg u)$ bits.

A prime example of succinct data structures are ordered rooted trees, where with n nodes we have $u \approx 4^n$. In 1989, Jacobson made a first step towards achieving this goal, by giving a data structure using $10n + o(n)$ bits, while supporting the most common navigational operations in $O(\lg n)$ time [19]. This was further improved to the optimal $2n + o(n)$ bits and optimal $O(1)$ navigation time by Munro and Raman [25]. Note that a conventional, pointer-based data structure for trees requires $\Theta(n \lg n)$ bits, which is off by a factor of $\lg n$ from the information-theoretical minimum.

Since the work of Munro and Raman, the research on succinct data structures has blossomed. We now have succinct data structures for bit-vectors [27], permutations [23], binary relations [2], dictionaries [26], suffix trees [29], to name just a few.

The practical value of those data structures has sometimes been disputed. However, as far as we know, in all cases where genuine attempts were made at

---

[1] Function lg denotes the binary logarithm throughout this paper.

M.S. Rahman and E. Tomita (Eds.): WALCOM 2015, LNCS 8973, pp. 65–76, 2015.

practical implementations, the results have mostly been successful [13,20,16, etc., to cite some recent papers presented in the algorithm engineering community]. Further examples of well-performing practical succinct tree implementations will be mentioned throughout this paper.

1.1 Our Contribution

We focus on the succinct representation of a very practical class of graphs: graphs that are "tree-like" in the sense that the number of edges, which can potentially be $\Theta(n^2)$ for an n-node graph, is much lower. We measure this tree-likeness by introducing two additional parameters: (1) k, the number of "additional" edges that have to be added to a spanning tree of the graph (note that $k = m - n + 1$ if m denotes the total number of edges), and (2) h, the number of nodes having more than one incoming edge (also called non-tree nodes in the following). This definition of tree-likeness is similar in flavor to the k-almost trees by Gurevich et al. [17], but in the latter the number of additional edges is counted separately for each biconnected component, with k being the maximum of these.

We think that our definition of tree-likeness encompasses a large range of instances arising in practice. One important example comes from computational biology, where one models the ancestral relationships between species by phylogenetic trees. However, sometimes there are also non-bifurcating specification events [18]. One approach to handle those events are phylogenetic networks, which have an underlying tree as a basis, but with added cross-edges to model the passing of genetic material that does not follow the tree.

Our first contribution (Sect. 3) is a theoretical formulation of a succinct data structure for graphs with the above mentioned parameters n, m, k, and h. It uses space at most $(2n + m) \lg 3 + h \lg n + k \lg h + o(m + k \lg h) + O(\lg \lg n)$ bits, which is close to the $2n + o(n)$ bits for succinct trees if k (and hence also m and h) is close to n. This should be compared to the $O((n + m) \lg n)$ bits that were needed if the graph was represented using a pointer-based data structure. Our second contribution is that we show that the data structure is amenable to a practical implementation (Sect. 4–5). We show that we can reduce the space from a conventional pointer-based representation by a factor of about 20, while the times for navigational operations (moving in either direction of the edges) increase by roughly the same factor; such a space-time tradeoff is typical for succinct data structures.

1.2 Further Theoretical Work on Succinct Graphs

Farzan and Munro [9] showed how to represent a general graph succinctly in $\lg \binom{n^2}{m}(1+o(1))$ bits of space, while supporting the operations supported both by adjacency lists and by adjacency matrices in optimal time. Other results exist for special types of graphs: separable graphs [5], planar graphs [25], pagenumber-k graphs [11], graphs of limited arboricity [21], and DAGs [8]. However, to the best of our knowledge, only the approach on separable graphs has been implemented so far [6]. Also, none of the approaches can navigate efficiently to the sources of the *incoming* edges (without doubling the space), as we do.

2 Preliminaries

In this section we introduce existing data structures that form the basis of our new succinct graph representation. All these results (hence also our new one) are in the word-RAM model of computation, where it is assumed that the machine consists of words of width w bits that can be manipulated in $O(1)$ time by a standard set of arithmetic and logical operations, and further that the problem size n is not larger than $O(2^w)$.

2.1 Succinct Data Structures

Let $S[0, n)$ be a *bit-string* of length n. We define the fundamental *rank-* and *select*-operations on S as follows: $\mathsf{rank}_1(S, i)$ gives the number of 1's in the prefix $S[0, i]$, and $\mathsf{select}_1(S, i)$ gives the position of the i'th 1 in S, reading S from left to right ($0 \leq i < n$). Operations $\mathsf{rank}_0(S, i)$ and $\mathsf{select}_0(S, i)$ are defined similarly for 0-bits. S can be represented in $n + o(n)$ bits such that rank- and select-operations are supported in $O(1)$ time [25].

These operations have been extended to sequences over larger alphabets, at the cost of slight slowdowns in the running times [14]: let $S[0, n)$ be a *string* over an alphabet Σ of size σ. Then S can be represented in $n \lg \sigma (1 + o(1))$ bits of space such that the operations $\mathsf{rank}_a(S, i)$ and $S[i]$ (accessing the i'th element) take $O(\lg \lg \sigma)$ time, and $\mathsf{select}_a(S, i)$ takes $O(1)$ time (all for arbitrary $a \in \Sigma$ and arbitrary $0 \leq i < n$). Note that by additionally storing S in plain form, the access-operation also takes $O(1)$ time, at the cost of duplicating the space. In some special cases the running times for the three operations is faster. For example, when the alphabet size is small enough such that $\sigma = w^{O(1)}$ for word size w, then Belazzougui and Navarro [3] proved that $O(1)$ time for all three operations is possible within $O(n \lg \sigma)$ bits of space.

2.2 The Level Order Unary Degree Sequence (LOUDS)

There are several ways to represent an ordered tree on n nodes using $2n$ bits [24, 4]; in this article, we focus on one of the oldest approaches, the *level order unary degree sequence* [19], which is obtained as follows (the reasons for preferring LOUDS over BPS [24] or DFUDS [4] will become evident when introducing the new data structure in Sect. 3). For convenience, we first augment the tree with an artificial *super-root* that is connected with the original root of the tree. Now initialize B as an empty bit-vector and traverse the nodes of the tree level by level (aka breadth-first). Whenever we see a node with k children during this level-order traversal, we append the bits $1^k 0$ to S, where 1^k denotes the juxtaposition of k 1-bits. See Fig. 1 for an example. In the LOUDS, each node is represented twice: once by a '1,' written when the node was seen as a *child*

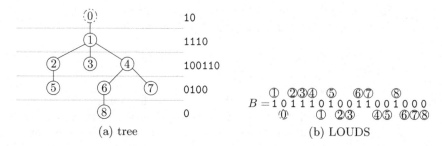

Fig. 1. An ordered tree (a) and its level order unary degree sequence (b)

during the level-order traversal, and once by a '0,' written when it was seen as a *parent*. The number of bits in B is $2n + 1$.

We identify the nodes with their level-order number, since both the 1- and the 0-bits appear in this order in B. It should be noted that all succinct data structures for trees [19, 24, 4, 10, 7] must have the freedom to fix a particular naming for the nodes; natural such namings are post- or pre-order [19, 24, 4], in-order [7], and level-order [19], as here.[2]

If we now augment B with data structures for rank and select (see Sect. 2.1), then the resulting space is $2n + o(n)$ bits, but basic navigational operations on the tree can be *simulated* in $O(1)$ time: for moving to the parent node of i $(1 \leq i \leq n)$, we jump to the position y of the i'th 1-bit in B by $y = \mathsf{select}_1(B, i)$, and then count the number j of 0's that appear before y in B by $j = \mathsf{rank}_0(B, y)$; j is then the level-order number of the parent of i. Conversely, listing the children of i works by jumping to the position x of the i'th 0-bit in B by $x = \mathsf{select}_0(B, i)$, and then iterating over the positions $x+1, x+2, \ldots$, as long as the corresponding bit is '1.' For each such position $x+k$ with $B[x+k] = 1$, the level-order numbers of i's children are $\mathsf{rank}_1(B, x) + k$, which can be simplified to $x - i + k + 1$.

3   New Data Structure

We now propose our new succinct data structure for tree-like graphs. Let G denote a directed graph. We use the following characteristics of G:

- n, the number of nodes in G,
- m, the number of edges in G,
- $c \leq n$, the number of strongly connected components with no incoming edge from a different strongly connected component,
- $k = m - n + 1$, the number of non-tree edges in G (the number of edges to be added to a spanning tree of G to obtain G), and
- $h \leq k$, the number of non-tree nodes in G (nodes with more than 1 incoming edge).

---

[2] If the naming is arbitrary (e.g., chosen by the user), then $n \lg n$ bits are inevitable, since *any* memory layout of the nodes has $n!$ possible namings.

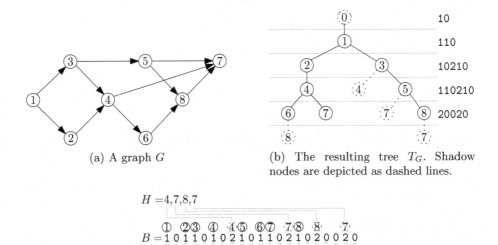

(a) A graph G

(b) The resulting tree T_G. Shadow nodes are depicted as dashed lines.

(c) The resulting ternary LOUDS B and the array H of non-tree edges

Fig. 2. Illustration of our new data structure. The nodes are numbered such that they correspond to the level-order numbers in the chosen BFT-tree.

For simplicity, assume for now that there exists a node r in G from which a path to every other node exists (i.e., $c = 1$). From r, perform a breadth-first traversal (BFT) of G. Let T_G^{BFT} denote the resulting BFT-tree. We augment T_G^{BFT} as follows: for each node w that is *inspected but not visited* during the BFT at node v (meaning that it has already been visited at an earlier point), we make a *copy* of w and append it as a child of v in the BFT-tree T_G^{BFT}. We call those nodes *shadow nodes*. Finally, we add a super-root to r, and call the resulting tree T_G, which has exactly $m+2$ nodes. See Fig. 2a and 2b for an example of G and T_G. If no such node r exists, we perform the BFT from c nodes $r_1, \ldots, r_c$ for every strongly connected component with no incoming edge from a different such component, and obtain a BFT-*forest*. All roots of this forest will be made children of the super-root. This adds at most c additional edges to T_G.

We now aim at representing the tree T_G space efficiently, similar to the LOUDS of Sect. 2.2. Since we need to distinguish between real nodes and shadow nodes, we cannot construct a *bit*-vector anymore. Instead, we construct B as a sequence of *trits*, namely values from $\{0, 1, 2\}$, as follows: again, B is initially empty, and we visit the nodes of T_G in level-order. For each visited node, the sequence appended to B is constructed as in the original LOUDS, but now using a '2' instead of a '1' for shadow nodes. The shadow nodes are *not* visited again during the level-order traversal and hence *not* represented by 0's.[3] We call the resulting

---

[3] Listing the shadow nodes by 0's would not harm, but does not yield any extra information; hence we can omit them.

Function `children(i)`: find the nodes directly reachable from i.

$x \leftarrow \text{select}_0(B, i) + 1;$ `// start of the list of i's children`
while $B[x] \neq 0$ **do**
    if $B[x] = 1$ **then** output $\text{rank}_1(B, x);$ `// actual node`
    else output $H[\text{rank}_2(B, x)];$ `// shadow node`
    $x \leftarrow x + 1;$
endw

Function `parents(i)`: find the nodes from which i is directly reachable.

---
output $\text{rank}_0(B, \text{select}_1(B, i));$ `// tree parent`
$j \leftarrow 1;$
$x \leftarrow \text{select}_i(H, j);$
while $x < k$ **do**
    output $\text{rank}_0(B, \text{select}_2(B, x + 1));$ `// non-tree parent`
    $j \leftarrow j + 1;$
    $x \leftarrow \text{select}_i(H, j);$
endw

---

trit-vector B the *ternary LOUDS*. The ternary LOUDS consists of $n + m + c + 1$ trits. See Fig. 2c for an example.

We also need an additional array H that lists the non-tree nodes in the order in which they appear in B. This array will be used for the navigational operations, as shown in Sect. 3.1. For the operations, besides accessing H, we will also need select-support on H. For this, we use the data structures mentioned in Sect. 2.1 [3,14].

3.1 Algorithms

The algorithms for listing the children and parents of a node are shown in Functions children(i) and parents(i). These functions follow the original LOUDS-functions as closely as possible. Listing the children just needs to make the distinction if there is a '1' or a '2' in the ternary LOUDS B; in the latter case, array H storing the shadow nodes needs to be accessed.

Listing the parents is only slightly more involved. First, the (only) tree parent can be obtained as in the original LOUDS. Then we iterate through the occurrences of i in H in a while-loop, using select-queries. For each occurrence found, we go to the corresponding '2' in B and count the number of '0's before that '2' as usual.

As in the original LOUDS, counting the number of children is faster than traversing them: simply calculate $\text{select}_0(B, i+1) - \text{select}_0(B, i) + 1$; this computes the desired result in $O(1)$ time.[4]

---

[4] For calculating the number of parents in $O(1)$ time, we would need to store those numbers explicitly for hybrid nodes; for all other nodes it is 1.

3.2   Space Analysis

The trit-vector B can be stored in $(n + m + c)(\lg 3 + o(1))$ bits [27], while supporting $O(1)$ access on its elements. Support for rank and select-queries needs additional $o(n + m)$ bits [25].

There are several ways to store H. Storing it in plain form uses $k \lg n$ bits. Using another $k \lg n(1 + o(1))$ bits, we can also support $\mathsf{select}_a(H, i)$-queries on H in constant time [14]. This sums up to $2k \lg n + o(k \lg n)$ bits.

On the other hand, since the number h of non-tree nodes can be much smaller than k (the number of non-tree edges), this can be improved with a little bit of more work: we store a translation table $T[0, h)$ such that $T[i]$ is the level order number of the i'th non-tree node. Then $H[0, k)$ can be implemented by a table $H'[0, k)$ that only stores values from $[0, h)$, such that $H[i] = T[H'[i]]$. The space for T and H' is $k \lg h + h \lg n$ bits. To also support select-queries on H within less than $k \lg n$ bits of space, we use the *indexable dictionaries* of Raman et al. [28]: store a bit vector $C[0, n)$ such that $C[i] = 1$ iff the i'th node in level order is a non-tree node. C can be stored in $h \lg n + o(h) + O(\lg \lg n)$ bits [28, Thm. 3.1], while supporting select- and partial rank-queries (only $\mathsf{rank}_1(C, i)$ with $C[i] = 1$, which is what we need here) in constant time. Now we only need to prepare H' for select-queries, this time using $k \lg h + o(k \lg h)$ bits. Queries $\mathsf{select}_a(H, i)$ can be answered by $\mathsf{select}_{\mathsf{rank}_1(C,a)}(H', i)$, so H can be discarded. Since the data structure of Raman et al. [28] automatically supports select-queries, we also do not need to store T in plain form anymore, since $T[i] = \mathsf{select}_1(C, i)$. Thus, the total space for H using this second approach is $h \lg n + k \lg h + o(h + k \lg h) + O(\lg \lg n)$ bits.

Summing up and simplifying ($c \le n$), the main theoretical result of this article can be formulated as follows:

Theorem 1. *A directed graph G with n nodes, m edges, and h non-tree nodes ($k = m - n + 1$ is the number of non-tree edges) can be represented in*

$$(2n + m) \lg 3 + h \lg n + k \lg h + o(m + k \lg h) + O(\lg \lg n)$$

bits such that listing the x incoming or y outgoing edges of any node can be done in $O(x)$ or $O(y)$ time, respectively. Counting the number of outgoing edges can be done in $O(1)$ time.

4   Implementation Details

We now give some details of our implementation of the data structure from Sect. 3, sometimes sacrificing theoretical worst-case guarantees for better results in practice.

4.1   Representing Trit-Vectors

We first explain how we store the trit sequence B such that constant time access, rank and select are supported. We group 5 trits together into one tryte, and store this tryte in a single byte. This results in space $\lceil (n+m+c)/5 \rceil \cdot 8 = \lceil 1.6\,(n+m+c) \rceil$ bits for B, which is only $\approx 1\%$ more than the optimal $\lceil (n+m+c)\lg 3 \rceil \approx \lceil 1.585\,(n+m+c) \rceil$ bits. The individual trits are reconstructed using Horner's method, in just one calculation.[5]

For rank and select on B, we use an approach similar to the *bit*-vectors of González et al. [15], but with a three-level scheme (instead of only 2), thus favoring space over time. This scheme basically stores rank-samples at increasing sample rates, and the fact that the bits are now intermingled with 2's does not cause any troubles. We used sample rates 25, 275, and 65 725 trits, respectively, which enable a fast byte-aligned layout in memory. On the smallest level we divided a 25-trit block into five trytes. Using the table lookup technique [22] on the trytes the calculation for rank on a 25-trit block is done in at most five steps with an overhead of $3^5 = 243$ bytes of space.

As in the original publication [15], select queries are solved by binary searches on rank-samples, again favoring space over time.

4.2   Other Data Structures

Instead of the complex representation of H as described in Sect. 3.2, needed for an efficient support of the parent-operation, we used a simpler array-based approach: we store the positions of 1-bits (in B) of the first occurrences of non-tree nodes in an array $P[0, h)$. (In the example of Fig. 2, we have $P = [5, 11, 14]$ for the non-tree nodes 4,7, and 8.) A second array $Q[0, k)$ lists the positions of the other occurrences of the non-tree nodes, in level order (In the example, $Q = [7; 13, 19; 16]$). A final third array $N[0, h)$ stores the starting positions of the non-tree nodes in Q (in the example, $N = [0, 1, 3]$). Then with a binary search on P (or a bit-vector marking the respective positions) we can find out if a node i has further shadow copies, and if so, list them using Q and N. Note that with these arrays, we can also efficiently list (in $O(1)$ time) the number of parents of non-tree nodes.

We also added a bit-vector $D = [0, n)$ with $D[i] = 1$ iff node i is a leaf node. This way, the question if a node has children can be quickly answered by just one look-up to D, omitting rank and select queries.

5   Practical Results

The aim of this section is to show the practicality of our approach on the example of phylogenetic networks. Such networks arise in computational biology. They are a generalization of the better known phylogenetic trees, which model the (hypothetic) ancestral relationships between species. In particular for fast reproducing organisms like bacteria, networks can better explain the observed data

---

[5] We did not investigate codes that exploit the fact that the distribution of the 0's, 1's, and 2's in B is not necessarily uniform. Some further space could be saved here, probably at the cost of increased access times. We leave this as a direction for future research.

than trees. Quoting Huson and Scornavacca [18], phylogenetic networks "may be more suitable for data sets where evolution involves significant amounts of reticulate events, such as hybridization, horizontal gene transfer, or recombination."

Since large real-life networks are not (yet) available, we chose to create them artificially for our tests. We did so by creating random tree-like graphs with 10% non-tree edges ($k = n/10$), by *directly* creating random trit-vectors of a given length, and randomly introducing k 2's to create non-tree edges. We further ensured that shadow nodes have different parents, and that all non-tree edges point only to nodes at the same height (in the BFS-tree), mirroring the structure of phylogentic networks (no interchange of genetic material with extinct species).

We compared our data structure to a conventional pointer-based data structure for graphs (where each node stores a list of its descendants, a pointer to an arbitrary father, and the number of its descendants). While there exist many implementations of succinct data structures for trees[6], we are not aware of any implementations for graphs, hence we did not compare our data structure to others.

Our machine was equipped with an Intel Core i7@2.2GHz and 8GB of RAM, running under Windows 7. We compiled the program for 32 bits, in order not to make the pointer-based representation unnecessarily large. All programs used only a single core of the CPU. We averaged the running times over 1 000 tests for $n = 10\,000$, over 100 tests for $100\,000 \leq n \leq 1\,000\,000$, over 15 tests for $n = 10\,000\,000$, and over 5 tests for $n = 100\,000\,000$.

Table 1. Comparison between a pointer based graph and our succinct LOUDS representation for graphs with 10% non-tree edges

| | space [MByte] | | time for children [μsec] | | time for parents [μsec] | |
n	LOUDS	pointer	LOUDS	pointer	LOUDS	pointer
10 000	0.0159	0.3654	0.3203	0.0295	0.3315	0.0129
100 000	0.1682	3.6533	0.3458	0.0311	0.3472	0.0130
1 000 000	1.6818	36.5433	0.3884	0.0332	0.3614	0.0136
10 000 000	18.8141	365.4453	0.3889	0.03374	0.3812	0.0138
100 000 000	188.1542	3 654.4394	0.4095	—	0.4198	—

Table 1 shows the sizes of the data structures and the average running times for the children- and parents-operations with either representation.[7] It can be seen that our data structure is consistently about 20–25 times smaller than the pointer-based structure, while the time for the operations increases by a factor

[6] For example, the well-known libraries for succinct data structures https://github.com/fclaude/libcds and https://github.com/simongog/sdsl both have well-tuned succinct tree implementations. Other sources are [1,12].

[7] For memory reasons, the running times of the pointer-based representation could not be measured for the last instances.

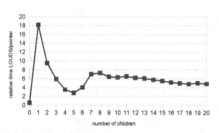

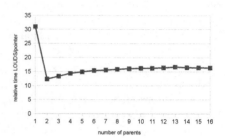

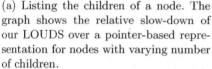

(a) Listing the children of a node. The graph shows the relative slow-down of our LOUDS over a pointer-based representation for nodes with varying number of children.

(b) The same as in (a), but now for listing the parents of a node

Fig. 3. Detailed evaluation of running times

of about 12 in case of the children-operation, and by a factor of about 25 in case of the parents-operation. Such trade-offs are typical in the world of succinct data structures.

To further evaluate our data structure, we more closely surveyed the children- and parents-operations in a graph with $1\,000\,000$ nodes and 10% of non-tree edges, in which a node has no more than 16 incoming edges. We executed both operations on every node in the graph and grouped the running times by the number of children and parents, respectively. The results are shown in Fig. 3. In (a), showing the results for the children-operations, several interesting points can be observed. First, for nodes with 0 children (a.k.a. leaves), our data structure is actually *faster* than the pointer-based representation (about twice as fast), because this operation can be answered by simply checking one bit in the bit-vector D, mentioned in Sect. 4. Second, for nodes with 5 children the slowdown is only about 3, then rises to a slowdown of about 7 for nodes with 8 children, and finally gradually levels off and seems to convert to a slowdown of about 5. We think that this can be explained by the different distributions of the *types* of the nodes listed in the children operation: while for tree-nodes the node numbers can be simply calculated from the LOUDS, for non-tree nodes this process involves further look-ups, e.g. to the H-array. Since we tested graphs with 10% non-tree edges, we think that at about 7–8 children/nodes this effect is most expressed. In (b) the parents operation on our LOUDS for nodes with one parent is around 30 times slower than the pointer representation. For a greater number of parents it is about 16 times slower. Our explanation is that at first a rank and select query is necessary to retrieve the first parent node, afterwards if the node has more than one parent the H-array is scanned. With our practical implementation of the H-array from Sect. 4 the select results are directly saved in the Q-array, hence there is no need for select queries anymore and a rank query seems to be around 16 times slower than a look-up.

6   Conclusions

We presented a framework and implementation for a new succinct data structure for "tree-like" graphs based on the LOUDS representation for trees. The practical evaluation confirmed that our succinct data structure achieves a significant space reduction. A trade-off between space and time can be observed, which is common in the world of succinct data structures.

References

1. Arroyuelo, D., Cánovas, R., Navarro, G., Sadakane, K.: Succinct trees in practice. In: Proc. ALENEX, pp. 84–97. SIAM (2010)
2. Barbay, J., Claude, F., Navarro, G.: Compact rich-functional binary relation representations. In: López-Ortiz, A. (ed.) LATIN 2010. LNCS, vol. 6034, pp. 170–183. Springer, Heidelberg (2010)
3. Belazzougui, D., Navarro, G.: New lower and upper bounds for representing sequences. In: Epstein, L., Ferragina, P. (eds.) ESA 2012. LNCS, vol. 7501, pp. 181–192. Springer, Heidelberg (2012)
4. Benoit, D., Demaine, E.D., Munro, J.I., Raman, R., Raman, V., Rao, S.S.: Representing trees of higher degree. Algorithmica 43(4), 275–292 (2005)
5. Blandford, D.K., Blelloch, G.E., Kash, I.A.: Compact representations of separable graphs. In: Proc. SODA, pp. 679–688. ACM/SIAM (2003)
6. Blandford, D.K., Blelloch, G.E., Kash, I.A.: An experimental analysis of a compact graph representation. In: ALENEX/ANALC, pp. 49–61. SIAM (2004)
7. Davoodi, P., Raman, R., Satti, S.R.: Succinct representations of binary trees for range minimum queries. In: Gudmundsson, J., Mestre, J., Viglas, T. (eds.) COCOON 2012. LNCS, vol. 7434, pp. 396–407. Springer, Heidelberg (2012)
8. Farzan, A., Fischer, J.: Compact representation of posets. In: Asano, T., Nakano, S.-i., Okamoto, Y., Watanabe, O. (eds.) ISAAC 2011. LNCS, vol. 7074, pp. 302–311. Springer, Heidelberg (2011)
9. Farzan, A., Munro, J.I.: Succinct representations of arbitrary graphs. In: Halperin, D., Mehlhorn, K. (eds.) ESA 2008. LNCS, vol. 5193, pp. 393–404. Springer, Heidelberg (2008)
10. Farzan, A., Munro, J.I.: A uniform approach towards succinct representation of trees. In: Gudmundsson, J. (ed.) SWAT 2008. LNCS, vol. 5124, pp. 173–184. Springer, Heidelberg (2008)
11. Gavoille, C., Hanusse, N.: On compact encoding of pagenumber k graphs. Discrete Mathematics & Theoretical Computer Science 10(3), 23–34 (2008)
12. Geary, R.F., Rahman, N., Raman, R., Raman, V.: A simple optimal representation for balanced parentheses. Theor. Comput. Sci. 368(3), 231–246 (2006)
13. Gog, S., Ohlebusch, E.: Fast and lightweight LCP-array construction algorithms. In: Proc. ALENEX, pp. 25–34. SIAM (2011)
14. Golynski, A., Munro, J.I., Rao, S.S.: Rank/select operations on large alphabets: a tool for text indexing. In: Proc. SODA, pp. 368–373. ACM/SIAM (2006)
15. González, R., Grabowski, S., Mäkinen, V., Navarro, G.: Practical implementation of rank and select queries. In: Poster Proceedings Volume of 4th Workshop on Efficient and Experimental Algorithms (WEA), Greece, pp. 27–38. CTI Press and Ellinika Grammata (2005)

16. Grossi, R., Ottaviano, G.: Design of practical succinct data structures for large data collections. In: Bonifaci, V., Demetrescu, C., Marchetti-Spaccamela, A. (eds.) SEA 2013. LNCS, vol. 7933, pp. 5–17. Springer, Heidelberg (2013)

17. Gurevich, Y., Stockmeyer, L., Vishkin, U.: Solving NP-hard problems on graphs that are almost trees and an application to facility location problems. J. ACM 31(3), 459–473 (1984)

18. Huson, D.H., Scornavacca, C.: A survey of combinatorial methods for phylogenetic networks. Genome Biology and Evolution 3, 23 (2011)

19. Jacobson, G.J.: Space-efficient static trees and graphs. In: Proc. FOCS, pp. 549–554. IEEE Computer Society (1989)

20. Joannou, S., Raman, R.: Dynamizing succinct tree representations. In: Klasing, R. (ed.) SEA 2012. LNCS, vol. 7276, pp. 224–235. Springer, Heidelberg (2012)

21. Kannan, S., Naor, M., Rudich, S.: Implicit representation of graphs. SIAM J. Discrete Math. 5(4), 596–603 (1992)

22. Munro, J.I.: Tables. In: Chandru, V., Vinay, V. (eds.) FSTTCS 1996. LNCS, vol. 1180, pp. 37–42. Springer, Heidelberg (1996)

23. Munro, J.I., Raman, R., Raman, V., Rao, S.S.: Succinct representations of permutations. In: Baeten, J.C.M., Lenstra, J.K., Parrow, J., Woeginger, G.J. (eds.) ICALP 2003. LNCS, vol. 2719, pp. 345–356. Springer, Heidelberg (2003)

24. Munro, J.I., Raman, V.: Succinct representation of balanced parentheses, static trees and planar graphs. In: Proc. FOCS, pp. 118–126. IEEE Computer Society (1997)

25. Munro, J.I., Raman, V.: Succinct representation of balanced parentheses and static trees. SIAM J. Comput. 31(3), 762–776 (2001)

26. Pagh, R.: Low redundancy in static dictionaries with constant query time. SIAM J. Comput. 31(2), 353–363 (2001)

27. Pătraşcu, M.: Succincter. In: Proc. FOCS, pp. 305–313. IEEE Computer Society (2008)

28. Raman, R., Raman, V., Rao, S.S.: Succinct indexable dictionaries with applications to encoding k-ary trees and multisets. ACM Transactions on Algorithms 3(4), Article No. 43 (2007)

29. Sadakane, K.: Compressed suffix trees with full functionality. Theory Comput. Syst 41(4), 589–607 (2007)

Forming Plurality at Minimum Cost

Wei-Yin Lin[1], Yen-Wei Wu[1], Hung-Lung Wang[4], and Kun-Mao Chao[1,2,3]

[1] Department of Computer Science and Information Engineering
[2] Graduate Institute of Biomedical Electronics and Bioinformatics
[3] Graduate Institute of Networking and Multimedia
National Taiwan University, Taipei, Taiwan 106
[4] Institute of Information and Decision Sciences
National Taipei University of Business, Taipei, Taiwan 100

Abstract. In this paper, we are concerned with a kind of spatial equilibria, the plurality points, in two-dimensional Euclidean space. Given a set of voters, each of which corresponds to a point, a plurality point is defined as a point closer to at least as many voters as any other point in the space. We generalize the definition by appending weights on the voters, and define the plurality point as a point $\triangle$ satisfying that the sum of weights of the voters closer to $\triangle$ is no less than that of the voters closer to any other point in the space. To remedy the issue of the non-existence of plurality points, we investigate the problem of eliminating some voters with minimum "cost" so that there is a plurality point with respect to the remaining voters. We show that the problem is NP-hard. Moreover, if all voters' weights are restricted to be equal, we show that the problem can be solved in $O(n^5 \log n)$ time, where n is the number of voters.

1 Introduction

A plurality point is an equilibrium location determined by election among voters located at different places. Given a set of points regarded as *voters*, a *plurality point* is a location which is closer to at least as many voters as any other location in the space [1]. There are multiple classifications for the location space, such as networks versus d-dimensional space, and continuous versus discrete space [7].

The importance of plurality points reveals in the process of election. The electoral process is often reduced to a multidimensional model of spatial competition for investigating the equilibrium [4, 11]. Such a representation model is known as the political spectrum, where each dimension is corresponding to an issue for discussion, and the value assigned to each voter shows the degree of support for that particular issue. For example, suppose that there are several citizens concerning two issues, one is environmental and the other one is economic. Each of them corresponds to a dimension of $\mathbb{R}^2$, and the coordinate of each citizen is his or her most preferred position in the issue space, where their preferences of both issues are quantified. To help candidates make decisions, the equilibrium can be found by computing plurality points. The model can be extended to $\mathbb{R}^d$

M.S. Rahman and E. Tomita (Eds.): WALCOM 2015, LNCS 8973, pp. 77–88, 2015.

when there are even more issues. In addition, finding plurality points is a natural mechanism to determine an adequate location to place facilities for public infrastructures.

Another equilibrium location which is strongly correlated to the plurality point is the *Condorcet point*. A Condorcet point is a location such that there is no other location closer to a strict majority of voters, i.e., no other location is closer to more than a half of the voters than the Condorcet point [6]. In Euclidean space, it has been shown that plurality points and Condorcet points are equivalent [12]. By considering it as the equilibrium point of social decisions under majority rule, some properties were proposed in two-dimensional space with different norms, as well as on a graph [10, 11]. There are also some prior results discussing the relationship between plurality points and Condorcet points in various normed spaces [5, 9]. It is known that the Condorcet point does not necessarily exist, and there are literatures dealing with this issue. In a network, Campos Rodríguez and Moreno Pérez [3] introduced a relaxation of the Condorcet point called the $\alpha\gamma$-*Condorcet points* and provided polynomial algorithms to solve the problems, in which two locations are considered indifferent for each voter if the difference of distances to them is within a positive threshold, and the proportion of voters needed to reject a location is different from one half.

In d-dimensional Euclidean space, the criterion of plurality was proposed mainly in the theory of spatial competition [11]. Most previous works appeared mainly in the literature of economics, politics and operations research [6, 7, 9, 11]. When $d > 2$, it has been shown that the plurality point is unique if not all voters are collinear, and it can be found in time $O(n^{d-1}\log n)$ if such a point exists [12], where n is the number of voters. However, the plurality point may not exist. Therefore we focus on providing an alternative solution by finding a subset of voters which admits a plurality point at minimum cost. The problem is generalized by assigning each voter a *weight* for voting and a *cost* to pay for being eliminated. Similar manipulation was also investigated in voting theory [2]. We show that this problem is NP-hard. For the special case where all voters are equally weighted, we provide a polynomial-time algorithm to solve this problem.

The rest of this paper is organized as follows. First, we formally define the problem and notation in Section 2.1, and some properties of plurality points are given in Section 2.2. In Section 3, we show that the problem investigated in this paper is NP-hard. In Section 4, we give a polynomial-time algorithm to solve the problem when voters are equally weighted. Finally, Section 5 concludes this paper by summarizing the main results and future work.

2   Preliminaries

2.1   Problem Definition

Let $V = \{v_1, v_2, \ldots, v_n\}$ be a finite set of voters, where each voter corresponds to a point in $\mathbb{R}^2$, and the coordinate of voter v_i is denoted by (x_i, y_i). For any point p, the coordinate of p is denoted by (x_p, y_p). The set V is also referred to as the corresponding set of points, and V is said to be a set of voters in $\mathbb{R}^2$. For each voter $v_i \in V$, a nonnegative weight $w(v_i)$ is given to specify how many votes that v_i holds. For any $U \subseteq V$, let $w(U) = \sum_{v_i \in U} w(v_i)$. We assume that no two voters in V correspond to a same point. A point $\triangle$ is said to be a *multidimensional median* of V if $\sum_{\{v \in V : x_v \le x_\triangle\}} w(v) \ge w(V)/2, \sum_{\{v \in V : x_v \ge x_\triangle\}} w(v) \ge w(V)/2, \sum_{\{v \in V : y_v \le y_\triangle\}} w(v) \ge w(V)/2$, and $\sum_{\{v \in V : y_v \ge y_\triangle\}} w(v) \ge w(V)/2$. We use $\mathcal{M}$ to denote the set of all multidimensional medians of V.

The distance $d(u,v)$ between two points u and v in $\mathbb{R}^2$ is measured by the ℓ_2-norm, i.e. $d(u,v) = \sqrt{(x_u - x_v)^2 + (y_u - y_v)^2}$. For two arbitrary points $\triangle_1$ and $\triangle_2$ in $\mathbb{R}^2$, we say that voter v_i *prefers* $\triangle_1$ to $\triangle_2$ if $d(v_i, \triangle_1) < d(v_i, \triangle_2)$. Define $[\triangle_1 \succ \triangle_2] = \{v_i \in V : d(v_i, \triangle_1) < d(v_i, \triangle_2)\}$ and $[\triangle_1 \sim \triangle_2] = V \setminus ([\triangle_1 \succ \triangle_2] \cup [\triangle_2 \succ \triangle_1])$. A point $\triangle^* \in \mathbb{R}^2$ is called a plurality point if $w([\triangle \succ \triangle^*]) \le w([\triangle^* \succ \triangle])$ holds for each $\triangle \in \mathbb{R}^2$.

We say that a set $D \subseteq V$ is a *feasible set* of V if $V \setminus D$ admits a plurality point. Let $\mathcal{F}(V)$ denote the set of all feasible sets of V. For each voter $v_i \in V$, a cost $c(v_i)$ is given to specify the price needed for sacrificing v_i in order to admit a plurality point. For any $U \subseteq V$, let $c(U) = \sum_{v_i \in U} c(v_i)$. We say that a set $D^* \in \mathcal{F}(V)$ is an *optimal feasible set* of V if $c(D^*) = \min_{D \in \mathcal{F}(V)} c(D)$. A formal problem definition is given below.

Problem 1 (The Minimum-Cost Plurality Problem). Given a finite set V of voters in $\mathbb{R}^2$ with nonnegative weights and costs, the minimum-cost plurality problem is to find an optimal feasible set D^* of V, where $D^* = \arg\min_{D \in \mathcal{F}(V)} c(D)$.

Since a nonempty set consisting of at most two voters admits a plurality point, it is clear that a feasible set sought in Problem 1 always exists. Notice that according to the definition of plurality points, one may modify an instance by inserting or deleting an arbitrary number of voters with weight zero such that the plurality points remain the same. Thus, in the discussion below, we assume that all voters are of positive weights.

2.2   Properties of Plurality Points

In this section, lemmas are given for finding the plurality points of V. Basically, there are some helpful observations proposed in our previous work [12]. However, those observations focus on the case that all voters are equally weighted. Lemmas 1–3 are a generalized version extended from [12]. The proofs of the lemmas in this section can be obtained by replacing the number of voters in the proofs in [12] with the sum of those voters' weights. We omit the proofs here due to the similarity.

We denote $\triangle_V$ as the set of all the plurality points of V. Given a line L, the plane $\mathbb{R}^2$ is partitioned into $L^+ \cup L^- \cup L$, where L^+ and L^- are two open halfplanes separated by L. Let $V_L^+ = V \cap L^+$, $V_L^- = V \cap L^-$, and $V_L = V \cap L$. For simplicity, we denote $w(V_L^+), w(V_L^-)$, and $w(V_L)$ by w_L^+, w_L^-, and w_L, respectively. For any $\triangle \in \mathbb{R}^2$, define $V_{L,\triangle}^{\nearrow} = \{v \in V_L : y_v > y_\triangle\} \cup \{v \in V_L : y_v = y_\triangle \text{ and } x_v > x_\triangle\}$, and $V_{L,\triangle}^{\swarrow} = \{v \in V_L : y_v < y_\triangle\} \cup \{v \in V_L : y_v = $

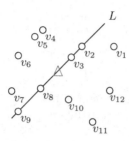

Fig. 1. Partitioning V into $V_L \cup V_L^+ \cup V_L^-$, where $V_L = \{v_2, v_3, v_8, v_9\}$, $V_L^+ = \{v_1, v_{10}, v_{11}, v_{12}\}$, $V_L^- = \{v_4, v_5, v_6, v_7\}$, $V_{L,\triangle}^{\nearrow} = \{v_2, v_3\}$, and $V_{L,\triangle}^{\searrow} = \{v_8, v_9\}$. If voters are of unit weight, then $w_L = w_L^+ = w_L^- = 4$ and $w_{L,\triangle}^{\nearrow} = w_{L,\triangle}^{\searrow} = 2$

$y_\triangle$ and $x_v < x_\triangle\}$. For simplicity, we denote $w(V_{L,\triangle}^{\nearrow})$ and $w(V_{L,\triangle}^{\searrow})$ by $w_{L,\triangle}^{\nearrow}$, and $w_{L,\triangle}^{\searrow}$, respectively. See Figure 1 for an illustration.

Lemma 1. *Let V be a set of voters in $\mathbb{R}^2$. If all voters in V are collinear, then*

(i) $\triangle_V = \mathcal{M}$;
(ii) $|\triangle_V| \geq 1$.

Lemma 2. *In $\mathbb{R}^2$, a point $\triangle$ is a plurality point if and only if for any line L passing through $\triangle$, $w_L^+ \leq w(V)/2$ and $w_L^- \leq w(V)/2$.*

Lemma 3. *Let V be a set of voters and $\triangle$ be a point in $\mathbb{R}^2$. If $\triangle \notin V$, then the following statements are equivalent:*

(i) $\triangle$ is a plurality point;
(ii) $w_L^+ = w_L^-$ for each line L passing through $\triangle$;
(iii) $w_{L,\triangle}^{\nearrow} = w_{L,\triangle}^{\searrow}$ for each line L passing through $\triangle$.

The lines passing through $\triangle$ and a voter $v_i \in V$ are called the *voting lines* of V with respect to $\triangle$. Note that the number of distinct voting lines of V with respect to $\triangle$ may not be exactly n, since some voters may be collinear with other voters and $\triangle$.

3   NP-Hardness of the Minimum-Cost Plurality Problem

In this section, we show that the minimum-cost plurality problem is NP-hard, even when the cost of all voters are equal. The decision version of this problem is given in Problem 2.

Problem 2 (The Unit-cost Plurality Problem (UCP)). Given a finite set V of weighted voters in $\mathbb{R}^2$ and an integer k with $0 \leq k < |V|$, each voter v is assigned with unit cost and a weight $w(v)$. UCP is to determine whether there is a subset V' of V such that $V \setminus V'$ admits a plurality point, and $c(V') \leq k$.

We show the NP-completeness of UCP by reducing the *maximum subset sum problem* (MSS) to it. The problem definition is given as below.

Problem 3 (The Maximum Subset Sum Problem (MSS)). Given a multiset of integers $A = \{a_1, a_2, \ldots, a_m\}$ and an integer k where $0 \leqslant k < m$, MSS is to determine if there is a subset A' of A such that $\sum_{a_i \in A \setminus A'} a_i = 0$ and $|A'| \leqslant k$.

Given a multiset of integers A, the *subset sum problem*, which is a well-known NP-complete problem, asks if there is a nonempty subset of A which sums to zero. Note that MSS is very similar to the subset sum problem, where the only difference in between is the size constraint. Hence, MSS is also NP-complete since we can answer the question of the subset sum problem throughout solving MSS by setting $k = m - 1$.

In the following, we present a reduction from MSS to UCP. Given an instance (A, k) of MSS, the reduction is done via transforming (A, k) into an instance $(\mathcal{R}(A), k)$ of UCP. We note here that MSS is NP-complete even for the case where $a \neq 0$, for all $a \in A$. Thus, for simplicity, we assume that there is no zero in A. Let $A = \{a_1, a_2, \ldots, a_m\}$. The instance $(\mathcal{R}(A), k)$, where $\mathcal{R}(A)$ consists of a set V of voters in $\mathbb{R}^2$ and a weight function $w \colon V \to \mathbb{R}^+$, satisfies the following three conditions.

- $V = \{v_1, \ldots, v_{3m}\}$.
- For $1 \leqslant i \leqslant m$, v_i corresponds to point $(\frac{a_i}{|a_i|} i, 0)$ with $w(v_i) = |a_i|$.
- For $m < i \leqslant 3m$, voter v_i is located on a unit circle such that $x_j = -x_{j+m}$, $y_j = -y_{j+m}$, $y_j > 0$, and $y_j / x_j \neq y_{j'} / x_{j'}$, for $m < j < j' \leqslant 2m$. The weight $w(v_i)$ is set to be $1 + \sum_{j=1}^m |a_j|$.

See Figure 2 for an illustration. Note that there are infinitely many rational coordinates satisfying the last condition. For example, we may define $x_j = \frac{2j+1}{2j^2+2j+1}$ and $y_j = \frac{2j^2+2j}{2j^2+2j+1}$, for $m < j \leqslant 2m$. With this fact, it is not difficult to verify that the construction of $\mathcal{R}(A)$ can be done in polynomial time. Moreover, $k < m < 3m = |V|$, with which we know $(\mathcal{R}(A), k)$ is an instance for UCP.

Lemma 4. *MSS returns true on (A, k) if and only if UCP returns true on $(\mathcal{R}(A), k)$.*

Proof. For necessity, suppose that there is a nonempty subset A' of A where $\sum_{a_i \in A \setminus A'} a_i = 0$ and $|A'| \leqslant k$. Let $V' = \{v_i \colon a_i \in A'\}$. We claim that $\triangle = (0,0)$ is a plurality point of $V \setminus V'$. Clearly, $c(V') = |V'| \leqslant k$. Since there is no voter on $\triangle$, according to Lemma 3, it suffices to show that each voting line L passing through $\triangle$ satisfies $w_{L,\triangle}^{\nearrow} = w_{L,\triangle}^{\swarrow}$. Let $V_1 = \{v_i \colon 1 \leqslant i \leqslant m, a_i > 0\}$, $V_2 = \{v_i \colon 1 \leqslant i \leqslant m, a_i < 0\}$, and $V_3 = \{v_i \colon m < i \leqslant 3m\}$. We categorize the voting lines passing through $\triangle$ into two types: (i) those passing through voters in $V_1 \cup V_2$, and (ii) those passing through voters in V_3. For voting lines of type (i), according to the reduction, there is only one such line L, which is the x-axis. Together with the fact that $\sum_{a_i \in A \setminus A'} a_i = 0$, we have $w_{L,\triangle}^{\nearrow} = w(V_1) = w(V_2) = w_{L,\triangle}^{\swarrow}$. For voting lines of type (ii), by the definition of voters in V_3, we have m such lines, on each of which there are exactly two voters. Denote these m voting lines by

L_j, $1 \leqslant j \leqslant m$, such that voters v_{m+j} and v_{2m+j} are on L_j. Since voters in V_3 are of the same weight, it follows that $w_{L_j, \triangle}^{\nearrow} = w(v_{m+j}) = w(v_{2m+j}) = w_{L_j, \triangle}^{\searrow}$.

For sufficiency, suppose that there is a subset V' of V such that $V \setminus V'$ admits a plurality point $\triangle$ with $|V'| \leqslant k$. Let $V_r = \{v_i : 1 \leqslant i \leqslant m, v_i \notin V', a_i > 0\}$ and $V_l = \{v_i : 1 \leqslant i \leqslant m, v_i \notin V', a_i < 0\}$.

First, we show that $\triangle$ lies within the unit circle centered at $(0,0)$. If $\triangle$ is outside the unit circle, let L be the line passing through $\triangle$ tangent to the circle. Without loss of generality let $(0,0) \in V_L^+$, then we have $w_L^+ > w(V \setminus V')/2$ due to the fact that $v_{m+1}, \ldots, v_{3m}$ lie on the same side of L. This contradicts the assumption that $\triangle$ is a plurality point according to Lemma 2.

Next, we show that $\triangle$ lies on the x-axis. Suppose to the contrary that $\triangle$ lies within the unit circle but not on the x-axis. According to the assumption that $k < m$, there is a voter $v_\ell \in V_r \cup V_l$. Without loss of generality, we assume that $v_\ell \in V_r$. Let L be the line passing through $(\ell, 0)$ and $\triangle$, and p be an intersection of L and the unit circle other than $(\ell, 0)$. If there is no voter on p, then obviously $w_{L,\triangle}^{\nearrow} \neq w_{L,\triangle}^{\searrow}$. Otherwise, there is a voter v_i lying on p satisfying $i > m$. Since $w(v_i) = 1 + \sum_{j=1}^{m} |a_j|$ and there is exactly one voter, v_ℓ, on $(\ell, 0)$, we have $w_{L,\triangle}^{\nearrow} \neq w_{L,\triangle}^{\searrow}$. By Lemma 3, $\triangle$ is not a plurality point of $V \setminus V'$. Therefore, $\triangle$ lies on the x-axis.

Let $A' = \{a_i : 1 \leqslant i \leqslant m, v_i \in V'\}$. According to the above arguments, one voting line, L, of $V \setminus V'$ with respect to $\triangle$ is the x-axis. By Lemma 3, we have $w_{L,\triangle}^{\nearrow} = w_{L,\triangle}^{\searrow}$, which implies $\sum_{i \in V_l} |a_i| = \sum_{i \in V_r} |a_i|$. It follows that $\sum_{A \setminus A'} a_i = 0$. Moreover, $|A'| \leqslant |V'| \leqslant k$, and the lemma is proved. □

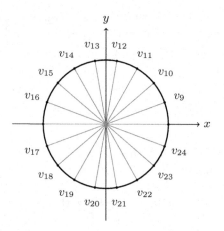

Fig. 2. An illustration of $2m$ voters lying on the unit circle, where $m = 8$. Note that voters $v_1, \ldots, v_8$ locate at the x-axis outside the unit circle.

According to the above polynomial-time reduction, we conclude this section with the following theorem.

Theorem 1. *UCP is NP-complete.*

4   The Case Where Voters Are Equally Weighted

Given a set V of voters, let D_V^* be an optimal feasible set of V, and V^* be $V \setminus D_V^*$. Since the minimum-cost plurality problem is NP-hard, we consider a special case where all voters are equally weighted in this section, and give a polynomial-time algorithm. For simplicity, all voters are assigned with unit weight and thus $w(V) = |V|$. Recall that $\triangle_V$ denotes the set of all plurality points of V. Two cases are discussed in order to find a feasible set with minimum cost: (i) $\triangle_{V^*} \cap V^* \neq \emptyset$; and (ii) $\triangle_{V^*} \cap V^* = \emptyset$. The set of possible locations that can be a plurality point for case (i) consists of n voters since $V^* \subseteq V$, while that for case (ii), as shown in Section 4.2, consists of $\binom{n}{4}$ voters. The strategy to compute an optimal feasible set is to treat all possible locations as the plurality points one by one, compute the corresponding feasible sets, and return the set of minimum cost as a solution.

4.1   The Case $\triangle_{V^*} \cap V^* \neq \emptyset$

When V^* consists of an odd number of voters, Lemma 5 shows that each line L passing through $\triangle^*$ should be *balanced*, that is, $w_{L,\triangle^*}^{\nearrow} = w_{L,\triangle^*}^{\nwarrow}$. However, the situation becomes more complicated when there are an even number of voters in V^* and is discussed in Lemmas 6, 7, and 8. Due to the space limitation, the proofs of these lemmas are omitted and will be given in the journal version.

Lemma 5. *Let U be a set of unit-weighted voters in $\mathbb{R}^2$. If $\triangle^* \in \triangle_U \cap U$ and $|U|$ is odd, then $w_{L,\triangle^*}^{\nearrow} = w_{L,\triangle^*}^{\nwarrow}$ for any line L passing through $\triangle^*$.*

Proof. Suppose to the contrary that there is a line L passing through $\triangle^*$ with $w_{L,\triangle^*}^{\nearrow} \geqslant w_{L,\triangle^*}^{\nwarrow} + 1$. Let L' and L'' be two lines resulting from slightly rotating L around $\triangle^*$ clockwise and counterclockwise, respectively, before they reach any other voter. See Figure 3 for an illustration. Since $\triangle^*$ is the only voter on L and $|U|$ is odd, we have $w_L^+ = w_L^- = (w(U) - 1)/2$. Without loss of generality let $U_{L,\triangle^*}^{\nearrow} \subseteq U_{L'}^-$ and $U_{L,\triangle^*}^{\nwarrow} \subseteq U_{L''}^+$. It follows that $w_{L''}^+ = w_L^+ + w_{L,\triangle^*}^{\nearrow} - w_{L,\triangle^*}^{\nwarrow} \geqslant (w(U) - 1)/2 + 1 > w(U)/2$, which contradicts the assumption that $\triangle^* \in \triangle_U$ according to Lemma 2. Hence, we have $w_{L,\triangle^*}^{\nearrow} = w_{L,\triangle^*}^{\nwarrow}$ for each line L passing through $\triangle^*$.   □

Lemma 6. *Let U be a set of unit-weighted voters in $\mathbb{R}^2$. If $\triangle^* \in \triangle_U \cap U$ and $|U|$ is even, then $|w_{L,\triangle^*}^{\nearrow} - w_{L,\triangle^*}^{\nwarrow}| \leqslant 1$ for any line L passing through $\triangle^*$.*

In the case where the number of voters is even, although it may be *unbalanced* for any line L passing through $\triangle^*$, i.e., L is not balanced, the bias generated by L could be adjusted by neighboring lines. To further elaborate this observation, we focus on the voting lines of V with respect to $\triangle^*$. The relationship between those adjacent voting lines is shown in the following lemma. Let $L_1, L_2, \ldots, L_r$ be all voting lines of V with respect to $\triangle^*$ sorted by their polar angles with respect to the x-axis in counterclockwise order. Two unbalanced voting lines L_i and L_j are said to be *adjacent* if L_t is balanced, for $i < t < j$. The following lemma states the relationship between two unbalanced voting lines.

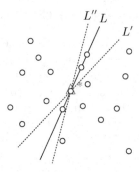

Fig. 3. Slightly rotating L in opposite directions to get L' and L''

Lemma 7 (alternating property). *Let U be a set of unit-weighted voters in $\mathbb{R}^2$. If $\triangle^* \in \triangle_U \cap U$ and $|U|$ is even, for any adjacent unbalanced voting lines L_i and L_j of U with respect $\triangle^*$, we have either*

- $w^{\nearrow}_{L_i,\triangle^*} > w^{\nwarrow}_{L_i,\triangle^*}$ *and* $w^{\nearrow}_{L_j,\triangle^*} < w^{\nwarrow}_{L_j,\triangle^*}$, or
- $w^{\nearrow}_{L_i,\triangle^*} < w^{\nwarrow}_{L_i,\triangle^*}$ *and* $w^{\nearrow}_{L_j,\triangle^*} > w^{\nwarrow}_{L_j,\triangle^*}$.

In fact, the *alternating property* can be extended to form a sufficient condition for $\triangle \in \triangle_U \cap U$ with $|U|$ being even. We summarize this main result in the following lemma.

Lemma 8. *Let U be a set of unit-weighted voters in $\mathbb{R}^2$ with $|U|$ being even. A point $\triangle \in U$ is a plurality point of U if and only if*

1. *the number of unbalanced voting lines of U with respect to $\triangle$ is odd, and*
2. *all the adjacent unbalanced voting lines of U w.r.t. $\triangle$ satisfy the alternating property.*

4.2 The Case $\triangle_{V^*} \cap V^* = \emptyset$

Let $\mathcal{I}_U$ be the set of intersection points of two lines segments $\overline{v_a v_b}$ and $\overline{v_c v_d}$, where v_a, v_b, v_c and v_d are four different voters in U.

Lemma 9. *Let U be a set of unit-weighted voters in $\mathbb{R}^2$. If $\triangle_U \cap U = \emptyset$, then any plurality point $\triangle^*$ of U is in $\mathcal{I}_U$.*

Proof. By Lemma 3, we have $w^{\nearrow}_{L,\triangle^*} = w^{\nwarrow}_{L,\triangle^*}$ for each line L passing through $\triangle^*$. Since all voters are equally weighted, we have $|U^{\nearrow}_{L,\triangle^*}| = |U^{\nwarrow}_{L,\triangle^*}|$ for each line L passing through $\triangle^*$. If there are two different voting lines passing through $\triangle^*$, then $\triangle^*$ is the intersection of the voting lines, formed by at least four voters in U, and we have $\triangle^* \in \mathcal{I}_U$. If there is at most one voting line passing through $\triangle^*$, then all voters are collinear. By Lemma 1, any multidimensional median is a plurality point of U, which contradicts that $\triangle_U \cap U = \emptyset$. $\square$

4.3   A Polynomial-Time Algorithm

For a given set V of voters, if a point $\triangle$ is known to be a plurality of V^*, one can apply Lemmas 3, 5, and 8 to determine V^*. Moreover, by Lemma 9, we have $\triangle \in V^* \cup \mathcal{I}_{V^*}$, which implies $\triangle \in V \cup \mathcal{I}_V$ immediately. Based on the observations, the idea of our algorithm is described as follows. First, each point $\triangle$ in $V \cup \mathcal{I}_V$ is taken as a candidate to be a plurality point. Second, we compute a "best possible" subset of V, whose removal makes $\triangle$ a plurality point. At last, the optimal solution V^* can be found from those "best possible" subsets. In more detail, for a given candidate $\triangle$, if the corresponding V^* satisfies $\triangle_{V^*} \cap V^* = \emptyset$ or $\triangle_{V^*} \cap V^* \neq \emptyset$ with $|V^*|$ being odd, we make sure that each voting line is balanced by deleting some voters appropriately via BALANCEDPLURALITY. For the case where $\triangle_{V^*} \cap V^* \neq \emptyset$ with $|V^*|$ being even, we introduce a dynamic programming algorithm called PLURALITYDP, which computes the corresponding subset by iteratively computing the scoring function of v_i. We then give a polynomial-time algorithm called UNITWEIGHTPLURALITY to solve the minimum-cost plurality problem when all voters are equally weighted.

Algorithm 1: BALANCEDPLURALITY($V = \{v_1, v_2, \ldots, v_n\}, \triangle$)

---

1   $D := \emptyset$;
2   Compute all the voting lines of V with respect to $\triangle$;
3   **foreach** *voting line* L **do**
4   **while** $w^{\nearrow}_{L,\triangle} > w^{\searrow}_{L,\triangle}$ **do**
5           $v_d := \arg\min_{v \in V^{\nearrow}_L} c(v)$;
6           $D := D \cup \{v_d\}$;
7   **while** $w^{\nearrow}_{L,\triangle} < w^{\searrow}_{L,\triangle}$ **do**
8           $v_d := \arg\min_{v \in V^{\searrow}_L} c(v)$;
9           $D := D \cup \{v_d\}$;
10 **return** D;

PLURALITYDP is developed based on Lemma 8. The idea is to add a voting line L_i at the ith iteration to compute the best situation so far for each of the following three cases: $w^{\nearrow}_{L_i,\triangle^*} > w^{\searrow}_{L_i,\triangle^*}$, $w^{\nearrow}_{L_i,\triangle^*} = w^{\searrow}_{L_i,\triangle^*}$, and $w^{\nearrow}_{L_i,\triangle^*} < w^{\searrow}_{L_i,\triangle^*}$. For any given $\triangle^*$, let $A_i = w^{\nearrow}_{L_i,\triangle^*}$ and $B_i = w^{\searrow}_{L_i,\triangle^*}$. We define the scoring functions used in the dynamic programming algorithm in the following.

The scoring functions $S^+_e(i)$, $S^-_e(i)$, $S^{\oplus}_e(i)$, $S^{\ominus}_e(i)$, $S^+_o(i)$, $S^-_o(i)$, $S^{\oplus}_o(i)$, and $S^{\ominus}_o(i)$ are defined as follows by denoting the superscript as the variable a and the subscript as the variable b, where $a \in \{+, -, \oplus, \ominus\}$ and $b \in \{e, o\}$, for $1 \leqslant i \leqslant k$. Suppose that $\triangle \in V$ and $L_1, L_2, \ldots, L_k$ are all voting lines of V w.r.t. $\triangle$. Let V_i be the set of all voters on $L_1, L_2, \ldots, L_i$, where $i \in \{1, \ldots, k\}$. For $a \in \{+, -\}$, we denote $D^a_b(i)$ as the set of all feasible set D of V_i with $\triangle$ being the plurality point satisfying $A_i > B_i$ ($A_i < B_i$) after deleting all voters in D if $a = +$ ($a = -$), and there are an even (odd) number of unbalanced voting lines of $V_i \setminus D$ w.r.t.

$\triangle$ if $b = e$ $(b = o)$. Similarly for $a \in \{\oplus, \ominus\}$, we denote $D_b^a(i)$ as the set of all feasible set D of V_i with $\triangle$ being the plurality point satisfying $A_i = B_i$ after deleting all voters in D, and for the greatest integer $h < i$ satisfying $A_h \neq B_h$, we have $A_h > B_h$ $(A_h < B_h)$ if $a = \oplus$ $(a = \ominus)$, and there are even (odd) number of unbalanced voting lines of $V_i \setminus D$ w.r.t. $\triangle$ if $b = e$ $(b = o)$. Hence, we define $S_b^a(i) = \min_{D \in D_b^a(i)} w(D)$ for any $a \in \{+, -, \oplus, \ominus\}$ and $b \in \{e, o\}$. Finally, we denote $D^\odot(i)$ as the set of all feasible set D of V_i with $\triangle$ being the plurality point satisfying $A_j = B_j$ after deleting all voters in D, $\forall j \in \{1, 2, \dots, i\}$. That is, all voting lines of $V_i \setminus D$ w.r.t. $\triangle$ are balanced, $\forall D \in D^\odot(i)$. Hence, we define $S^\odot(i) = \min_{D \in D^\odot(i)} w(D)$.

For simplicity, the following recursive relation of scoring functions are presented for the case where all voters are assigned with unit cost. We demonstrate the modification for a general case in the upcoming paragraph. By the alternating property, for $i = 2, \dots, k$, we have

$$S_e^+(i) = |A_i - B_i - 1| + \min\{S_o^-(i-1), S_o^\ominus(i-1)\},$$
$$S_e^-(i) = |A_i - B_i + 1| + \min\{S_o^+(i-1), S_o^\oplus(i-1)\},$$
$$S_e^\oplus(i) = |A_i - B_i| + \min\{S_e^+(i-1), S_e^\oplus(i-1)\},$$
$$S_e^\ominus(i) = |A_i - B_i| + \min\{S_e^-(i-1), S_e^\ominus(i-1)\},$$
$$S_o^+(i) = |A_i - B_i - 1| + \min\{S_e^-(i-1), S_e^\ominus(i-1), S^\odot(i-1)\},$$
$$S_o^-(i) = |A_i - B_i + 1| + \min\{S_e^+(i-1), S_e^\oplus(i-1), S^\odot(i-1)\},$$
$$S_o^\oplus(i) = |A_i - B_i| + \min\{S_o^+(i-1), S_o^\oplus(i-1)\},$$
$$S_o^\ominus(i) = |A_i - B_i| + \min\{S_o^-(i-1), S_o^\ominus(i-1)\}, \text{and}$$
$$S^\odot(i) = |A_i - B_i| + \min\{S^\odot(i-1)\},$$

with initial conditions $S_e^+(1) = S_e^-(1) = S_e^\oplus(1) = S_e^\ominus(1) = S_o^\oplus(1) = S_o^\ominus(1) = \infty$, $S_o^+(1) = |A_1 - B_1 - 1|$, $S_o^-(1) = |A_1 - B_1 + 1|$, and $S^\odot(1) = |A_1 - B_1|$.

The absolute value added in each formula is an abbreviation for different cases. For example, suppose that we want to get $A_i = B_i + 1$ after deletion, then $A_i - B_i - 1$ voters on L_i should be eliminated if $A_i > B_i$, and $B_i - A_i + 1$ voters on L_i should be eliminated if $B_i > A_i$. Since $|A_i - B_i - 1| = A_i - B_i - 1$ if $A_i > B_i$ and $|A_i - B_i - 1| = B_i - A_i + 1$ if $B_i > A_i$, we set $|A_i - B_i - 1|$ to combine these two values. In addition, for the case where the costs are different, by replacing the added term of absolute value $f_i = |A_i - B_i - 1|$ to $f_i' = \min_{V' \subseteq \hat{V}_i, |V'| = f_i} c(V')$, where $\hat{V}_i = V_i \setminus (\bigcup_{j=1,\dots,i-1} V_j)$, we obtain the recursive function for the general problem of non-unit cost. Other cases can be derived with the same argument.

Hence, to find an optimal feasible set when all voters are equally weighted, the objective is to compute $\min\{S_o^+(k), S_o^-(k), S_o^\oplus(k), S_o^\ominus(k)\}$. We name this dynamic programming procedure as PLURALITYDP$(V, \triangle)$, which returns an optimal feasible set D^* such that $\triangle \in \triangle_{V^*} \cap V^*$ and $|V^*|$ is even.

Theorem 2. *Let V be a set of voters in $\mathbb{R}^2$. The minimum-cost plurality problem can be solved in $O(n^5 \log n)$ time when all voters are equally weighted, where $n = |V|$.*

Algorithm 2: UNITWEIGHTPLURALITY($V = \{v_1, v_2, \ldots, v_n\}$)

---

1 $D^* := V$;
2 **if** *all voters are collinear* **then**
3     $D^* := \emptyset$;
4 **else**
5   Compute $\mathcal{I}_V$;
6   **foreach** $\triangle \in \mathcal{I}_V$ **do**
7         $D := $ BALANCEDPLURALITY$(V, \triangle)$;
8         $D := D \cup (\{\triangle\} \cap V)$;
9   **if** $c(D) \leqslant c(D^*)$ **then**
10             $D^* := D$;
11   **foreach** $v_i \in V$ **do**
12         $D := $ BALANCEDPLURALITY(V, v_i);
13   **if** $c(D) \leqslant c(D^*)$ **then**
14             $D^* := D$;
15         $D := $ PLURALITYDP(V, v_i);
16   **if** $c(D) \leqslant c(D^*)$ **then**
17             $D^* := D$;
18 **return** D^*;

Proof. The correctness of Algorithm 2 follows from Lemmas 3 and 5–9. For the time complexity, let $\triangle^*$ be the plurality point of V^*. Lines 5–10, lines 12–14, and lines 15–17 manipulate the cases where $\triangle_{V^*} \cap V^* = \emptyset$, $\triangle_{V^*} \cap V^*$ with $|V^*|$ being odd, and $\triangle_{V^*} \cap V^*$ with $|V^*|$ being even, respectively. All the manipulations need the information about the voting lines of V with respect to $\triangle^*$, and such information can be computed in $O(n \log n)$ time via sorting the slopes of the n lines, each of which passes through $\triangle^*$ and a point in V. Moreover, based on the information given above, PLURALITYDP takes $O(n)$ time. Therefore, the overall time complexity is $O(n^5 \log n)$ since the most time-consuming loop, lines 5–10, repeated $|\mathcal{I}_V| = O(n^4)$ times. $\square$

5   Concluding Remarks

In this paper, we deal with the problem where no plurality point exists with respect to a given instance and try to find one via eliminating some voters with minimum cost. In $\mathbb{R}^2$, the problem is NP-hard, whereas it is polynomial-time solvable when all voters' weights are equal. We note here that all the procedures given above can be extended immediately to $\mathbb{R}^d$, for $d \geqslant 3$, except PLURALITYDP. It remains open whether the problem is polynomial-time solvable in $\mathbb{R}^3$ for the case where voters' weights are equal.

Acknowledgements. The authors would like to thank the anonymous reviewers for helpful comments. Wei-Yin Lin, Yen-Wei Wu, and Kun-Mao Chao were supported in part by MOST grants 101-2221-E-002-063-MY3 and 103-2221-E-002-157-MY3, and Hung-Lung Wang was supported in part by MOST grant 103-2221-E-141-004 from the Ministry of Science and Technology, Taiwan.

References

1. Bandelt, H.-J.: Networks with Condorcet solutions. European Journal of Operational Research 20(3), 314–326 (1985)
2. Bartholdi, J.J., Tovey, C.A., Trick, M.A.: How hard is it to control an election? Mathematical and Computer Modeling 16(8–9), 27–40 (1992)
3. Campos Rodríguez, C.M., Moreno Pérez, J.A.: Relaxation of the Condorcet and Simpson conditions in voting location. European Journal of Operational Research 145(3), 673–683 (2003)
4. Davis, O.A., Hinich, M.J., Ordeshook, P.C.: An expository development of a mathematical model of the electoral process. The American Political Science Review 64(2), 426–448 (1970)
5. Durier, R.: Continuous location theory under majority rule. Mathematics of Operations Research 14(2), 258–274 (1989)
6. Hansen, P., Thisse, J.-F.: Outcomes of voting and planning: Condorcet, Weber and Rawls locations. Journal of Public Economics 16(1), 1–15 (1981)
7. Kress, D., Pesch, E.: Sequential competitive location on networks. European Journal of Operational Research 217(3), 483–499 (2012)
8. Labbé, M.: Outcomes of voting and planning in single facility location problems. European Journal of Operational Research 20(3), 299–313 (1985)
9. McKelvey, R.D., Wendell, R.E.: Voting equilibria in multidimensional choice spaces. Mathematics of Operations Research 1(2), 144–158 (1976)
10. Wendell, R.E., McKelvey, R.D.: New perspectives in competitive location theory. European Journal of Operational Research 6(2), 174–182 (1981)
11. Wendell, R.E., Thorson, S.J.: Some generalizations of social decisions under majority rule. Econometrica 42(5), 893–912 (1974)
12. Wu, Y.-W., Lin, W.-Y., Wang, H.-L., Chao, K.-M.: Computing plurality points and Condorcet points in Euclidean space. In: Proceedings of the International Symposium on Algorithms and Computation, pp. 688–698 (2013)

Approximate Distance Oracle in $O(n^2)$ Time and $O(n)$ Space for Chordal Graphs

Gaurav Singh, N.S. Narayanaswamy, and G. Ramakrishna

Department of Computer Science and Engineering,
Indian Institute of Technology Madras, India
{gsingh,swamy,grama}@cse.iitm.ac.in

Abstract. We preprocess a given unweighted chordal graph G on n vertices in $O(n^2)$ time to build a data structure of $O(n)$ size such that any subsequent distance query can be answered in constant time with a bounded constant factor error. In particular, for each pair of vertices $u_i, u_j \in V(G), 1 \leq i, j \leq n$, we take constant time to output a distance value $d_{ij} \leq 2d_G(u_i, u_j) + 8$ using our data structure, where d_G is the distance between u_i and u_j in G. In contrast, for the closely related APSP problem on chordal graphs, the current best algorithm runs in $O(n^{2.373})$ time. Our improvement comes from a relationship that we discover between the graph distance and minimum hitting sets of cliques on certain paths in a clique tree associated with a chordal graph. We design an efficient data structure which additively approximates (error of $+3$) these minimum hitting sets of cliques for all the paths in the clique tree. This data structure is then integrated with an efficient data structure which answers LCA queries in rooted trees to yield our distance oracle for the given chordal graph.

1 Introduction

The All Pairs Shortest Path (APSP) problem is one of the most fundamental problems in graph theory. For a given graph $G = (V, E)$, where V is the set of vertices and E is the set of edges with $|V| = n$ and $|E| = m$, the *all pairs shortest path* problem on G is to compute the shortest paths between all the pairs of vertices of G. A *distance oracle* is a data structure that stores the APSP information and can respond to distance queries efficiently. The time to set up this oracle, the amount of space it uses, the query response time, and the update time are among the most well-studied problems in the area of Distance Oracles. Many questions about this data structure still remain open and it is of immense practical relevance. Surprisingly, despite the fact that APSP is one of the oldest and fundamental problems in computer science, there does not exist any algorithm that can solve APSP in truly sub cubic time, i.e., $O(n^{3-\epsilon})$ time for some $\epsilon > 0$. This has motivated researchers to design sub cubic time algorithms that can approximately answer distances between any pair of vertices. In this paper, we restrict our attention on *chordal graphs* with an intent to explore how the underlying graph structure can play an important role in setting up

M.S. Rahman and E. Tomita (Eds.): WALCOM 2015, LNCS 8973, pp. 89–100, 2015.

efficient distance oracles. We give an algorithm to preprocess a chordal graph G in $O(n^2)$ time to setup an approximate distance oracle that occupies $O(n)$ space. Subsequently, this distance oracle answers the distance query between u_i and u_j in constant time, and the given answer is at most $2d_G(u_i, u_j) + 8$, where d_G is the distance between u_i and u_j in G.

Related Work. A classical algorithm for solving APSP in general graphs is given by Floyd and Warshall [1]. It outputs an $n \times n$ matrix that contains the shortest distances between every pair of vertices in $\Theta(n^3)$ time. Seidel [2] has given an algorithm to solve APSP problem in $O(M(n) \log n)$ time, where $M(n)$ is the time to multiply two $n \times n$ matrices. The current best known algorithm for finding $M(n)$ runs in $O(n^{2.373})$ time [3]. Han et al. [4] have given an algorithm to solve APSP problem for chordal graphs in $O(M(n))$ time. Gavoille et al. [5] have designed a distance oracle of size $O(n\log^2 n)$ for chordal graphs, such that their distance oracle can answer the distance queries in constant time with an error of $+1$. The distance oracle designed in this paper is better than the oracle presented by Gavoille et al. [5] in terms of the size of the oracle, but we loose the battle when it comes to the approximation factor of the answer returned by the oracle. APSP can be solved in $O(n^2)$ time for some specific subclasses of chordal graphs. Ravi et al. [6] have given an algorithm to solve APSP optimally in $O(n^2)$ time for *interval graphs*. In the same year, Radhakrishnan et al. [7] gave algorithm to solve APSP problem for *k-trees* in $O(n^2)$ time. One year later, Dahlhaus [8] devised algorithms to solve APSP problem for *strongly chordal graphs*, *distance hereditary graphs*, *path graphs* and *permutation graphs* in $O(n^2)$ time.

Thorup et al. [9] proved that for any integer $k \geq 1$, a general graph can be preprocessed in $O(kmn^{1/k})$ time to produce a distance oracle that occupies $O(n^{1+1/k})$ space. Subsequent distance queries can be answered approximately in $O(k)$ time with an approximation factor of $2k - 1$. Cohen et al. [10] have designed an algorithm for solving APSP approximately on weighted undirected graphs. Their algorithm takes $\tilde{O}(n^2)$ time to build a data structure of size $O(n^2)$. This data structure is capable of answering distance queries in constant time within a stretch of 3. Baswana et al. [11] proved that general unweighted graphs can be preprocessed to set up a distance oracle of size $O(kn^{1+1/k})$ in $O(\min(n^2, kmn^{1+1/k}))$ time to answer distance queries with stretch factor of $2k - 1$.

Relevance of Our Results. In Table 1 we contrast our results with the existing results. Thorup et al. [9] have proved that any distance oracle for general graphs, which gives distances with stretch factor strictly less then $2k+1$, requires $\Omega(n^{1+1/k})$ space. In particular, if stretch is strictly less than 3, then the size of oracle must be $\Omega(n^2)$. However, for chordal graphs with diameter at least 10, although our stretch is strictly less than 3 yet the size of data structure is only $O(n)$. Secondly, algorithms designed by both Seidel [2] and Han et al. [4] compute the square of the input graph G as a subroutine. Given a graph $G = (V, E)$, the square of the graph is defined as $G^2 = (V, E')$ where $(u, v) \in E'$ if and only if $1 \leq d_G(u, v) \leq 2$. Methods known for computing G^2 for a general graph or a chordal graph require $O(n^{2.373})$ time. Therefore, computation of G^2 acts as a bottleneck for solving APSP on chordal and general graphs. Han et al.[4] also proved that computing G^2 for *split graphs*, which is a subclass of chordal graphs, is as hard as the problem for general graphs. In this paper, we exploit structural properties of chordal graphs to bypass computation of G^2 and achieve an $O(n^2)$ running time for building distance oracle.

Table 1. Exact and approximate distance oracles for different classes of graphs

Graphs	Processing time	Size	Stretch	Query time
General unweighted graphs (Seidel [2])	$O(n^{2.373} \log n)$	$O(n^2)$	1	$O(1)$
General weighted graphs (Thorup et al. [9])	$O(kmn^{1/k})$	$O(n^{1+1/k})$	$2k-1$	$O(k)$
General weighted graphs (Baswana et al. [11])	$O(\min(n^2, kmn^{1+1/k}))$	$O(n^{1+1/k})$	$2k-1$	$O(k)$
Unweighted Chordal graphs (Han et al. [4])	$O(n^{2.373})$	$O(n^2)$	1	$O(1)$
Unweighted Chordal graphs (this paper)	$\boldsymbol{O(n^2)}$	$\boldsymbol{O(n)}$	**(2,8)**	$\boldsymbol{O(1)}$
Interval graphs (Ravi et al. [6])	$O(n^2)$	$O(n^2)$	1	$O(1)$
Strongly chordal graphs (Dahlhaus [8])	$O(n^2)$	$O(n^2)$	1	$O(1)$
k trees (Radhakrishnan et al. [7])	$O(kn^2)$	$O(n^2)$	1	$O(1)$

Outline of Our Algorithm. Given a chordal graph G, we first construct a clique tree T of G in linear time [12]. For each vertex $u_i \in V(G)$, we identify a maximal clique C_i in the clique tree T such that $u_i \in C_i$. To approximate the distance between u_i and u_j in G, we compute the size of a hitting set S which hits all the cliques in the path joining C_i and C_j in the clique tree T. The distance that is output is twice the value of $|S|$. We prove that the size of S is upper bounded by the number of vertices in any shortest path between u_i and u_j. Also, we prove that $d_G(u_i, u_j) \leq 2|S|$. We first construct a linear size data structure that can compute the size of the hitting set S for any pair of cliques in clique tree T in constant time. Finally we use this data structure to answer distance queries in constant time. Our main result is the following theorem.

Theorem 1. *Given a chordal graph G, we can be preprocess G in $O(n^2)$ time and $O(n^2)$ space to build a data structure of $O(n)$ size such that for any query pair $u_i, u_j \in V(G)$, a distance response upper bounded by $2d_G(u_i, u_j) + 8$ can be given in constant time.*

Outline of the Paper. In Section 2, we present preliminaries and basic notations used throughout this paper. In Section 3, we prove some novel properties related to shortest paths in the graph and hitting sets of certain maximal cliques in the associated clique tree of chordal graphs. We present the algorithms to construct distance oracle and to answer the distance queries in constant time in Section 4. Finally, Section 5 concludes the paper.

2   Preliminaries

Chordal Graphs. A *chord* of a cycle is an edge between two non adjacent vertices of the cycle. A graph $G = (V, E)$ is *chordal* if every cycle of length greater than three has a chord. The book on perfect graphs by Golumbic [13] is an excellent reference for chordal graphs. Let K_G be the set of maximal cliques of a chordal graph G and $K_G(v)$ be the set of maximal cliques of G containing v. It is well known that there exists a tree T, called the *clique tree* of G, whose vertices correspond bijectively to the elements of K_G; and for each $v \in V(G)$, the nodes of T corresponding to $K_G(v)$ induce a subtree of T. This property is called as *induced-subtree property* [14]. The closely related *clique intersection property* [14] is as follows: let v be present in two maximal cliques C_1 and C_2 in G, and let u_1 and u_2 be the corresponding tree nodes, respectively. Let u be a node on the path between u_1 and u_2 in T, and let C be the maximal clique in G corresponding to u. Then $v \in C$.

Hitting Set. For a set K of cliques in G, the *hitting set* S of K is a subset of $V(G)$ that has a non-empty intersection with each clique in K.

Least Common Ancestors. Let T' be a rooted tree on n vertices. Harel et al. [15], Scheiber et al. [16] and Powell [17] have given three different algorithms to preprocess T' in $O(n)$ time to build a data structure D of $O(n)$ size. Subsequent queries of the least common ancestor of two vertices can be answered in constant time. We will use the algorithm presented in [16] as a subroutine in our paper.

Abuse of Notation. Throughout, we assume that $G = (V, E)$ is a connected, unweighted chordal graph with $|V| = n$ and $|E| = m$ and T is a clique tree of G. Henceforth, by the word *clique*, we imply maximal clique in G Further, when we refer to a clique of a clique tree T, we actually mean the clique in G that a node in T corresponds to. For a vertex $u \in V(G)$, we use T_u to refer to the set of cliques in T that contain u. In other words, T_u is the subtree of T whose nodes correspond to $K_G(u)$. This notational abuse makes the writing, and we believe the reading, of the paper easier. Finally, when we refer to a vertex $u_i \in V(G)$, we mean that the subscript i satisfies $1 \leq i \leq n$. This is crucial because the index i uniquely identifies the vertex in the algorithm.

3   Distance Information from Clique Trees

We now show the connection between the distance in the graph between u and v and the hitting set of the cliques on the path connecting T_u and T_v in T.

Lemma 1. *Let $u, v \in V(G)$ such that $d_G(u, v) > 1$, and let $S \subseteq V$ be a minimum set of vertices that hits all the cliques in the path in T joining T_u and T_v. Then $|S| \leq d_G(u, v) - 1$, i.e., $|S|$ is upper bounded by the number of internal vertices in any shortest path joining u and v in G.*

Proof. Let P_{uv} be a shortest path between u and v in G. Clearly, no two edges of P_{uv} can be present in one maximal clique as this would contradict the premise that P_{uv} is the shortest path. Furthermore, each edge of P_{uv} is in a maximal clique. Let σ be the path joining T_u and T_v in T. First, we prove that any edge $e \in P_{uv}$ is present in some clique of σ. Let $\{e_1, e_2, \ldots, e_{|P_{uv}|}\}$ be the set of edges present in P_{uv} such that e_k is adjacent to e_{k-1} and e_{k+1}. For contradiction, let all the edges of P_{uv} except e_k be present in some clique of σ. Also, let C_k be a clique that contains e_k and $C_k \notin \sigma$. Since T is a tree (connected), there must exist a path between C_k and σ. Let $C'_k \in \sigma$, such that among all the cliques of σ, C'_k is at the minimum distance from C_k. Clearly, C'_k is present in the path joining C_{k-1} and C_k as well as in the path joining C_{k+1} and C_k, where C_{k-1} and C_{k+1} are cliques of σ which contain e_{k-1} and e_{k+1} respectively. Therefore, by clique intersection property C'_k contains the vertex common to e_k and e_{k-1}. Similarly, C'_k contains the vertex common to e_k and e_{k+1}, i.e., C'_k contains both the end points of e_k. Hence, e_k is also present in $C'_k \in \sigma$.

Let C_i and C_j be two cliques of σ, and e_i and e_j be two adjacent edges of P_{uv}, such that $e_i \in C_i$ and $e_j \in C_j$. By the clique intersection property, all the cliques in the path joining C_i and C_j contain the common vertex of e_i and e_j. Therefore, for any clique $C \in \sigma$, either C contains an edge of P_{uv} or it is present in the path joining some C_i and C_j, and in this case, C contains the common vertex of the edges e_i and e_j. Therefore, all the cliques of σ are definitely hit by the internal vertices of P_{uv} which are $d_G(u, v) - 1$ in number, i.e., $|S| \leq d_G(u, v) - 1$. Hence the lemma. □

Corollary 1. *Let u and v be two vertices in G, and C_u and C_v be two maximal cliques of T containing u and v, respectively, and let S be a minimum hitting set that hits all the cliques in the path joining C_u and C_v in T. Then $|S| \leq d_G(u, v) + 1$.*

Proof. If u and v are non-adjacent, then from Lemma 1 we know that the size of the hitting set of the path joining T_u and T_v is upper bounded by $d_G(u, v) - 1$. To this hitting set, if we add the vertices u and v, we get a hitting set of all the cliques on the paths from C_u to C_v. Therefore, $|S| \leq d_G(u, v) + 1$. But if u and v are adjacent in G, then there must exist a clique C_{uv} that contains both u and v. If C_{uv} is in the path joining C_u and C_v, then the path joining C_u and C_{uv} is hit by u and similarly the path joining C_{uv} and C_v is hit by v. Therefore, the size of the minimum hitting set for the path joining C_u and C_v is 2. If C_u is in the path joining C_{uv} and C_v, then the entire path joining C_u and C_v is hit by v (because the entire path joining C_{uv} and C_v is hit by v). Therefore, in this case the size of the hitting set is 1. Similarly, if C_v is in the path joining C_{uv} and C_u, then the size of the hitting set for the path joining C_u and C_v is 1.

By the above three cases, it is clear that if u and v are adjacent in G, even then $|S| \leq d_G(u, v) + 1$.   □

Theorem 2. *Let $S \subseteq V$ be a set of vertices that hits all the cliques in the path joining two cliques C_u and C_v in T, then there exists a path P between u and v in G, such that the size of P is upper bounded by $2|S|$, i.e., $|P| \leq 2|S|$.*

Proof. Let C_x and C_y be two adjacent cliques of the path joining C_u and C_v in T. Also, let x and y be the vertices of S which hit C_x and C_y respectively. Clearly, the distance between x and y is at most 2 because C_x and C_y are cliques and $C_x \cap C_y \neq \phi$ (intersection of adjacent cliques is non-empty in a clique tree). Therefore, if S is a hitting set of the path joining C_u and C_v in T, then, from the following steps, there exists a path from u to v of length at most $2|S|$ in G via the vertices of the hitting set S. The factor 2 is worst possible and is achieved in the case when S is an independent set.

We consider the elements of S as being ordered according to the order of occurrence on the path from C_u to C_v. Since any two consecutive cliques on the path from C_u to C_v share at least one common vertex, two consecutive vertices in the hitting set S can be connected by a path of length at most 2 via this common vertex. The hitting set vertices are now joined by a path of length at most $2|S| - 2$ in G. Further, by adding edges to u and v from the ends of this path, there is a path of length at most $2|S|$ from u to v in G. Hence the distance between u and v in G is at most $2|S|$.   □

We now show that a minimum hitting set of cliques in the path connecting T_u and T_v can be found efficiently.

Structure of a Minimum Hitting Set. Let $P = [C_1, C_2, \ldots, C_b]$ be the path joining T_u and T_v in T, and $u \in C_1, v \in C_b$. We reiterate that C_1 and C_b are the only cliques of P in T_u and T_v, respectively. In this part, it is useful to visualize this path as a sequence of cliques. Therefore a prefix of the path P is a prefix of the corresponding sequence. Let S be a minimum hitting set of the cliques in P. Let v_m be a vertex in C_1 such that it occurs in the maximum number of consecutive cliques in P. Consequently, there is a minimum hitting set S of cliques in P such that $v_m \in S$. Indeed, this is true because if we have a minimum hitting set S' that does not satisfy this property, then we can replace $x \in S' \cap C_1$ by v_m, and we obtain a minimum hitting set satisfying the desired property. This gives us an algorithm to find a minimum hitting set:

Data Structure for a Minimum Hitting Set: We now propose a data structure to partition the cliques in P into sub-paths $IS_1, IS_2, \ldots, IS_r$ such that for $1 \leq i \leq r$, sub path IS_i is the *maximum length prefix* of cliques in $P \setminus \{IS_1, \ldots, IS_{i-1}\}$ that have a common element. Apart from constructing this partition, we also construct a hitting set S of the cliques in P. S is initialized to be an empty set. To compute this maximum length prefix, we start with the empty prefix and a set M of common elements. M is initialized to contain all the vertices of the first clique of the sequence. Let C_j be the clique considered in the j-th iteration. Clearly, it is the first clique in the remaining sequence.

If $M \cap C_j = \phi$, then we have found the current prefix as the maximum length prefix. Let $v \in M \cap C_{j-1}$ be an arbitrary vertex. The hitting set S is updated to $S \cup \{v\}$. Now, M is then reset to be the set of all vertices in C_j. If $M \cap C_j$ is non-empty then C_j is added to the current prefix and M is updated to be $M \cap C_j$. With this modified M, we move to the next iteration till all the cliques in P are considered.

Theorem 3. *The* hitting set S *computed above is a minimum hitting set of cliques in* P. *Further, it can be computed in* $O(n^2)$ *time.*

Proof. The optimality is clear from the structure of a canonical minimum hitting set. The running time is also straightforward, as M in the data structure is upper bounded by the size of the maximum clique, and its intersection with another clique C is computed. Intersection between two subsets M and C of n element universe can be computed in $O(n)$ time by representing the sets as bit vectors. A chordal graph on n vertices has at most n maximal cliques, therefore, the set intersection will be done at most n times. Therefore, it follows that the hitting set can be computed in $O(n^2)$ time. Hence the theorem.   □

This theorem plays a crucial role in an algorithm in the next section in which we find the hitting set of the cliques in each path from the root clique in the rooted clique tree. The algorithm is presented in Algorithm 1, and also is a formal presentation of the algorithm analyzed in Theorem 3 for the case of the hitting set of cliques on the path between two cliques.

Lemma 2. *Let C_k be a clique on the path P connecting cliques C_1 and C_2 in the clique tree T. Let P_1 be the path from C_1 to C_k, and P_2 be the path from C_k to C_2. Let S, S_1 and S_2 be the minimum hitting sets for P, P_1, and P_2 respectively, then $|S| \le |S_1| + |S_2| \le |S| + 1$.*

Proof. Clearly, $S_1 \cup S_2$ is a hitting set for the cliques in P. Therefore, $|S| \le |S_1| + |S_2|$. In S, consider those elements that hit C_k. There can be at most 2 elements that hit C_k owing to the structure of a minimum hitting set. Let H_1 and H_2 be those elements of S that hit only cliques in P_1 (i.e. $P_1 \setminus C_k$) and P_2 (i.e. $P_2 \setminus C_k$), respectively. Further, by adding to H_1 one vertex of S, which is in C_k and covers P_1, we get a hitting set of cliques in P_1. Similiarly, by adding one vertex of S, which is in C_k and covers P_2, we get a hitting set of cliques in P_2. Therefore, $|S_1| \le |H_1| + 1$ and $|S_2| \le |H_2| + 1$. Also, $|S| \ge |H_1| + |H_2| + 1$. Therefore, it follows that $|S_1| + |S_2| \le |H_1| + |H_2| + 2 \le |S| + 1$.   □

Lemma 2 is used to find an efficient and good approximation to the hitting set of cliques in many paths in a clique tree. The efficiency is gained by using the exact hitting set of cliques in some paths in the clique tree to calculate approximate hitting sets of cliques in other paths.

4   Approximate Distance Oracle for Chordal Graphs

In this section, we present the details of the data structures involved in setting up the distance oracle. The preprocessing algorithm receives a chordal graph $G = (V, E)$ as an input and sets up the distance oracle for G. It first constructs a clique tree T of the chordal graph G. This is achieved using a known linear time algorithm [12] to construct a clique tree of a chordal graph. Each clique in the clique tree is assumed to be a maximal clique in the graph G. Indeed, if a clique in the clique tree is not a maximal clique, then we can convert it into a maximal clique in linear time.

Vertex to Maximal Clique Mapping: This is very crucial, as the idea is to approximate the distance between two vertices u_i, u_j in the graph by the hitting set of the maximal cliques in the path connecting the corresponding maximal cliques in T. Towards this end, an n-element array is used. In this array, the i-th element corresponds to u_i and it points to a maximal clique of clique tree T that contains u_i. This is achieved by inspecting each clique of T in an arbitrary order, and whenever the array entry corresponding to a vertex u_i is empty and is encountered in the current clique C_j, the array entry is updated to point to C_j. This again takes $O(n^2)$ time in the worst case, though it can be tightly written as $O(\omega(G)|K_G|)$ where $\omega(G)$ is the size of the maximum clique and K_G is the set of maximal cliques of G. We refer to this array as $\mathcal{H}$. To populate the array $\mathcal{H}$, the algorithm inspects each clique of the clique tree T. The size of a clique is bounded by n and also $|K_G| \leq n$. Hence, it takes $O(n^2)$ time to populate $\mathcal{H}$. Clearly, the array $\mathcal{H}$ requires $O(n)$ space only.

Hitting Set Oracle for a Rooted Clique Tree: We now present an algorithm to calculate the hitting set of cliques on all the paths originating from a single vertex C_r of the clique tree, where C_r is the arbitrarily assigned root of the clique tree. The algorithm is presented by way of a function definition $DFS(C, X, l)$ in Algorithm 1. This information is then used to calculate the approximate distance information efficiently for a distance query.

Lemma 3. *On termination of the function call $DFS(C_r, C_r, 1)$, for each $1 \leq i \leq n$, the array $HS[i]$ has the value of the minimum hitting set of the cliques in the path from C_r to clique C_i in the clique tree T.*

Proof. The correctness proof works based on Theorem 3. The proof is a direct application of Theorem 3 if the tree rooted at C_r is a path. In case the tree is not a path, let us focus on one particular node C_i in the tree. When the DFS traversal visits C_i, owing to the careful use of recursion, we can consider just the path from C_r to C_i in our analysis. Now we appeal to Theorem 3 to conclude that $HS[i]$ contains the minimum hitting set of the cliques on the path from C_r to C_i.   □

The array HS has one element per maximal clique in G. Therefore, it has at most n elements in it. Since a DFS traversal is used to populate HS using an $O(n)$ time set intersection performed at each node, it follows that the entries of

---

Algorithm 1. DFS(C,X,l)- Used to Set up the Hitting Set Oracle

---

1: Let i be the index of maximal clique C.
2: $X \leftarrow C \cap X$
3: **if** $X \neq \phi$ **then**
4:     HS[i] $= l$
    /* no increase in the size of hitting set to hit C */
5: **else**
6:     HS[i] $= l + 1$
    /* increase in the size of hitting set to hit C */
7: **end if**
8: **if** C is a leaf node in T **then**
9:     return
10: **else**
11:     **for** each $C' \in$ CHILDREN(C) **do**
    /* Find hitting sets of the cliques on the paths to the descendants of C */
12:       **if** $X \neq \phi$ **then**
13:         DFS(C',X,l)
14:       **else**
15:         $X \leftarrow C'$
16:         DFS(C',X,$l + 1$)
17:       **end if**
18:     **end for**
19: **end if**

HS are calculated in $O(n^2)$ time. Further, the clique tree itself requires $O(n^2)$ space and therefore, the setup time and space of the algorithm is $O(n^2)$.

Setting Up the Rooted T for LCA Queries: To be able to answer distance queries efficiently, our aim is to use the hitting set of the cliques on the path between a pair of cliques in T. For this, we will be using Lemma 2. In particular, for the distance between u and v, we propose to use the hitting set of the cliques on the path between $C_u, C_v \in T$. To compute the value of this hitting set, we will take the least common ancestor C of C_u and C_v in the rooted tree, and compute the hitting set of the cliques on the path from C and C_u, and similarly between C and C_v. Using these hitting set values, we apply Lemma 2. Therefore, it is crucial that we can perform LCA queries efficiently. Towards this end, we preprocess the rooted tree T, using algorithm presented in [16], to build a data structure D of linear size in linear time. Using D, we can compute the least common ancestor of any two cliques C_u and C_v in T in constant time.

As a consequence of the above results in Lemma 3 and the LCA algorithm in [16], we have the following theorem about the time to construct our distance oracle and the space occupied by it.

Theorem 4. *Given a chordal graph G, the data structure, HS, $\mathcal{H}$ and D use $O(n)$ space, and as described above, can be constructed in $O(n^2)$ time and $O(n^2)$ space.*

4.1 $O(1)$ Time Response to Distance Queries

The query consists of $u_i, u_j \in V$ as input and the response is a value d_{ij}, which is the approximate distance between u_i and u_j in G. Algorithm 2 uses the oracle set up earlier in this section. Recall that $\mathcal{H}[i]$ gives the index of a maximal clique in the clique tree that contains vertex u_i. Similarly, $HS[i]$ gives the size of the minimum hitting set of the maximal cliques on path connecting C_r and C_i. Finally, D can respond to LCA queries on T in constant time.

The most important benefit of using the hitting sets of the cliques on a path between two cliques in T is that for any pair of cliques this can be computed within an additive error of 3. Using this approximate hitting set value, we can then provide an estimate of the distance, as shown in Algorithm 2.

Algorithm 2. $Query(u_i, u_j)$ – uses the Oracle $\mathcal{H}, HS, D$ as in Theorem 4

```
 1: p = H[i];
 2: q = H[j];
    /* these are indices of two maximal cliques containing u_i and u_j */
 3: if p == r (or q == r) then
    /* r be the index of root clique C_r */
 4:     h_pq = HS[q] (or h_pq = HS[p])
    /* one of the two queried vertices is in root clique C_r */
 5: else
 6:     Query D with p and q.
 7:     Let k be the least common ancestor of p and q in T.
 8:     if p == k (or q == k) then
 9:         h_pq = (HS[q] − HS[p] + 1) (or h_pq = (HS[p] − HS[q] + 1))
        /* r, p, q are in a single path */
10:     else
11:         if k == r then
12:             h_pq = HS[q] + HS[p]
13:         else
14:             h_pq = HS[q] + HS[p] − 2HS[k] + 2
15:         end if
16:     end if
17: end if
18: return d_ij = 2 * h_pq
    /* Approximate Distance Returned */
```

Lemma 4. *For two cliques C_p and C_q in T, let S_{pq} denote a minimum hitting set of the cliques on the path connecting C_p and C_q in T. Then, value h_{pq} calculated in Algorithm 2 is such that $h_{pq} \leq |S_{pq}| + 3$.*

Proof. The proof follows a case analysis as in the algorithm 2. If C_p is same as C_r, then $h_{pq} = HS[q] = |S_{rq}| = |S_{pq}|$. The case when C_q is same as C_r, is similar with $h_{pq} = HS[p] = |S_{rp}| = |S_{pq}|$. In the case when p, q, r, lie on a single path in T, it follows that the LCA of p, q is p or q. Let us assume that it is p for the analysis, and the case of the LCA being q is symmetric. From Lemma 2, we know that $HS[q] \leq HS[p] + |S_{pq}| \leq HS[q] + 1$. Therefore, $HS[q] - HS[p] \leq |S_{pq}| \leq HS[q] - HS[p] + 1 = h_{pq}$. Therefore, $h_{pq} - 1 \leq |S_{pq}| \leq h_{pq}$, and it follows that $h_{pq} \leq |S_{pq}| + 1$. In the case when r is the LCA of p and q, from Lemma 2, we know that $|S_{pq}| \leq HS[p] + HS[q] = h_{pq} \leq |S_{pq}| + 1$. Finally, in the case when $k \neq r$ is the LCA of p and q, we have three inequalities that together give tight bounds on $|S_{pq}|$. Again, from Lemma 2 we know that $HS[p] \leq |S_{pk}| + HS[k] \leq HS[p] + 1$, and $HS[q] \leq |S_{qk}| + HS[k] \leq HS[q] + 1$. Further, we also know that $|S_{pq}| \leq |S_{pk}| + |S_{kq}| \leq |S_{pq}| + 1$. From these three inequalities, it follows $h_{pq} = HS[p] + HS[q] - 2HS[k] + 2 \leq |S_{pk}| + |S_{qk}| + 2 \leq |S_{pq}| + 3$. We have now placed a bound on h_{pq} in all the cases, and the lemma is proved. □

Lemma 5. *For any two vertices $u_i, u_j \in V(G)$, the value d_{ij} output by Algorithm 2 is such that $d_{ij} \leq 2d_G[u_i, u_j] + 8$. Further, there is a path in G of length at most d_{ij}.*

Proof. We know that $d_{ij} = 2h_{pq}$. From Lemma 4, $h_{pq} \leq |S_{pq}| + 3$. Further, from Corollary 1, we know that $|S_{pq}| \leq d_G(u_i, u_j) + 1$. Therefore, it follows that $d_{ij} \leq 2(d_G(u_i, u_j) + 1 + 3) = 2d_G(u_i, u_j) + 8$. Since there is a hitting set of size at most h_{pq} that hits the cliques on the path from C_p to C_q, from Theorem 2 it follows that there is a path of length $d_{ij} = 2h_{pq}$. Hence the lemma. □

We have now proved all the properties of our distance oracle for chordal graphs and these complete the proof of Theorem 1.

5   Conclusion

We have presented a linear size data structure to answer the distance queries of an unweighted chordal graph. More precisely, the data structure designed in this paper gives a response d for the distance query between u and v in constant time, where $d_G(u, v) \leq d \leq 2d_G(u, v) + 8$. Furthermore, by maintaining the approximate hitting sets as well, along with the whole clique tree, we can modify the distance oracle in this paper to use $O(n^2)$ space, and respond to distance queries with paths of length at most $2d_G(u, v) + 8$. The query time is proportional to the length of the path being output.

References

1. Cormen, T.H., Leiserson, C.E., Rivest, R.L., Stein, C., et al.: Introduction to algorithms, vol. 2. MIT Press, Cambridge (2001)
2. Seidel, R.: On the all-pairs-shortest-path problem in unweighted undirected graphs. Journal of Computer and System Sciences 51(3), 400–403 (1995)
3. Williams, V.V.: Multiplying matrices faster than coppersmith-winograd. In: Proceedings of the Forty-fourth Annual ACM Symposium on Theory of Computing, pp. 887–898. ACM (2012)

4. Han, K., Sekharan, C.N., Sridhar, R.: Unified all-pairs shortest path algorithms in the chordal hierarchy. Discrete Applied Mathematics 77(1), 59–71 (1997)
5. Gavoille, C., Katz, M., Katz, N.A., Paul, C., Peleg, D.: Approximate distance labeling schemes. In: Meyer auf der Heide, F. (ed.) ESA 2001. LNCS, vol. 2161, pp. 476–487. Springer, Heidelberg (2001)
6. Ravi, R., Marathe, M.V., Pandu Rangan, C.: An optimal algorithm to solve the all-pair shortest path problem on interval graphs. Networks 22(1), 21–35 (1992)
7. Radhakrishnan, V., Hunt, H., Stearns, R.: On Solving Systems of Linear Equations and Path Problems for Bounded Treewidth Graphs. State University of New York at Albany, Department of Computer Science (1992)
8. Dahlhaus, E.: Optimal (parallel) algorithms for the all-to-all vertices distance problem for certain graph classes. In: Mayr, E.W. (ed.) WG 1992. LNCS, vol. 657, pp. 60–69. Springer, Heidelberg (1993)
9. Thorup, M., Zwick, U.: Approximate distance oracles. Journal of the ACM (JACM) 52(1), 1–24 (2005)
10. Cohen, E., Zwick, U.: All-pairs small-stretch paths. In: Proceedings of the eighth annual ACM-SIAM Symposium on Discrete algorithms, pp. 93–102. Society for Industrial and Applied Mathematics (1997)
11. Baswana, S., Kavitha, T.: Faster algorithms for approximate distance oracles and all-pairs small stretch paths. In: 47th Annual IEEE Symposium on Foundations of Computer Science, FOCS 2006, pp. 591–602. IEEE (2006)
12. Shibata, Y.: On the tree representation of chordal graphs. Journal of Graph Theory 12(3), 421–428 (1988)
13. Golumbic, M.C.: Algorithmic graph theory and perfect graphs, vol. 2. Elsevier (2004)
14. Blair, J.R., Peyton, B.: An introduction to chordal graphs and clique trees. In: Graph Theory and Sparse Matrix Computation, pp. 1–29. Springer (1993)
15. Harel, D., Tarjan, R.E.: Fast algorithms for finding nearest common ancestors. SIAM Journal on Computing 13(2), 338–355 (1984)
16. Schieber, B., Vishkin, U.: On finding lowest common ancestors: simplification and parallelization. SIAM Journal on Computing 17(6), 1253–1262 (1988)
17. Powell, P.: A further improved LCA algorithm. Technical report TR90-01. University of Minnesota, Institute of Technology, Computer Science Department (1990)

Straight-Path Queries in Trajectory Data*

Mark de Berg and Ali D. Mehrabi

Department of Computer Science
TU Eindhoven

Abstract. Inspired by sports analysis, we study data structures for storing a trajectory representing the movement of a player during a game, such that the following queries can be answered: Given two positions s and t, report all sub-trajectories in which the player moved in a more or less straight line from s to t. We consider two measures of straightness, namely *dilation* and *direction deviation*, and present efficient construction algorithms for our data structures, and analyze their performance.

We also present an $O(n^{1.5+\varepsilon})$ algorithm that, given a trajectory P and a threshold τ, finds a simplification of P with a minimum number of vertices such that each edge in the simplification replaces a sub-trajectory of length at most τ times the length of the edge. This significantly improves the fastest known algorithm for the problem.

1 Introduction

Background. Video analysis is nowadays an important tool for sports coaches. Traditionally, video analysis is done manually: someone watches a video of a match and annotates the video with various types of events—goals or points being scored, changes of ball possession, and so on. However, manual analysis is labor intensive and annotating all league matches of an entire season would be very time-consuming and expensive. Therefore there has been considerable interest in automating parts of the process. A basic step in an automated analysis is to extract the movements of the players and the ball from the video. Nowadays this can be done quite accurately, giving us for each player a *trajectory*: a sequence of the player's location at regular time steps. The availability of high-quality trajectories enables the use of geometric algorithms and data structures. In this paper we study two problems in this area.

The first problem is an indexing problem, related to the following query a coach may wish to ask: show me all video fragments in which player X runs in a more or less straight line from a certain position s on the field to another position t. We thus need a data structure storing a collection of trajectories (corresponding to the movements of player X in all matches) such that we can efficiently answer *straight-path queries*: given a directed query segment st, report all subtrajectories starting near s and going in a more or less straight line to a point near t. (We will define the problem more formally below.)

---

* Research by MdB and AM was supported by the Netherlands Organization for Scientific Research (NWO) under projects 024.002.003 and 612.001.118, respectively.

M.S. Rahman and E. Tomita (Eds.): WALCOM 2015, LNCS 8973, pp. 101–112, 2015.

The second problem we study is a simplification problem: given a trajectory P, compute a simplification P' with a minimum number of vertices under the condition that P' is sufficiently similar to P. Here we require (as is usually done) that the vertices of P' form a subset of the vertices of P. Computing such a simplification is useful to reduce storage requirements, and also to smooth out irregularities in the data due to small errors in the reported locations.

Straight-path queries: related work. The focus of our work is on data structures that come with proven guarantees on the query time but also on the quality of the reported results. For the latter we need to define when a subtrajectory is sufficiently similar to the query segment st. We are aware of only one such result, obtained by De Berg et al. [4]. They show how to store a trajectory P of n vertices such that, given a query segment st and a threshold Δ, one can find all subtrajectories of P whose so-called *Fréchet distance* to st is at most Δ. However, their work has several drawbacks. First of all, in addition to all the correct subtrajectories their data structure may report additional subtrajectories whose Fréchet distance to st can be up to a factor $2+3\sqrt{2}$ times larger than Δ. Second, their data structure is a complicated multi-level structure which is difficult to implement and unlikely to be efficient in practice. Finally, they only show how to (approximately) count the subtrajectories—it is unclear how to actually report them in an efficient manner. Gudmundsson and Smid [8] recently studied a more general version of the problem, but their solution only works for c-packed paths and only reports a single subpath (and is rather involved).

There are also a few non-algorithmic papers related to ours, such as the work of Shim et al. [11]. However, they do not have any guarantees on the performance of their solution. Our main goal is thus to develop a data structure for straight-path queries that has with provable performance guarantees and that is simple enough to be effective in our soccer scenario.

Straight-path queries: our approach and results. We take the following practical approach. We partition the soccer field into a grid of square cells (the cell size can be set by the user). To specify a query the coach indicates a starting cell C_s and a target cell C_t, and the data structure should report all subtrajectories where the player moved in a more or less straight line from C_s to C_t. We still have to define what it means when "a player moves from C_s to C_t in a more or less straight line". Let s be the point where the player's trajectory P exits C_s and let t be the point where it enters C_t. Then we want the subtrajectory from s to t—we denote this subtrajectory by $P[s,t]$—to be similar to the segment st. We study two different definitions for this similarity.

- The first option is to use the so-called *dilation* of $P[s,t]$, which is defined as $|P[s,t]|/|st|$, where $|\cdot|$ denotes the Euclidean length of a path or segment. We now say that the player moves in a more or less straight line from C_s to C_t when the dilation of $P[s,t]$ is at most some (predetermined) threshold $\tau \geqslant 1$. In other words, $P[s,t]$ can be at most a factor τ longer than the segment st.
- The second option is to require that the player always moves in more or less the same direction along $P[s,t]$. We define the *direction deviation* of a

trajectory to be the maximum angle between any two (directed) segments on the trajectory. We then say that the player moves in a more or less straight line from C_s to C_t when the direction deviation of $P[s,t]$ is at most some (predetermined) threshold $\alpha < \pi/2$. We call such a subtrajectory α-*straight*.

Our first data structure for straight-path queries is a look-up table that stores, for all pairs of grid cells C_s, C_t, the set $S(C_s, C_t)$ of straight C_s-to-C_t subtrajectories (according the chosen definition of straightness). Thus a query can be answered in $O(1+A)$ time by a look-up table, where A is the number of reported subtrajectories. Our contributions for this simple data structure are (i) efficient algorithms to compute all sets $S(C_s, C_t)$, and (ii) a theoretical and experimental analysis of the size of the data structures. Due to space limitations we defer the experimental analysis and several proofs to the full version.

Because the worst-case size of our first data structure is large, we also present a data structure that uses much less storage. This data structure can be used when the straightness measure is the direction deviation. A drawback is that, in addition to the correct subtrajectories, the data structure may also report some additional α-straight subtrajectories that start near C_s. We analyze the maximum possible error—that is, how far from C_s the reported subtrajectories may start—theoretically (and, in the full version, experimentally).

The minimum-vertex path-simplification problem. In path-simplification problems the goal is to compute, for a given trajectory P, a trajectory Q with fewer vertices than P that is sufficiently similar to P. We study a variant called the *minimum-vertex path-simplification (MVPS) problem*, introduced by Gudmundsson et al. [7]. In the MVPS problem we want to find a minimum-size subset Q of vertices of P such that for any two consecutive vertices p_i and p_j in Q we have $|P[p_i, p_j]| \leqslant \tau |p_i p_j|$. (In other words, we are only allowed to use a shortcut $p_i p_j$ when the dilation of $P[p_i, p_j]$ is at most τ.) Gudmundsson et al. [7] solve this problem in $O(n^2)$ time exactly and they give an $(1+\varepsilon)$-approximation algorithm that runs in time $O(n \log n + n/\varepsilon)$. We present a dynamic-programming algorithm with expected running time $O(n^{1.5+\varepsilon})$, for any $\varepsilon > 0$, thus significantly improving the running time of their exact algorithm.

2   The Data Structures

For simplicity of presentation we assume we are given a single trajectory P with n vertices, denoted by $v_0, \ldots, v_{n-1}$; it is trivial to extend the results to multiple trajectories. We further assume that the grid $\mathcal{G}$ we use to partition the soccer field is a square grid with $m \times m$ cells. Recall that $P[p, p']$ denotes the subtrajectory from p to p'. We say that $P[p, p']$ is a *C-to-C' subtrajectory* if p lies on the boundary of cell C and p' lies on the boundary of cell C' and $P[p, p']$ does not intersect C and C' except at p and p'. For two points $p, p' \in P$ we write $p \prec p'$ when p comes before p' in the order along P.

2.1   A Look-up Table for Straight-Path Queries

As explained in the introduction, our first data structure is a look-up table that stores for every pair of grid cells C, C' the set $S(C, C')$ of all C-to-C' subtrajectories that are considered straight with respect to the given measure of straightness (dilation or direction deviation) and parameter (τ or α). More precisely, for each such subtrajectory $P[p, p']$ we store its starting point p and endpoint p'. The main questions are then: (i) how do we construct the sets $S(C, C')$ efficiently, and (ii) what is the maximum size of the data structure, that is, how large can $\sum_{C,C'} |S(C, C')|$ be.[1] Next we answer these questions for the two straightness measures that we use.

In the sequel we call a point where P crosses from one cell into the next a *transition point*. (To deal correctly with degenerate situations we define each cell to be closed on the bottom and to the left, and open on the top and to the right. Thus vertical edges belong to the cell lying to their right and horizontal edges belong to the cell above; vertices belong to the cell to their top-right.) A transition point is an *exit point* for the cell being exited, and an *entry point* for the cell being entered. We denote the sequence of transition points by $p_0, p_1, \ldots$, where the transition points are ordered along P. We denote the cell from which P exits at p_i by $C_{\text{exit}}(p_i)$, and the cell being entered by $C_{\text{entry}}(p_i)$.

Direction Deviation. We first describe how to compute the sets $S(C, C')$ when direction deviation is used as straightness measure. Let α be the given straightness parameter, where we assume $0 \leqslant \alpha < \pi/2$. Note that α-straightness is a monotone criterion: if a subtrajectory $P[p, q]$ is α-straight, then any subtrajectory $P[p', q']$ with $p \prec p' \prec q' \prec q$ is also α-straight. Thus we can follow the following strategy: we walk along P from start to finish, and at each transition point p_j we walk back along P to report all α-straight subtrajectories of the form $P[p_i, p_j]$, where p_i is a transition point with $i < j$. Because α-straightness is a monotone criterion, we can stop the backwards walk as soon as we encounter a transition point p_i for which $P[p_i, p_j]$ is not α-straight. A problem with this approach is that if P has many consecutive vertices inside the same cell then we spend a lot of time walking back through that cell, which can cause a high running time. We thus have to proceed more carefully.

We model directions as points on the unit circle S^1. A subtrajectory $P[p_i, p_j]$ is α-straight if and only if the smallest circular interval of S^1 that contains all points corresponding to the directions of the edges of the subtrajectory has length α—see Fig. 1. Our algorithm now works as follows. As we walk along P we compute for each consecutive pair of transition points p_j, p_{j+1} the smallest circular interval $I(p_j, p_{j+1})$ containing all directions of the subtrajectory $P[p_j, p_{j+1}]$, if this interval has length at most α; if the interval has length greater than α then $I(p_j, p_{j+1})$ is defined to be NIL. (The smallest interval is uniquely defined when it has length at most α, since $\alpha < \pi$.) Note that if $P[p_j, p_{j+1}]$ is a single

---

[1] By hashing techniques we can make sure we only store information for pairs C, C' such that $S(C, C') \neq \emptyset$, so the amount of storage is indeed $O(\sum_{C,C'} |S(C, C')|)$.

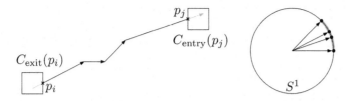

Fig. 1. A subtrajectory and the points on S^1 corresponding to its edges. The smallest circular interval containing the points is shown in grey.

segment then $I(p_j, p_{j+1})$ degenerates to a point on S^1. At each transition point p_i we walk backwards from transition point to transition point (thus skipping over vertices of P) as long as the subtrajectories are α-straight. To check this we maintain the smallest circular interval I^* that contains all intervals $I(p_j, p_{j+1})$ that we encountered in the backwards walk.

We ignored one aspect so far: the fact that $P[p_i, p_j]$ is α-straight is not sufficient for the subtrajectory to be reported. We also need $P[p_i, p_j]$ to be a valid $C_{\text{exit}}(p_i)$-to-$C_{\text{entry}}(p_j)$ subtrajectory. This is violated if $P[p_i, p_j]$ intersects $C_{\text{exit}}(p_i)$ or $C_{\text{entry}}(p_j)$ at some point other than p_i or p_j. To make sure our algorithm is output sensitive, we have to avoid reporting $P[p_i, p_j]$ in this case. This can be done by removing a transition point p_i from our list of transition points as soon as we encounter another transition point $p_{i'}$ with $C_{\text{exit}}(p_{i'}) = C_{\text{exit}}(p_i)$. This ensures that when we report $P[p_i, p_j]$, then $P[p_i, p_j]$ does not intersect $C_{\text{exit}}(p_i)$ except at p_i. When we remove p_i, we have to "merge" the intervals $I(p_{i-1}, p_i)$ and $I(p_i, p_{i+1})$ into a new interval $I(p_{i-1}, p_{i+1})$. To facilitate this we maintain an ordered list $\mathcal{L}$ of all encountered transition points that have not been deleted yet, and with each transition point p_i in $\mathcal{L}$ we store an interval $I(p_i)$ which is the smallest circular interval containing all directions of the subtrajectory $P[p_{i'}, p_i]$ (or NIL if this interval has length more than α), where $p_{i'}$ is the predecessor of p_i in $\mathcal{L}$. To avoid reporting a subtrajectory $P[p_i, p_j]$ that intersects $C_{\text{entry}}(p_j)$ at some point other than p_j, we can simply stop our backwards walk when we encounter a transition point p_i with $C_{\text{exit}}(p_i) = C_{\text{entry}}(p_j)$.

Algorithm 1 describes our algorithm in more detail. Subroutine *SmallestInterval*$(P[p_i, p_j])$ outputs the smallest circular interval containing all edge directions of $P[p_{j-1}, p_j]$ or, when this interval has length more than α, it outputs NIL. Similarly, for two circular intervals I_1, I_2 the subroutine *Merge*(I_1, I_2) outputs the smallest circular interval containing I_1 and I_2 or, when this interval has length more than α, it outputs NIL.

Since walking back from a transition point p_j takes time $O(1 + k_j)$, where k_j is the total number of reported subtrajectories, we get the following theorem.

Theorem 1. *Let P be a trajectory with n vertices in a domain that is an $m \times m$ grid, and let α be a constant with $0 \leqslant \alpha < \pi/2$. Then we can compute all sets $S(C, C')$ of α-straight subtrajectories of P in $O(n + k)$ time, where k is the total size of all sets.*

Algorithm 1. *FindStraightSubtrajectories*(P, α).

1. Set $j := 0$ and create an empty list $\mathcal{L}$ for storing transition points.
2. Walk along P from v_0 to v_{n-1}, tracing the trajectory through the grid. Whenever P crosses from one cell into another, do the following:
    (i) Create a transition point p_j at the crossing point. If $j = 0$ then skip to Step (v), otherwise continue with Step (ii).
   (ii) Set $I(p_{j-1}, p_j) := SmallestInterval(P[p_{j-1}, p_j])$
  (iii) $I^* := I(p_{j-1}, p_j)$; let *ptr* point to the end of $\mathcal{L}$.
        while $I^* \neq$ NIL and *ptr* $\neq$ NIL
        do Let p_i be the transition point *ptr* points to.
            if $C_{\text{exit}}(p_i) = C_{\text{entry}}(p_j)$
            then *ptr* := NIL
            else Report $P[p_i, p_j]$ as an α-straight $C_{\text{exit}}(p_i)$-to-$C_{\text{entry}}(p_j)$ subtrajectory. If $I(p_i) =$ NIL then $I^* :=$ NIL, else $I^* := Merge(I^*, I(p_i))$. Move *ptr* backwards (to its predecessor).
   (iv) If there is a transition point p_i in $\mathcal{L}$ with $C_{\text{exit}}(p_i) = C_{\text{exit}}(p_j)$—we can test this in $O(1)$ time by maintaining some extra information—then we remove p_i from $\mathcal{L}$ and we set $I(p_{i'}) := Merge(I(p_i), I(p_{i'}))$, where $p_{i'}$ is the successor of p_i in $\mathcal{L}$.
    (v) Append p_j to $\mathcal{L}$ with $I(p_j) := I(p_{j-1}, p_j)$, and set $j := j + 1$.

Analysis of the number of α-straight trajectories. Next we prove bounds on $\sum_{C,C'} |S(C, C')|$, the total number of α-straight trajectories. In the analysis we make the assumption that the average length of the segments $v_i v_{i+1}$ of P is at most the edge length of the grid cells. With our grid cells having size of 1m × 1m, for instance, and a sampling rate of 20Hz this is clearly a realistic assumption. For our soccer application, from now on we assume without loss of generality that the cells in the grid $\mathcal{G}$ have unit size, and that the average length of the segments $v_i v_{i+1}$ in P is at most 1. We call such a trajectory a *short-edge trajectory*. The following theorem states the main result for this case. Observe that in practice α would be chosen fairly small, in which case $\cos(\alpha/2)$ will be close to 1. The bound in the next theorem then becomes $O(nm)$.

Theorem 2. *Let P be a short-edge trajectory with n vertices within an $m \times m$ unit grid, and let $0 \leqslant \alpha < \pi/2$. For a pair C, C' of grid cells, let $S(C, C')$ be the collection of all C-to-C' α-straight subtrajectories. Then $\sum_{C,C'} |S(C, C')| = O(\min\{n^2, nm^2, nm/\cos(\alpha/2)\})$, and this bound is tight in the worst case.*

To prove Theorem 2 we first bound the length of any α-straight subtrajectory.

Lemma 1. *The length of any α-straight trajectory P' in an $m \times m$ unit grid is at most $\sqrt{2}m/\cos(\alpha/2)$.*

Proof. Since the directions of all edges in P' differ by at most α, there is a direction $\boldsymbol{d}$ such that any edge in P' makes an angle at most $\alpha/2$ with $\boldsymbol{d}$. Let ℓ be a line with direction $\boldsymbol{d}$, and project all edges of P' orthogonally onto ℓ. Note that if the projection of some edge s_i of P' onto ℓ has length x_i, then $|s_i| \leqslant x_i/\cos(\alpha/2)$. Furthermore, since $\alpha < \pi/2$, the projections of these segments have disjoint interiors. Hence, $|P'| = \sum_i |s_i| \leqslant \sum_i x_i / \cos(\alpha/2) = |pp'| / \cos(\alpha/2)$, where p and p' are the projections of the start and endpoint of P' onto ℓ. Because P' lies inside an $m \times m$ grid, we have $|pp'| \leqslant \sqrt{2}m$. $\qquad\square$

Using Lemma 1, we can now prove the upper bound from Theorem 2. Indeed, since the average length of the segments is at most 1, the total length of P is at most n. Because a trajectory of length L visits $O(L+1)$ grid cells, this implies that the total number of transition points of P is $O(n)$, which gives an $O(n^2)$ upper bound on the total number of subtrajectories between transition points (irrespective of whether they are α-straight or not). It also implies an upper bound of $O(nm^2)$, since there are only m^2 distinct cells C to start a C-to-$C_{\text{entry}}(p_j)$ subtrajectory for a fixed transition point p_j. On the other hand, for each transition point p_j, the length of the longest α-straight subtrajectory ending at p_j is at most $\sqrt{2}m/\cos(\alpha/2)$ by Lemma 1. Using the fact that a trajectory of length L visits $O(L+1)$ grid cells, we can therefore bound the number of α-straight subtrajectory ending at p_j by $O(m/\cos(\alpha/2))$, which gives an $O(nm/\cos(\alpha/2))$ on the total number of α-straight subtrajectories. Combining this with the $O(n^2)$ and $O(nm^2)$ bounds, we obtain the claimed upper bound. In the full version we show a matching lower bound, thus completing the proof of Theorem 2.

Remark. If the trajectory is not restricted to be short-edge then the upper bound of Theorem 2 becomes $O(\min\{n^2 s^2, nm^2 s, nms/\cos(\alpha/2)\})$, where s denotes the length of the longest segment in the trajectory, and this bound is tight in the worst case.

Dilation. We now turn our attention to dilation as a measure of straightness. We denote the dilation of $P[p, p']$ by $\text{dil}(P[p, p'])$. Compared to direction deviation as straightness measure, dilation is more difficult to handle, because it is not a monotone criterion: if $\text{dil}(P[p, p']) \leqslant \tau$ then a subtrajectory $P[p', q']$ with $p \prec p' \prec q' \prec q$ may still have dilation larger than τ. Next we show that we can nevertheless get an output-sensitive algorithm.

The idea of our solution is that, as we walk along P, we store the transition points in a suitable data structure $\mathcal{D}$. This data structure allows us to perform a query with the current transition point p_j to find all transition points p_i such that $P[p_i, p_j]$ is a $C_{\text{exit}}(p_i)$-to-$C_{\text{entry}}(p_j)$ subtrajectory of dilation at most τ. To this end we associate to each transition point p_i a point $\psi_i := (x_i, y_i, d_i)$ in $\mathbb{R}^3$, where $d_i = |P[v_0, p_i]|$ and x_i and y_i are the x- and y-coordinate of p_i, respectively. Note that the values d_i can be computed in constant time as we walk on P, if we maintain the total length of the traversed part of P. Now when we arrive at transition point p_j, we are looking for all transition points p_i such that $\text{dil}(P[p_i, p_j]) \leqslant \tau$, that is, such that $(d_j - d_i)/\sqrt{(x_j - x_i)^2 + (y_j - y_i)^2} \leqslant \tau$. (A similar idea was used by Agarwal *et al.* [2].) Thus if we define the range $\Gamma_\tau(p_j)$ in $\mathbb{R}^3$ for p_j as

$$\Gamma_\tau(p_j) := \{(x, y, d) : \tau(x_j - x)^2 + \tau(y_j - y)^2 - (d_j - d)^2 \geqslant 0\}.$$

then we are looking for all points p_i such that $\psi_i \in \Gamma_\tau(p_j)$.

Algorithm 2. *FindSmallDilationSubtrajectories*(P, τ).

---

1. Set $j := 0$ and initialize an empty data structure $\mathcal{D}$.
2. Walk along P from v_0 to v_{n-1}, tracing the trajectory through the grid. Whenever P crosses from one cell into another, do the following:
   (i) Create a transition point p_j at the crossing point.
   (ii) If $j > 0$ then query the data structure $\mathcal{D}$ to find all transition points p_i such that $\psi_i \in \Gamma_\tau(p_j)$ and $i > j'$, where $p_{j'}$ is the most recent transition point with $C_{\text{entry}}(p_{j'}) = C_{\text{entry}}(p_j)$. (If there is no such transition point, then $j' = -1$.) For each such transition point p_i, report $P[p_i, p_j]$ as a $C_{\text{exit}}(p_i)$-to-$C_{\text{entry}}(p_j)$ subtrajectory with dilation at most τ.
   (iii) Insert $\psi_j := (x_j, y_j, d_j)$ into $\mathcal{D}$. If $\mathcal{D}$ already stored a transition point p_i with $C_{\text{exit}}(p_i) = C_{\text{exit}}(p_j)$ then we delete the corresponding point ψ_i from $\mathcal{D}$.
   (iv) Set $j := j + 1$.

As before, there is one other aspect to be taken into account: we are only allowed to report a subtrajectory $P[p_i, p_j]$ when it does not intersect $C_{\text{exit}}(p_i)$ except at p_i and it does not intersect $C_{\text{entry}}(p_j)$ except at p_j. The former is guaranteed by deleting a transition point p_i from $\mathcal{D}$ when we encounter a transition point $p_{i'}$ with $i' > i$ such that $C_{\text{exit}}(p_{i'}) = C_{\text{exit}}(p_i)$. The latter is guaranteed by refining our query: when we arrive at transition point p_j we find the most recent transition point $p_{j'}$ with $C_{\text{entry}}(p_{j'}) = C_{\text{entry}}(p_j)$—we can find this point (if it exists) in $O(1)$ time if we maintain a pointer from each grid cell to its most recent entry point—and then we only search for exit points p_i with $i > j'$. Algorithm 2 describes this in more detail.

It remains to describe a data structure $\mathcal{D}$ for answering the following queries:

> Given a query point p_j and an index j', report the points p_i such that $\psi_i \in \Gamma_\tau(p_j)$ (in other words, with $\text{dil}(P[p_i, p_j]) \leqslant \tau$) and $i > j'$.   (∗)

First we focus on the condition $\psi_i \in \Gamma_\tau(p_j)$. The range $\Gamma_\tau(p_j)$ is a semi-algebraic set in $\mathbb{R}^3$. Hence, we can use the range-searching data structure of Agarwal *et al.* [1], which uses $O(n^{1+\varepsilon})$ storage and expected preprocessing (for any fixed $\varepsilon > 0$) and has query time $O(n^{2/3} + k)$, where k is the number of reported points. We can improve the query time if we allow more preprocessing, using standard techniques. For instance, we can obtain logarithmic query time using $O(n^3)$ preprocessing. To this end we map every point p_i to an algebraic surface $\Sigma_\tau(p_i)$ in $\mathbb{R}^3$, defined as

$$\Sigma_\tau(p_i) := \{(x, y, d) : \tau(x - x_i)^2 + \tau(y - y_i)^2 - (d - d_i)^2 = 0\}.$$

Now, whether or not $\text{dil}(P[p_i, p_j]) \leqslant \tau$ is determined by on which side of $\Sigma_\tau(p_i)$ the point p_j lies. Thus we can find all points p_i such that $\text{dil}(P[p_i, p_j]) \leqslant \tau$ by performing point location with p_j in the arrangement defined by the surfaces $\{\Sigma_\tau(p_i) : i < j\}$. The latter can be solved in $O(\log n)$ time after $O(n^{3+\varepsilon})$ preprocessing [5]. Unfortunately, we cannot afford cubic preprocessing. However, we can combine our first data structure with the cubic-storage solution in a standard manner [3, Exercise 16.16] to obtain a trade-off between storage and query time. In particular, for any s with $n \leqslant s \leqslant n^3$ we can construct a data structure using $O(s^{1+\varepsilon})$ expected preprocessing so that a query can be answered in time $O(n^{1+\varepsilon}/s^{1/3})$.

Recall that we need to extend the data structure such that when we do a query for entry point p_j we only report subtrajectory $P[p_i, p_j]$ when $i > j'$, where j' is defined as in Algorithm 2. This can be done by adding a so-called *range restriction* to the data structure [12]. We also need our data structure to be dynamic, that is, we need to be able to do insertions and deletions. This can be done by applying the logarithmic method [9] in combination with *weak deletions*. By applying these techniques we can obtain, for any fixed $\varepsilon > 0$, a data structure in which queries take $O(n^{1+\varepsilon}/s^{1/3})$ time and updates take $s^{1+\varepsilon}/n$ expected time. We now choose $s = n\sqrt{n}$ to balance the query time and insertion time. Putting everything together, we obtain the following result.

Theorem 3. *Let P be a trajectory with n vertices in a domain that is an $m \times m$ grid, and let $\tau \geqslant 1$ be a constant. Then, for any fixed $\varepsilon > 0$, we can compute all sets $S(C, C')$ of subtrajectories of P with dilation at most τ in expected time $O(n^{1.5+\varepsilon} + k)$, where k is the total size of all sets.*

Analysis of the number of subtrajectories with dilation at most τ. As before we assume the grid consists of unit-size cells, and we make the realistic assumption that we are dealing with short-edge trajectories. The proof of the following theorem is similar to the proof of Theorem 2.

Theorem 4. *Let P be a short-edge trajectory with n vertices within an $m \times m$ unit grid, and let $\tau \geqslant 1$. For a pair C, C' of grid cells, let $S(C, C')$ be the collection of all C-to-C' subtrajectories of dilation at most τ. Then $\sum_{C,C'} |S(C, C')| = O(\min\{n^2, nm^2, \tau nm\})$, and this bound is tight in the worst case.*

2.2   A More Space-Efficient Alternative

Explicitly storing all sets $S(C, C')$ of straight subtrajectories gives fast and accurate queries, but it is costly in terms of storage. Below we present a much more space-efficient alternative. This comes at the cost of slightly slower queries times and the fact we may also report some subtrajectories that pass near to the starting cell C_s of the query (rather than starting exactly at C_s). The alternative solution works for direction deviation as straightness measure.

Let P be the given n-vertex trajectory inside an $m \times m$ grid $\mathcal{G}$, and let α be a given straightness threshold with $0 \leqslant \alpha < \pi/2$. Recall that direction deviation is a monotone criterion, so for any entry point p_j there is a point $p \prec p_j$ such that $P[p', p_j]$ is α-straight for all $p \preceq p' \prec p_j$ and $P[p', p_j]$ is not α-straight for any $p' \prec p$. We call $P[p, p_j]$ the *longest α-straight subtrajectory* for p_j. For a cell $C \in \mathcal{G}$, let $L(C)$ denote the set of all longest α-straight subtrajectories of P

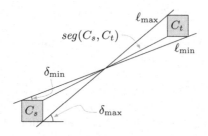

Fig. 2. Definition of $seg(C_s, C_t)$, $\delta_{\min}$ and $\delta_{\max}$

ending at some entry point on the boundary of C. For each cell C, we store the set $L(C)$ in a priority search tree[2] PST$[C]$, as explained next.

Consider a cell C, an entry point p_j of C and the longest α-straight subtrajectory $P[p, p_j] \in L(C)$. We associate a 2-dimensional point $\chi(p_j)$ with this subtrajectory, as follows. Let $\phi(pp_j)$ be the counterclockwise angle that the directed segment pp_j makes with the positive x-axis. Then the point $\chi(p_j)$ is defined as $\chi(p_j) := (\phi(pp_j), |pp_j|)$. This gives us a set $X(C) := \{\chi(p_j) : P[p, p_j] \in L(C)\}$ of points in $\mathbb{R}^2$, which we store in PST$[C]$. Recall that a short-edge trajectory induces $O(n)$ transition points. Since for each transition point p_j we only store the longest subtrajectory $P[p, p_j]$ ending at p_j, we have the following lemma.

Lemma 2. *Let P be a short-edge trajectory with n vertices. Then the total amount of storage needed for all priority search trees PST$[C]$ is $O(n)$.*

We now explain how we answer a straight-path query with starting cell C_s and target cell C_t. To simplify the presentation, we assume that C_t lies to north-east of C_s; the other cases can be handled in a symmetrical manner.

Let $seg(C_s, C_t)$ denote a shortest directed line segment connecting C_s to C_t. Let $\ell_{\min}$ denote the common tangent of C_s and C_t with minimum slope, and $\ell_{\max}$ denote the common tangent of C_s and C_t with maximum slope. Finally, let $\delta_{\min}$ and $\delta_{\max}$ denote the angles that $\ell_{\min}$ and $\ell_{\max}$ make with the positive x-axis; see Figure 2. We then now perform a semi-infinite range query on the priority search tree PST$[C_t]$ with the semi-infinite range $R(C_s, C_t)$ defined as

$$R(C_s, C_t) := [\delta_{\min} - \alpha : \delta_{\max} + \alpha] \times [|s(C_s, C_t)| : \infty).$$

Thus, intuitively, we report a longest subtrajectory $P[p, p_j]$ if the direction of pp_j is similar to the direction of $seg(C_s, C_t)$ and pp_j is at least as long as the minimum distance between C_s and C_t. The following lemma states that the subtrajectories we report include all subtrajectories from C_s to C_t, and that any subtrajectory we report passes near C_s.

Lemma 3. *(i) Let $P[p_i, p_j]$ be an α-straight C_s-to-C_t subtrajectory. Then $\chi(p_j)$, the point stored for $P[p_i, p_j]$ in PST$[C_t]$, lies in the range $R(C_s, C_t)$. (ii) Let*

---

[2] A priority search tree [3, Section 10.2] stores a planar point set such that all points in a semi-infinite range $[x_1 : x_2] \times [y : \infty)$ can be reported efficiently.

$P[p, p_j]$ be a subtrajectory in $L(C_t)$ such that $\chi(p_j) \in R(C_s, C_t)$. Then the distance from $P[p, p_j]$ to C_s is at most $O(1 + \sin(2\alpha) \cdot |seg(C_s, C_t)|)$.

Notice that the error in (ii)—the distance from $P[p, p_j]$ to C_s—is a constant number of cells plus a fraction of $|seg(C_s, C_t)|$ that tends to zero as α tends to zero. Thus the error does not depend on $|pp_j|$, which is desirable since $|pp_j|$ can be large compared to $|seg(C_s, C_t)|$. It should be noted that the subtrajectory we report is guaranteed to pass near C_s, but does not necessarily start near C_s.

3   Distance-Preserving Path Simplification

Let $P = (p_1, p_2, \ldots, p_n)$ be a path with n vertices and $\tau \geqslant 1$ be a real number. A path $Q = (p_{i_1}, p_{i_2}, \ldots, p_{i_k})$, with $1 = i_1 < i_2 < \cdots < i_k = n$, is called a τ-distance-preserving approximation of P if $\mathrm{dil}(P[p_{i_t}, p_{i_{t+1}}]) \leqslant \tau$ for all $1 \leqslant t < k$. Gudmundsson *et al.* [7] introduced the *Minimum Vertex Path Simplification (MVPS) problem*: given a path P and a threshold $\tau \geqslant 1$, compute a τ-distance-preserving approximation of P having the minimum number of vertices. Their algorithm for the problem runs in $O(n^2)$ time. Our approach to improve this time bound uses dynamic programming. For any $1 \leqslant j \leqslant n$, let $M[j]$ denote the minimum number of vertices on any τ-distance-preserving approximation of $P[p_j, p_n]$. Define $S(j) := \left\{ p_i : i > j \text{ and } \mathrm{dil}(P[p_i, p_j]) \leqslant \tau \right\}$. Then $M[j]$ satisfies

$$M[j] = \begin{cases} 1 & \text{if } j = n \\ 1 + \min\left\{ M[i] \; : \; p_i \in S(j) \right\} & \text{if } j < n \end{cases}$$

Explicitly checking each point $p_i \in S(j)$ when computing $M[j]$ will lead to a quadratic algorithm. To speed up the algorithm we will augment the data structure from Section 2.1 so that, given a query index i, we can compute $\min\{M[i] : p_i \in S(j)\}$ quickly.

The data structure. Observe that the condition $p_i \in S(j)$ is essentially the same as the condition (∗) on page 108. Hence, we can report all points in $S(j)$ using the data structure described on page 108–109. However, we only want to report the point p_i with the minimum $M[i]$ value. To this end we have to augment the data structure with extra information.

The data structure is a two-level tree[3] whose first level is a binary search tree $\mathcal{T}$ storing all the vertices on the path based on their indices. This level enables us to restrict the attention to points p_i with $i > j$. With each node v of the first-level tree $\mathcal{T}$, we have an associated tree $\mathcal{T}'_v$ that allows us to select all points p_i with $\mathrm{dil}(P[p_i, p_j]) \leqslant \tau$. This selection comes in the form of a number of nodes w whose canonical subsets together form a disjoint partition of $S(j)$. Thus we augment the data structure as follows: with each node w in any of the associated trees, we store the value $M_w := \min\{M[i] : p_i \in S_w\}$, where S_w is

---

[3] For the current application we do not need insertions and deletions, so we do not need to apply the logarithmic method.

the canonical subset associated to w. (When $M[i]$ is not known yet, it is defined as $+\infty$.) When we now perform a query with a point p_j, we search our data structure to select a set of nodes w whose canonical subsets form $S(j)$, and we take the smallest M_w-value among the selected nodes.

The algorithm. Now the algorithm simply works as follows. We construct the data structure described above, where each M_w-value is initialized to $+\infty$, and we initialize $M[n] := 1$. We then compute each value $M[j]$, for $j = n-1, \ldots, 1$, by performing a query with p_j as described above. After having computed $M[j]$ we set $M_w := \min(M_w, M[j])$ for each node w whose canonical subset contains p_j.

We have seen before that the data structure can be constructed in $O(n^{1.5+\varepsilon})$ time and that each query takes $O(n^{0.5+\varepsilon})$ time. Since any point p_j is stored in $O(n^{0.5+\varepsilon})$ canonical subsets, updating the M_w-values at each step can be done in $O(n^{0.5+\varepsilon})$ as well. Hence, we obtain the following result.

Theorem 5. *The MVPS problem can be solved in $O(n^{1.5+\varepsilon})$ expected time.*

References

1. Agarwal, P.K., Matoušek, J., Sharir, M.: On range searching with semialgebraic sets II. SIAM J. Comput. 42, 2039–2062 (2013)
2. Agarwal, P.K., Klein, R., Knauer, C., Langerman, S., Morin, P., Sharir, M., Soss, M.A.: Computing the detour and spanning ratio of paths, trees, and cycles in 2D and 3D. Discr. Comput. Geom. 39, 17–37 (2008)
3. de. Berg, M., Cheong, O., van Kreveld, M., Overmars, M.: Computational Geometry: Algorithms and Applications, 3rd edn. Springer (2008)
4. de. Berg, M., Cook, A.F., Gudmundsson, J.: Fast Fréchet queries. Comput. Geom. Theory Appl. 46(6), 747–755 (2013)
5. Chazelle, B., Edelsbrunner, H., Guibas, L.J., Sharir, M.: A singly exponential stratification scheme for real semi-algebraic varieties and its applications. Theor. Comput. Sci. 84(1), 77–105 (1991)
6. Chazelle, B., Sharir, M., Welzl, E.: Quasi-optimal upper bounds for simplex range searching and new zone theorems. Algorithmica 8(5), 407–429 (1992)
7. Gudmundsson, J., Narasimhan, G., Smid, M.: Distance-preserving approximations of polygonal paths. Comput. Geom. Theory Appl. 36(3), 183–196 (2007)
8. Gudmundsson, J., Smid, M.: Fréchet queries in geometric trees. In: Bodlaender, H.L., Italiano, G.F. (eds.) ESA 2013. LNCS, vol. 8125, pp. 565–576. Springer, Heidelberg (2013)
9. Overmars, M. (ed.): The Design of Dynamic Data Structures. LNCS, vol. 156. Springer, Heidelberg (1983)
10. Pettersen, S.A., Johansen, D., Johansen, H., Berg-Johansen, V., Gaddam, V.R., Mortensen, A., Langseth, R., Griwodz, C., Stensland, H.K., Halvorsen, P.: Soccer video and player position dataset. In: Proc. 5th Int. Conf. Multi. Syst., pp. 18–23 (2014)
11. Shim, C.-B., Chang, J.-W., Kim, Y.-C.: Trajectory-based video retrieval for multimedia information systems. In: Yakhno, T. (ed.) ADVIS 2004. LNCS, vol. 3261, pp. 372–382. Springer, Heidelberg (2004)
12. Willard, D.E., Luecker, G.S.: Adding range restriction capability to dynamic data structures. J. ACM 32, 597–617 (1985)

Folding a Paper Strip to Minimize Thickness*

Erik D. Demaine[1,**], David Eppstein[2], Adam Hesterberg[3], Hiro Ito[4],
Anna Lubiw[5], Ryuhei Uehara[6], and Yushi Uno[7]

[1] Computer Science and Artificial Intelligence Lab,
Massachusetts Institute of Technology, USA
edemaine@mit.edu

[2] Computer Science Department, University of California, Irvine, USA
eppstein@uci.edu

[3] Department of Mathematics, Massachusetts Institute of Technology, USA
achester@mit.edu

[4] School of Informatics and Engineering,
University of Electro-Communications, Japan
itohiro@uec.ac.jp

[5] David R. Cheriton School of Computer Science, University of Waterloo, Canada
alubiw@uwaterloo.ca

[6] School of Information Science, Japan Advanced Institute of Science and
Technology, Japan
uehara@jaist.ac.jp

[7] Graduate School of Science, Osaka Prefecture University, Japan
uno@mi.s.osakafu-u.ac.jp

Abstract. In this paper, we study how to fold a specified origami crease
pattern in order to minimize the impact of paper thickness. Specifically,
origami designs are often expressed by a mountain-valley pattern (plane
graph of creases with relative fold orientations), but in general this specification is consistent with exponentially many possible folded states. We
analyze the complexity of finding the best consistent folded state according
to two metrics: minimizing the total number of layers in the folded state (so
that a "flat folding" is indeed close to flat), and minimizing the total amount
of paper required to execute the folding (where "thicker" creases consume
more paper). We prove both problems strongly NP-complete even for 1D
folding. On the other hand, we prove the first problem fixed-parameter
tractable in 1D with respect to the number of layers.

* This research was performed in part at the 29th Bellairs Winter Workshop on Computational Geometry.

** Erik Demaine was supported in part by NSF ODISSEI grant EFRI-1240383 and
NSF Expedition grant CCF-1138967. David Eppstein was supported in part by NSF
grant 1228639 and ONR grant N00014-08-1-1015. Adam Hesterberg was supported
in part by DoD, Air Force Office of Scientific Research, National Defense Science
and Engineering Graduate (NDSEG) Fellowship, 32 CFR 168a. Hiro Ito was sup-
ported in part by JSPS KAKENHI Grant Number 24650006 and MEXT KAKENHI
Grant Number 24106003. Anna Lubiw was supported in part by NSERC. Ryuhei
Uehara was supported in part by JSPS KAKENHI Grant Number 23500013 and
MEXT KAKENHI Grant Number 24106004. Yushi Uno was supported in part by
KAKENHI Grant numbers 23500022 and 25106508.

1   Introduction

Most results in computational origami design assume an idealized, zero-thickness piece of paper. This approach has been highly successful, revolutionizing artistic origami over the past few decades. Surprisingly complex origami designs are possible to fold with real paper thanks in part to thin and strong paper (such as made by Origamido Studio) and perhaps also to some unstated and unproved properties of existing design algorithms.

This paper is one of the few attempts to model and optimize the effect of positive paper thickness. Specifically, we consider an origami design specified by a *mountain-valley pattern* (a crease pattern plus a mountain-or-valley assignment for each crease), which in practice is a common specification for complex origami designs. Such patterns only partly specify a folded state, which also consists of an *overlap order* among regions of paper. In general, there can be exponentially many overlap orders consistent with a given mountain-valley pattern. Furthermore, it is NP-hard to decide flat foldability of a mountain-valley pattern, or to find a valid flat folded state (overlap order) given the promise of flat foldability [3]. But for 1D pieces of paper, the same problems are polynomially solvable [1,4], opening the door for optimizing the effects of paper thickness among the exponentially many possible flat folded states—the topic of this paper.

Preceding Research. One of the first mathematical studies about paper thickness is also primarily about 1D paper. Britney Gallivan [5], as a high school student, modeled and analyzed the effect of repeatedly folding a positive-thickness piece of paper in half. Specifically, she observed that creases consume a length of paper proportional to the number of layers they must "wrap around", and thereby computed the total length of paper (relative to the paper thickness) required to fold in half n times. She then set the world record by folding a 4000-foot-long piece of (toilet) paper in half twelve times, experimentally confirming her model and analysis.

Motivated by Gallivan's model, Uehara [7] defined the *stretch* at a crease to be the number of layers of paper in the folded state that lie between the two paper segments hinged at the crease. We will follow the terminology of Umesato et al. [9] who later replaced the term "stretch" with *crease width*, which we adopt here. Both papers considered the case of a strip of paper with *equally spaced* creases but an arbitrary mountain-valley assignment. When the mountain-valley assignment is uniformly random, its expected number of consistent folded states is $\Theta(1.65^n)$ [8]. Uehara [7] asked whether it is NP-hard, for a given mountain-valley assignment, to minimize the maximum crease width or to minimize the total crease width (summed over all creases). Umesato et al. [9] showed that the first problem is indeed NP-hard, while the second problem is fixed-parameter tractable. Also, there is a related study for a different model, which tries to compact orthogonal graph drawings to use minimum number of rows [2].

Models. We consider the problem of minimizing crease width in the more general situation where the creases are not equally spaced along the strip of paper. This more general case has some significant differences with the equally spaced case. For one thing, if the creases are equally spaced, all mountain-valley patterns can be folded flat by repeatedly folding from the rightmost end; in contrast, in the general case, some mountain-valley patterns (and even some crease patterns) have no consistent flat folded state that avoids self-intersection. Flat foldability of a mountain-valley pattern can be checked in linear time [1] [4, Sec. 12.1], but it requires a nontrivial algorithm.

For creases that are not equally spaced, the notion of crease width must also be defined more precisely, because it is not so clear how to count the layers of paper between two segments at a crease. For example, in Fig. 1, although no layers of paper come all the way to touch the three creases on the left, we want the sum of their crease widths to be 100.

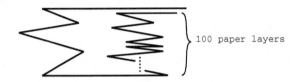

Fig. 1. How can we count the paper layers?

We consider a folded state to be an assignment of the segments to horizontal *levels* at integer y coordinates, with the creases becoming vertical segments of variable lengths. See Fig. 2 and the formal definition below. Then the *crease width* at a crease is simply the number of levels in between the levels of the two segments of paper joined by the crease. That is, it is one less than the length of the vertical segment assigned to the crease. This definition naturally generalizes the previous definition for equally spaced creases. Analogous to Uehara's open problems [7], we will study the problems of minimizing the maximum crease width and minimizing the total crease width for a given mountain-valley pattern. The total crease width corresponds to the extra length of paper needed to fold the paper strip using paper of positive thickness, naturally generalizing Gallivan's work[1] [5].

In the setting where creases need not be equally spaced, there is another sensible measure of thickness: the *height* of the folded state is the total number of levels. The height is always $n + 1$ for n equally spaced creases, but in our setting different folds of the same crease pattern can have different heights. Figure 2 shows how the three measures can differ. Of course, the maximum crease width is always less than the height.

Contributions. Our main results (Section 3) are NP-hardness of the problem of minimizing height and the problem of minimizing the total crease width. See Table 1. In addition, we show in Section 4 that the problem of minimizing

---

[1] Although we assume orthogonal bends in this paper, while Gallivan measures turns as circular arcs, this changes the length by only a constant factor. Gallivan's model seems to correspond better to practice.

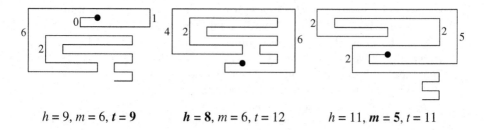

$$h = 9, m = 6, t = 9 \qquad h = 8, m = 6, t = 12 \qquad h = 11, m = 5, t = 11$$

Fig. 2. Three different folded states of the crease pattern VMVMVVMMMM (ending at the dot). The crease width of each crease is given beside its corresponding vertical segment. Each folding is better than the other two in one of the three measures, where h is the height, m is the maximum crease width, and t is the total crease width.

height is fixed-parameter tractable, by giving a dynamic programming algorithm that runs in $O(2^{O(k \log k)}n)$ time, where k is the minimum height. This dynamic program can be adapted to minimize maximum crease width or total crease width for foldings of bounded height, with the same time complexity as measured in terms of the height bound. Table 1 summarizes related results.

Table 1. Complexity of minimizing thickness, by model, for the case of equally spaced creases and for the general case

thickness measure	eq. spaced creases	general creases
height	trivial	NP-hard (this paper)
		FPT wrt. min height (this paper)
max crease width	NP-hard [9]	$\implies$ NP-hard [9]
total crease width	open	NP-hard (this paper)

2   Preliminaries

We model a *paper strip* as a one-dimensional line segment. It is rigid except at *creases* $p_1, p_2, \ldots, p_n$ on it; that is, we are allowed to fold only at these crease points. For notational convenience, the two ends of the paper strip are denoted by p_0 and p_{n+1}. We are additionally given a *mountain-valley string* $s = s_1 s_2 \cdots s_n$ in $\{M, V\}^n$. In the *initial state* the paper strip is placed on the x-axis, with each crease p_i at a given coordinate x_i. Without loss of generality, we assume that $x_0 = 0 < x_1 < \cdots < x_n < x_{n+1}$. Sometimes we will normalize so $x_{n+1} = 1$. We may consider the paper strip as a sequence of $n+1$ *segments* S_i of length $x_{i+1} - x_i$ delimited by the creases p_i and p_{i+1} for each $i \in \{0, 1, \ldots, n\}$. We fold the strip through two dimensions, so we distinguish the *top* side of the strip (the positive y side) and the *bottom* side of the strip (the negative y side). Each crease's letter determines how we can fold it: when it is M (resp. V), the two paper segments sharing the crease are folded in the direction such that their bottom sides (resp. top sides) are close to touching (although they may not necessarily touch if they have other paper layers between them).

Following Demaine and O'Rourke [4] we define a *flat folding* (or *folded state*) via the relative stacking order of collocated layers of paper. We begin with x_0 at the origin, and the first segment lying in the positive x-axis. The lengths of the segments determine where each segment lies along the x-axis (because they zig-zag). Suppose that point p_i is mapped to x-coordinate $f(p_i)$. The mountain-valley assignment determines for each segment S_i whether S_i lies above or below S_{i+1}. We extend this to specify the relative vertical order of any two segments that overlap horizontally. This defines a *folded state* so long as the vertical ordering of segments is transitive and *non-crossing*. More formally:

1. if segments S_i and S_{i+1} are joined by a crease at x-coordinate $f(p_i)$ then for any segment S that extends to the left and the right of $f(p_i)$, either $S < S_i, S_{i+1}$ or $S > S_i, S_{i+1}$,

2. if segments S_i and S_{i+1} are joined by a crease at x-coordinate $f(p_i)$, segments S_j and S_{j+1} are joined by a crease at the same x-coordinate $f(p_j) = f(p_i)$, and all 4 segments extend to the same side of the crease, then the two creases do not *interleave*, i.e., we do not have $A < B < A' < B'$ where A and A' are one of the pairs joined at a crease and B and B' are the other pair.

When the x_i's are not equally spaced, the paper strip cannot necessarily be folded flat with the given mountain-valley assignment. For example, segments of lengths 2, 1, 2 do not allow the assignment VV. There is a linear time algorithm to test whether an assignment has a flat folding [4].

In order to define crease width, we will use an enhanced notion of folded states: a *leveled folded state* is an assignment of the segments to *levels* from the set $\{1, 2, \ldots\}$ such that the resulting vertical ordering of segments is a valid folded state. See Fig. 2. We can draw a leveled folded state as a rectilinear path of alternating horizontal and vertical segments, where the horizontal segments are the given ones, and the vertical segments (which represent the creases) have variable lengths.

Clearly a leveled folded state provides a folded state, but in the reverse direction, a folded state may correspond to many leveled folded states. However, for the measures we are concerned with, we can efficiently compute the best leveled folded state corresponding to any folded state.

The *height* of a leveled folded state is the number of levels used. Given a folded state, the minimum height of any corresponding leveled folded state can be computed efficiently, since it is the length of a longest chain in the partial order defined on the segments in the folded state.

The *crease width* of a crease in a leveled folded state is the number of levels in-between the two segments joined at the crease. We are interested in minimizing the maximum crease width and in minimizing the total crease width, i.e., the sum of the crease widths of all the creases. In both cases, given a folded state, we can compute the best corresponding leveled folded state using linear programming.

A mountain-valley string that alternates $MVMVMV\ldots$ is called a *pleat*. For equally-spaced creases, the legal folded state is unique (up to reversal of the paper) if and only if s is a pleat [7,8].

In this paper, we consider three versions of minimizing thickness in a flat folding. For all three problems we have the following instance in common:

INSTANCE: A paper strip P, with creases $p_1, \ldots, p_n$ at positions $x_1, \ldots, x_n$ with a mountain-valley string $s \in \{M, V\}^n$, and a natural number k.

The questions of the three problems are as follows:

MinHeight: Is there a leveled folded state of height at most k?

MinMaxCW: Is there a leveled folded state with maximum crease width at most k?

MinSumCW: Is there a leveled folded state with total crease width at most k?

3   NP-completeness

In this section, we show NP-completeness of the MinHeight and MinSumCW problems. We remind the reader that the pleat folding has a unique folded state [7,8]. We borrow some useful ideas from [9].

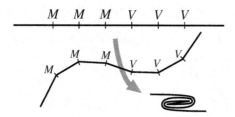

Fig. 3. The unique flat folding of the string MMMVVV

Observation 1. *Let n be a positive integer, P be a strip with creases $p_1, \ldots, p_{2n}$, and s be a mountain-valley string $M^n V^n$. We suppose that the paper segments are of equal length except a longer one at each end. Precisely, we have $|S_i| = |S_j| < |S_0| = |S_{2n}|$ for all i, j with $0 < i, j < 2n$, where $|S_i|$ denotes the length of the segment S_i, Then the legal folded state with respect to s is unique up to reversal of the paper. Precisely, the legal folded state has the segments in vertical order $S_0, S_{2n-1}, S_2, S_{2n-3}, \ldots, S_{2i}, S_{2(n-i)-1}, \ldots, S_1, S_{2n}$ or the reverse.*

A simple example is given in Fig. 3. We call this unique folded state the *spiral folding* of size $2n$.

Our hardness proofs reduce from 3-PARTITION, defined as follows.

3-PARTITION (cf. [6])
Instance: A finite multiset $A = \{a_1, a_2, \ldots, a_{3m}\}$ of $3m$ positive integers.
Define $B = \sum_{j=1}^{3m} a_j/m$. We may assume each a_j satisfies $B/4 < a_j < B/2$.
Question: Can A be partitioned into m disjoint sets $A^{(1)}, A^{(2)}, \ldots, A^{(m)}$
such that $\sum A^{(i)} = B$ for every i with $1 \le i \le m$?

It is well-known that 3-PARTITION is strongly NP-complete, i.e., it is NP-hard
even if the input is written in unary notation [6]. Our reductions are based on a
similar reduction of Umesato et al. [9].

Theorem 2. *The* MinHeight *problem for paper folding height is NP-complete.*

Proof. It is easy to see that the problem is in NP. To prove hardness, we reduce
from 3-PARTITION.

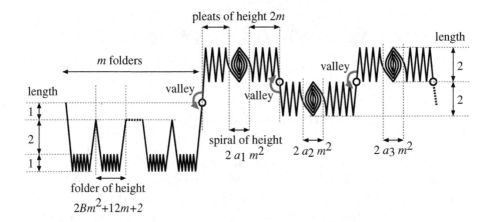

Fig. 4. Outline of the reduction for Theorem 2. Note that this figure and the next one
are sideways compared to previous figures, so height is horizontal.

Given an instance $\{a_1, a_2, \ldots, a_{3m}\}$ of 3-PARTITION, we construct a corresponding paper strip P as follows (Fig. 4). The left part of P is folded into m
folders, where each folder is a pleat consisting of $2Bm^2 + 12m$ *short* segments of
length 1 between two segments of length 3, except for the very first and last long
segments, which have length 4.[2] The right part of P contains $3m$ gadgets, where
the ith gadget represents the integer a_i. The ith gadget consists of one spiral
of height $2a_i m^2$ between two $2m$ pleats. Each line segment in the gadget has
length 2 except for the one end segment which has length 3. This construction
can be carried out in polynomial time.

---

[2] In the reduction in [9], this folder consists of just two segments.

By Observation 1, each spiral folds uniquely, and also we know that each pleat folds uniquely [7,8]. Therefore, the folders and gadgets fold uniquely. Figure 4 shows the unique combination of these foldings before folding at the *joints*, depicted by white circles. Once the joints are valley folded, the folding will no longer be unique.

The intuition is that the pleats of each gadget give us the freedom to place the spiral of each gadget in any folder. The heights of the spirals ensure that the packing of spirals into folders acts like 3-PARTITION. More precisely, we show:

Claim. An instance (A, B) of 3-PARTITION has a solution if and only if the paper strip P can be folded with height at most $2Bm^3 + 12m^2 + 2m$.

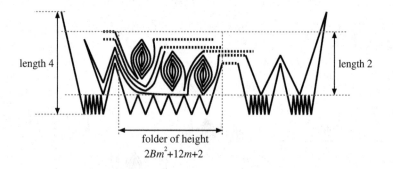

Fig. 5. Putting spirals into a folder

To prove the claim, first suppose that the 3-PARTITION instance $\{a_1, a_2, \ldots, a_{3m}\}$ has a solution, say, $A^{(1)}, A^{(2)}, \ldots, A^{(m)}$. Then we have $A^{(i)} \subset A$, $|A^{(i)}| = 3$, $\sum A^{(i)} = B$ for each i in $\{1, 2, \ldots, m\}$, and $A = \bigcup_{i=1}^{m} A^{(i)}$. For the three items in $A^{(i)}$, we put the three corresponding spirals into the ith folder; see Fig. 5. Because the items sum to B, the total height of the spirals is $2Bm^2$. Each gadget uses $2(m-1)$ of the $4m$ total pleats to position its spiral, leaving $2(m+1)$ pleats which we put in the folder of the spiral, for a total of $6(m+1)$. The $3m-3$ other gadgets also place two pleats in this spiral, just passing through, for a total of $6m-6$. Thus each folder has at most $2Bm^2 + 12m$ layers added and, because it already had $2Bm^2+12m$ short pleat segments, its final height is $2Bm^2+12m+2$ (including the two long segments). Therefore the total height of the folded state is $2Bm^3 + 12m^2 + 2m$ as desired.

Next suppose that the paper strip P can be folded with height at most $k = 2Bm^3+12m^2+2m$. There are m folders each with height at least $2Bm^2+12m+2$. Therefore, each folder must have height exactly $2Bm^2 + 12m + 2$ and the number of levels inside the folder is $2Bm^2+12m$. Furthermore, the spirals must be folded into the folders. We claim that the spirals in each folder must have total height at most $2Bm^2$. For, if the spirals in one of the folders have total height more than $2Bm^2$, then they have height at least $2(B+1)m^2 = 2Bm^2 + 2m^2$, which is greater than $2Bm^2 + 12m$ if $2m^2 > 12m$, i.e., if $m > 6$ (which we may assume without loss of generality). In particular, each folder must have at most three spirals: because each a_j is greater than $B/4$, each spiral has height larger than $Bm^2/2$, so four spirals would have height larger than $2Bm^2$. Because the $3m$ spirals are partitioned among m folders, exactly three spirals are placed in each folder, and their total height of at most $2Bm^2$ corresponds to three elements of sum at most (and thus exactly) B. Therefore we can construct a solution to the 3-PARTITION instance.   □

Theorem 3. *The* MinSumCW *problem is NP-complete.*

Proof. This reduction from 3-PARTITION is a modification to the reduction to MinHeight in the proof of Theorem 2; refer to Figures 6 and 7. We introduce a deep "molar" at both ends of each gadget, which must fit into deep "gums" at either end of the folders. Specifically, for $z = m^4$, each gum has $2z + 4m$ pleats, and each molar in the ith gadget has $2z + 4(m - i)$ pleats. In the intended folded state, the left molars nest inside each other (smaller/later inside larger/earlier) within the left gum, and similarly for the right molars into the right gum. In this case, every molar and every gum remains at its minimum possible height given by its pleats.

The heights of the molars guarantee that, in any legal folding, every molar ends up in a gum. If, in any of the m gadgets, the right molar folds into the left gum, then the left molar of that gadget also folds into the left gum, so the left gum has height at least $4z$ in the folded state, $2z - 4m$ more than its minimum height. This increase in height translates into an equal increase in the total crease width (because the number of creases remains fixed). Because $z = m^4$, this increase will dominate the total crease width. Therefore every folding with a right molar in the left gum, or with a left molar in the right gum, has total crease width larger than the intended folded state.

This argument guarantees that, in any solution folding to the MinSumCW instance, each gadget has its left molar in the left gum and its right molar in the right gum. In this case, the height of each gadget is the height of its spiral plus the height of all the folders, which will be minimized precisely when the folders do not grow in height. The total crease width of a gadget differs from its height by a fixed amount (the number of creases), so we arrive at the same minimization problem. Thus the proof reduces to the MinHeight construction.   □

4   Fixed-Parameter Tractability

In this section, we show the following theorem.

Theorem 4. *Testing whether a strip with n folds has a folded state with height at most k can be done in time $O(2^{O(k \log k)} n)$.*

Proof. We use a dynamic programming algorithm that sweeps from left to right across the line onto which the strip is folded, stopping at each of the points on the

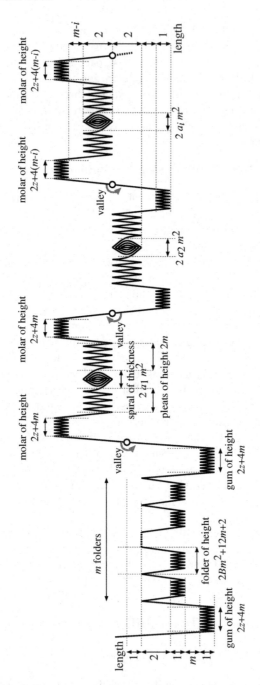

Fig. 6. Outline of the reduction. Note that height is vertical.

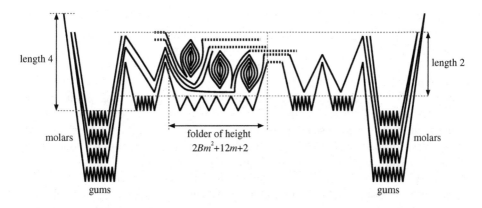

Fig. 7. Putting spirals into folders and molars into gums

line where a strip endpoint or a crease (fold point) is placed. At each point of the line between two stopping points, there can be at most k segments of the strip, for otherwise the height would necessarily be larger than k and we could terminate the algorithm, returning that the height is not less than or equal to k. We define a *level assignment* for a point p between two stopping points to be a function a from input segments that overlap p to distinct integer levels from 1 to k. The number of possible level assignments for any point is therefore at most k^k.

Let $\varepsilon > 0$ be smaller than the distance between any two stopping points. At each stopping point p of the algorithm, we will have a set A of allowed level assignments a^- for the point $p - \varepsilon$; initially (for the leftmost point of the folded input strip) A will contain the unique level assignment for the empty set of segments. For each combination of a level assignment a^- in A for the point $p - \varepsilon$ and an arbitrary level assignment a^+ for the point $p + \varepsilon$, we check whether there is a valid folding of the part of the strip between $p - \varepsilon$ and $p + \varepsilon$ that matches this level assignment. To do so, we check the following four conditions that capture the *noncrossing* conditions defined in Section 2:

- If a segment s extends to both sides of p without being folded at p, then it has the same level on both sides. That is, $a^-(s) = a^+(s)$.
- For each two input folds at p that connect pairs of segments that overlap $p - \varepsilon$, the levels of these pairs of segments are nested or disjoint. That is, if we have a fold connecting segments s_0 and s_1, and another fold connecting segments s_2 and s_3, then $[a^-(s_0), a^-(s_1)]$ and $[a^-(s_2), a^-(s_3)]$ are either disjoint intervals or one of these two intervals contains the other.
- For each two input folds at p that connect pairs of segments that overlap $p + \varepsilon$, the levels of these pairs of segments are nested or disjoint. This is a symmetric condition to the previous one, using a^+ instead of a^-.
- For each fold at p, connecting segments s_0 and s_1, and for each input segment s_2 that crosses p without being folded there, the interval of levels occupied by the fold should not contain the level of s_2. That is, if the two segments s_0 and s_1 extend to the left of p, then the interval $[a^-(s_0), a^-(s_1)]$ should not contain

$a^-(s_2)$. If the two segments extend to the right of p, then we have the same condition using a^+ instead of a^-.

If the pair (a^-, a^+) passes all these tests, we include a^+ in the set of valid level assignments for $p + \varepsilon$, which we will then use at the next stopping point of the algorithm.

If, at the end of this process, we reach the rightmost stopping point with a nonempty set of valid level assignments (necessarily consisting of the unique level assignment for the empty set of segments) then a folding of height k exists. The folding itself may be recovered by storing, for each level assignment a^+ considered by the algorithm, one of the level assignments a^- such that $a^- \in A$ and (a^-, a^+) passed all the tests above. Then, backtracking through these pointers, from the rightmost stopping point back to the leftmost one, will give a sequence of level assignments such that each consecutive pair is valid, which describes a consistent folding of the entire input strip.

The time for the algorithm is the number of stopping points multiplied by the number of pairs of level assignments for each stopping point and the time to test each pair of level assignments. This is $O(2^{O(k \log k)} n)$, as stated. □

5   Conclusion

In this paper, we considered three problems MinHeight, MinMaxCW and Min-SumCW for 1D strip folding, and showed some intractable results. We have some interesting open questions. Although we gave an FPT algorithm for MinHeight, it is not clear if the other two problems have FPT algorithms. Extending our models to 2D foldings would also be interesting.

References

1. Arkin, E.M., Bender, M.A., Demaine, E.D., Demaine, M.L., Mitchell, J.S.B., Sethia, S., Skiena, S.S.: When can you fold a map? Computational Geometry: Theory and Applications 29(1), 23–46 (2004)
2. Bannister, M.J., Eppstein, D., Simons, J.A.: Inapproximability of orthogonal compaction. Journal of Graph Algorithms and Applications 16, 651–673 (2012)
3. Bern, M., Hayes, B.: The complexity of flat origami. In: Proceedings of the 7th Annual ACM-SIAM Symposium on Discrete Algorithms, pp. 175–183 (1996)
4. Demaine, E.D., O'Rourke, J.: Geometric Folding Algorithms: Linkages, Origami, Polyhedra. Cambridge University Press (2007)
5. Gallivan, B.: Folding paper in half 12 times: An 'Impossible Challenge' Solved and Explained. Historical Society of Pomona Valley (2002)
6. Garey, M.R., Johnson, D.S.: Computers and Intractability: A Guide to the Theory of NP-Completeness. W. H. Freeman & Co. (1979)
7. Uehara, R.: On stretch minimization problem on unit strip paper. In: 22nd Canadian Conference on Computational Geometry (CCCG), pp. 223–226 (2010)
8. Uehara, R.: Stamp foldings with a given mountain-valley assignment. In: Origami[5]: Proceedings of the 5th International Meeting of Origami Science, Mathematics, and Education, pp. 585–597. AK Peters/CRC Press (2011)
9. Umesato, T., Saitoh, T., Uehara, R., Ito, H., Okamoto, Y.: The complexity of the stamp folding problem. Theoretical Computer Science 497, 13–19 (2013)

An Almost Optimal Algorithm for Voronoi Diagrams of Non-disjoint Line Segments*

(Extended Abstract)

Sang Won Bae

Department of Computer Science, Kyonggi University, Suwon, South Korea
swbae@kgu.ac.kr

Abstract. This paper presents an almost optimal algorithm that computes the Voronoi diagram of a set S of n line segments that may intersect or cross each other. If there are k intersections among the input segments in S, our algorithm takes $O(n\alpha(n)\log n+k)$ time, where $\alpha(\cdot)$ denotes the functional inverse of the Ackermann function. The best known running time prior to this work was $O((n + k)\log n)$. Since the lower bound of the problem is shown to be $\Omega(n\log n + k)$ in the worst case, our algorithm is worst-case optimal for $k = \Omega(n\alpha(n)\log n)$, and is only a factor of $\alpha(n)$ away from the lower bound. For the purpose, we also present an improved algorithm that computes the medial axis or the Voronoi diagram of a polygon with holes.

1 Introduction

There is no doubt that the Voronoi diagram is one of the most fundamental and the most well studied structures in computational geometry. Voronoi diagrams and their variations play an important role not only in computer science but also many other fields in engineering and sciences, finding a lot of applications. For a comprehensive survey, we refer to Aurenhammer and Klein [2] or to a book by Okabe et al. [14].

In this paper, we are interested in the Voronoi diagram of line segments in the plane. As one of the most popular variants of the ordinary Voronoi diagram, the line segment Voronoi diagram has been extensively studied in the computational geometry community, finding lots of applications in computer graphics, pattern recognition, motion planning, shape representation, and NC machining [8,10,13]. For the set of line segments that are disjoint or may intersect only at their endpoints, a variety of optimal $O(n\log n)$-time algorithms that compute the diagram are known. For examples, Kirkpatric [10], Lee [13], and Yap [17] presented divide-and-conquer algorithms, Fortune [7] presented a plane sweep algorithm, and a pure abstract approach to Voronoi diagrams by Klein [11] is also applied to yield an optimal time algorithm [12].

---

* This research was supported by Basic Science Research Program through the National Research Foundation of Korea (NRF) funded by the Ministry of Science, ICT & Future Planning (2013R1A1A1A05006927).

M.S. Rahman and E. Tomita (Eds.): WALCOM 2015, LNCS 8973, pp. 125–136, 2015.

However, only few research considers line segments that may intersect or cross each other freely. Let S be a set of n arbitrary line segments in the plane and k be the number of intersecting pairs of the segments in S. In fact, one can easily achieve the time bound $O((n+k)\log n)$ for computing the Voronoi diagram of S as follows: first, specify all the intersection points among the segments in S and consider the set S' of sub-segments obtained by cutting the original segments in S at the intersection points. Then, S' consists of at most $n + 2k$ line segments that can intersect only at endpoints, so we can apply any of the above existing algorithms. On the other hand, by a reduction from the problem of computing all the intersections of given line segments, it is not difficult to see that the lower bound of the problem of computing the Voronoi diagram of line segments is $\Omega(n\log n + k)$ in the worst case.

In this paper, we present an almost optimal algorithm that computes the Voronoi diagram of line segments. Our algorithm takes $O(n\alpha(n)\log n + k)$ time, where $\alpha(n)$ denotes the functional inverse of the Ackermann function. Since the lower bound is shown to be $\Omega(n\log n + k)$, our algorithm is only a factor of $\alpha(n)$ away from the optimal running time, and is optimal for large $k = \Omega(n\alpha(n)\log n)$. To our best knowledge, prior to our result, there was no known algorithm better than the simple $O((n + k)\log n)$-time algorithm.

In order to achieve our main result, we also consider an interesting special case where S forms a polygon. In this case, the Voronoi diagram of S is closely related to the *medial axis* of S [6, 13]. When S forms a simple polygon, it is known that its Voronoi diagram and medial axis can be computed in linear time by Chin et al. [6]. In this work, we extend their result into more general form of polygons, namely, *weakly simple polygons* and *polygonal domains*. In particular, we devise an $O(m\log(m + t) + t)$-time algorithm that computes the Voronoi diagram or the medial axis of a given polygonal domain, where m denotes the total number of vertices of its holes and t denotes the number of vertices of its outer boundary. Note that our algorithm is strictly faster than any $O(n\log n)$ time algorithm when t is relatively larger than m. We exploit this algorithm for a polygonal domain as a subroutine of the $O(n\alpha(n)\log n + k)$-time algorithm that computes the Voronoi diagram of non-disjoint line segments.

The remaining of the paper is organized as follows: After introducing some preliminaries in Section 2, we present out algorithm that computes the Voronoi diagram of a polygonal domain in Section 3. Then, Section 4 is devoted to describe and analyze our algorithm that computes the Voronoi diagram of line segments.

Due to space limit, some proofs are omitted, but will be found in a full version of the paper.

2   Preliminaries

Throughout the paper, we use following notations: For a subset $A \subset \mathbb{R}^2$, we denote by ∂A the boundary of A with the standard topology and by $\overline{pq}$ the line segment joining two points $p \in \mathbb{R}^2$ and $q \in \mathbb{R}^2$.

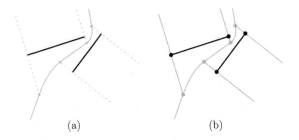

Fig. 1. (a) The Voronoi diagram of two line segments. (b) The Voronoi diagram of six sites: two open line segments and four of their endpoints. Sites are depicted as black segments or dots, and Voronoi edges and vertices are as gray segments and dots.

2.1   Voronoi Diagrams of Line Segments

Let S be a set of n line segments in the plane $\mathbb{R}^2$, and let k denote the number of intersection points among the segments in S.

The *Voronoi diagram* $\mathsf{VD}(S)$ of S is a subdivision of the plane $\mathbb{R}^2$ into *Voronoi regions* $R(s, S)$, defined to be

$$R(s, S) := \bigcap_{s' \in S, s' \neq s} \{x \in \mathbb{R}^2 \mid d(x, s) < d(x, s')\},$$

where $d(x, s)$ denotes the Euclidean distance from point x to segment s.

As done in the literature [1, 17], we regard each segment $s \in S$ as three distinct *sites*: the two endpoints of s and the relative interior of s. We thus assume that the set S is implicitly the set of points and open line segments that form n closed line segments. See Fig. 1. Note that each Voronoi edge of $\mathsf{VD}(S)$ is then either a straight or parabolic segment. The Voronoi vertices of $\mathsf{VD}(S)$ are distinguished into two kinds: those of one kind are proper vertices which are equidistant points from three distinct sites, while those of the other kind are simply the intersection points of S at which at least three Voronoi edges meet. Fig. 2 illustrates the Voronoi diagram of an example set of line segments.

When we are interested in the diagram $\mathsf{VD}(S)$ inside a compact region $A \subseteq \mathbb{R}^2$, we shall write $\mathsf{VD}_A(S)$ to denote the subdivision of A induced by $\mathsf{VD}(S)$. In other words, $\mathsf{VD}_A(S)$ is obtained by intersecting the diagram $\mathsf{VD}(S)$ itself with A.

The following fact is well known as the *star-shape* or *weak star-shape property* of Voronoi regions of a point or an open line segment.

Lemma 1. *If $s \in S$ is a point, then $R(s, S)$ is star-shaped with respect to s. If $s \in S$ is an open line segment, then $R(s, S)$ is weakly star-shaped in the sense that for any $x \in R(s, S)$, the segment $\overline{xs_x}$ is totally contained in $R(s, S)$, where s_x is the perpendicular foot from x to segment s.* □

Note that this fundamental observation implies the monotonicity of each Voronoi region $R(s, S)$ of any site $s \in S$ with respect to the site s; that is, $R(s, S)$ is

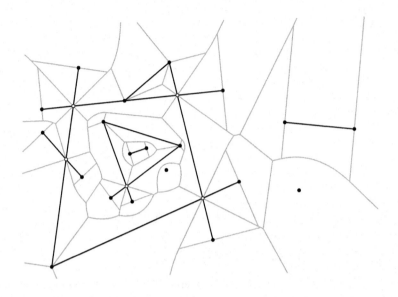

Fig. 2. The Voronoi diagram of non-disjoint line segments

rotationally monotone around s if s is a point, or is monotone in the direction of s if s is a segment.

As pointed out in Section 1, cutting segments in S at every intersection point and applying any existing optimal algorithm, including Yap [17], Klein [11], and Fortune [7], yields the following immediate upper bound.

Lemma 2. *Let S be a set of n line segments with k pairwise intersections. Then, the Voronoi diagram* $\mathsf{VD}(S)$ *of S has $O(n+k)$ combinatorial complexity and can be computed in $O((n+k)\log n)$ time.* ☐

On the other hand, we show a lower bound on computing $\mathsf{VD}(S)$ in terms of n and k by a reduction from the problem of finding the intersection points of line segments, of which the lower bound is known to be $\Omega(n\log n + k)$ in the worst case in the algebraic decision tree model [5].

Lemma 3. *Any algorithm that correctly computes the Voronoi diagram* $\mathsf{VD}(S)$ *of S takes $\Omega(n\log n + k)$ time in the worst case in the algebraic decision tree model.*

Proof. This easily follows from the fact that all the intersection points among the segments in S also appear as Voronoi vertices of $\mathsf{VD}(S)$. Therefore, after computing the Voronoi diagram $\mathsf{VD}(S)$, it suffices to traverse all the Voronoi vertices of $\mathsf{VD}(S)$. Since any algorithm that computes all the intersection points among n line segments takes $\Omega(n\log n + k)$ time in the worst case in the algebraic decision tree model [5], the same lower bound applies to the problem of computing $\mathsf{VD}(S)$. ☐

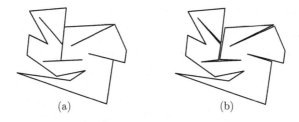

Fig. 3. (a) A weakly simple polygon and (b) a simple polygon obtained by perturbation

2.2 Weakly Simple Polygons

A *polygon* in the plane is a closed curve that is piecewise linear. A polygon can be defined by a cyclic sequence of points in $\mathbb{R}^2$, called *vertices*, that are connected by line segments, called *edges*. A polygon is called *simple* if and only if all of its vertices are distinct and its edges intersect only at their common endpoints. Note that if a polygon is not simple, then its vertices do not have to be distinct and its edges may overlap or cross each other.

A *weakly simple polygon* is, informally, a polygon without "self-crossing." Intuitively, one may think that a weakly simple polygon is allowed to have vertices and edges "touched." Although the definition using the term "self-crossing" is common in the literature, it is turned to be imprecise as a formal definition. See Chang et al. [3] for discussions on this subject. In this work, we adopt their definition of weakly simple polygons.

Definition 1 (Chang et al. [3]). *A polygon P is* weakly simple *if and only if P has at most two vertices or P can be made simple by an arbitrarily small perturbation of its vertices.*

By definition, we can find a simple polygon that is arbitrarily close to any given weakly simple polygon by a perturbation of its vertices; see Fig. 3. Note that a line segment or a point itself is considered a degenerate form of weakly simple polygons, and that the interior of a weakly simple polygon may be an empty set. By convention, any (weakly) simple polygon is also considered as a compact subset containing its interior.

3 Voronoi Diagram of a Polygon

In this section, we consider some special cases of the set S of line segments, namely, when S forms a weakly simple polygon or a polygonal domain having polygonal holes.

3.1 Voronoi Diagram of a Weakly Simple Polygon

Let P be a weakly simple polygon with n vertices. The *Voronoi diagram of a (weakly) simple polygon P* is the Voronoi diagram $\mathsf{VD}_P(P)$ of the vertices and edges of P inside the polygon P. We denote by $\mathsf{VD}^-(P) = \mathsf{VD}_P(P)$ the Voronoi diagram of a polygon P.

If P is a simple polygon, $\mathsf{VD}^-(P)$ can be computed in linear time by Chin et al. [6]. By the definition of weakly simple polygons, it is not difficult to show that our case can also be handled by the algorithm of Chin et al. [6] after a perturbation of its vertices.

Lemma 4. *The Voronoi diagram* $\mathsf{VD}^-(P)$ *of a weakly simple polygon* P *with* n *vertices can be computed in* $O(n)$ *time.*

The *medial axis* $\mathsf{MA}(P)$ of a polygon P is the set of points x in P such that x has at least two distinct points that are closest from x among all points on ∂P. Remark that the medial axis $\mathsf{MA}(P)$ of a weakly simple polygon P can also be computed in linear time since $\mathsf{MA}(P)$ is a subset of $\mathsf{VD}^-(P)$ [6].

3.2   Voronoi Diagram of a Polygon with Holes

We now extend our interests to a polygon with polygonal holes, called a *polygonal domain*, also known as a *multiply-connected polygon* [16] or a *pocket* [8] in the literature. A polygonal domain $D \subset \mathbb{R}^2$ is a polygonal region that may have one or more *holes*. If D has h holes, then its boundary ∂D consists of $h+1$ pairwise disjoint (weakly) simple polygons $P, Q_1, Q_2, \ldots, Q_h$ such that P contains all the other $Q_1, \ldots, Q_h$ in its interior. We call P the *outer boundary* of D, and $Q_1, \ldots, Q_h$ the *holes* of D. In the literature, the outer boundary and the holes of polygonal domains are supposed to be simple polygons [8,16]. However, in this paper, we relax the condition for P and the Q_i to be weakly simple polygons.

Let $\mathsf{VD}^-(D) := \mathsf{VD}_D(P \cup Q_1 \cup \cdots \cup Q_h)$ be the Voronoi diagram of a given polygonal domain D. Since the edges of D intersect only at their endpoints, that is, the vertices of D, we can compute $\mathsf{VD}^-(D)$ in $O(n \log n)$ time using existing algorithms [7,11,17], where n denotes the total number vertices of D. In the following, we achieve an improved time bound by introducing more parameters: let m denote the total number of vertices of the holes $Q_1, \ldots, Q_h$ of D and t denote the number of vertices of the outer boundary P of D. Note that $n = m+t$.

Our strategy is as follows: Let $Q := Q_1 \cup \cdots \cup Q_h$ be the set of all the vertices and the edges that form the holes $Q_1, \ldots, Q_h$. We first compute the Voronoi diagram $\mathsf{VD}^-(P)$ of the outer boundary P, and the Voronoi diagram $\mathsf{VD}(Q)$ of the edges of the holes. We then merge the two diagrams into our target $\mathsf{VD}^-(D)$. Recall that $\mathsf{VD}^-(P)$ can be computed in $O(t)$ time by Lemma 4 and $\mathsf{VD}(Q)$ in $O(m \log m)$ time since P is a weakly simple polygon and Q is the set of line segments that may intersect only at their endpoints.

Consider the vertices and the edges of $\mathsf{VD}^-(D) = \mathsf{VD}_D(P \cup Q)$ to be computed. We make a general position assumption that no point $x \in D$ is equidistant from four distinct sites in $P \cup Q$; one can remove this assumption by a standard perturbation technique [6]. Some of them come from $\mathsf{VD}^-(P)$ or from $\mathsf{VD}(Q)$, and the others cannot be found in either of the two diagrams $\mathsf{VD}^-(P)$ and $\mathsf{VD}(Q)$, which are the set of equidistant points from two closest sites $s \in P$ and $s' \in Q$.

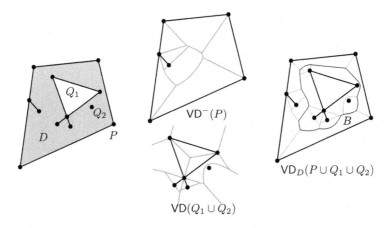

Fig. 4. How to merge $\mathsf{VD}^-(P)$ and $\mathsf{VD}(Q)$ to obtain $\mathsf{VD}_D(P \cup Q)$, for an example polygonal domain D having its outer boundary P and two holes Q_1, Q_2. The outer boundary P and the holes Q_1, Q_2 are weakly simple polygons with 7 vertices, 7 vertices, and one vertex, respectively. The merge curve $\beta = B$ is depicted as black solid line.

We denote the union of all such edges and vertices of $\mathsf{VD}_D(P \cup Q)$ by B. The set $B \subset D$ can be described as the set of points x such that $d(x, P) = d(x, Q)$, where $d(x, P) = \min_{s \in P} d(x, s)$ and $d(x, Q) = \min_{s' \in Q} d(x, s')$. By definition, B properly divides the interior of D into two regions, one closer to P and the other closer to Q. The following lemma describes how B looks topologically. For a simple closed curve $C \subset \mathbb{R}^2$, we denote by C^- the *interior* of C, that is, the region bounded by C, and by $C^+ := \mathbb{R}^2 \setminus (C^- \cup C)$ the *exterior* of C.

Lemma 5. *Let $\beta_1, \ldots, \beta_b$ be the connected components of B. We then have:*

(i) Each component β_i for $i = 1, \ldots, b$ is a simple closed curve in D.
(ii) Every hole of D is contained in the interior β_i^- of some β_i.
(iii) Each β_i contains at least one hole of D in its interior β_i^-.
(iv) No two distinct curves β_i and β_j are nested, that is, $\beta_i \not\subset \beta_j^-$ and $\beta_j \not\subset \beta_i^-$.

We call such a closed curve $\beta \subseteq B$, described in Lemma 5, a *merge curve*. Lemma 5 tells us that B consists of one or more merge curves and each hole Q_i is contained in a unique merge curve. Once all the merge curves in B are specified, one can easily compute the target diagram $\mathsf{VD}_D(P \cup Q)$ by cutting and gluing $\mathsf{VD}^-(P)$ and $\mathsf{VD}(Q)$ along the merge curves B in time linear to the combinatorial complexity of the final output $\mathsf{VD}_D(P \cup Q)$, which is bounded by $O(n)$. Hence, we now focus on how to efficiently find every merge curve $\beta \subseteq B$.

Following lemmas will be helpful for further discussions.

Lemma 6. *Let $\beta \subseteq B$ be any merge curve, and $s \in P$ and $s' \in Q$ be any two sites. Then, $\beta^+ \cap R(s, P)$ and $\beta^- \cap R(s', Q)$ are monotone with respect to s and s', respectively. Also, if $\beta^+ \cap R(s, P)$ is nonempty, then s lies in β^+; if $\beta^- \cap R(s', Q)$ is nonempty, then s' lies in β^-.*

Algorithm. MERGETWOVDs

1: **while** there is Q_i that is not bounded by a merge curve traced before **do**
2:    Let $\beta \subseteq B$ be the merge curve such that $Q_i \subset \beta^-$.
3:    Find a point $z \in \beta$.
4:    Trace β from z.
5:    Identify all Q_j bounded by β, that is, $Q_j \subset \beta^-$.
6: **end while**
7: Cut and glue $\mathsf{VD}^-(P)$ and $\mathsf{VD}(Q)$ along all merge curves.
8: **return** the resulting diagram as $\mathsf{VD}_D(P \cup Q) = \mathsf{VD}^-(D)$.

Fig. 5. Algorithm for merging $\mathsf{VD}^-(P)$ and $\mathsf{VD}(Q)$ into $\mathsf{VD}_D(P \cup Q)$

Lemma 7. *For any merge curve $\beta \subseteq B$, if its interior β^- intersects $R(s, Q)$ for some $s \in Q_j$, then the hole Q_j is contained in β^-.*

Now, suppose that $\mathsf{VD}^-(P)$ and $\mathsf{VD}(Q)$ have been computed correctly, and an optimal point location structure on $\mathsf{VD}^-(P)$ has been built. Such a structure supports a point location query on $\mathsf{VD}^-(P)$ in $O(\log t)$ time, and can be constructed in $O(t)$ time by Kirkpatrick [9] after triangulating each Voronoi region of $\mathsf{VD}^-(P)$ in $O(t)$ time [4]. Our merge algorithm is then described in Fig. 5.

Note that after a merge curve β is traced and identified, one can easily find out which hole lies in its interior β^- by traversing the regions of $\mathsf{VD}(Q)$ intersected by β^- by Lemma 7. From now on, we thus describe and analyze (1) how to find a point $z \in \beta$ and (2) how to trace the merge curve β in more details.

Finding a Point $z \in \beta$. Assume that we are to find a point z on a merge curve β, which contains Q_i in its interior β^-. For the purpose, we first pick any vertex q of Q_i and find a point $p \in P$ that is the closest from q among all points in P. This can be done by a point location query on $\mathsf{VD}^-(P)$. We then walk along $\overline{qp}$ from q towards p until we meet a point z on β.

Our strategy is promising due to the following lemma.

Lemma 8. *The segment $\overline{qp}$ intersects β exactly in a single point.*

More precisely, we walk on the diagram $\mathsf{VD}(Q)$ along the segment $\overline{qp}$ from q towards p. While walking in a Voronoi region $R(s', Q)$, the function $f(x) := d(x, Q) = d(x, s')$ can be explicitly described for $x \in \overline{qp} \cap R(s', Q)$. We check whether the equation $f(x) = d(x, p)$ has a solution $z \in \overline{qp} \cap R(s', Q)$. If so, we are done; otherwise, we proceed to the next region of $\mathsf{VD}(Q)$ along $\overline{qp}$. Lemma 8 guarantees that this procedure will terminate with a point $z \in \beta$.

Tracing β from $z \in \beta$. Let $s \in P$ and $s' \in Q$ be the sites such that $z \in R(s, P)$ and $z \in R(s', Q)$. This is automatically identified once $z = \overline{qp} \cap \beta$ is found in the above step. Then, z must lie on the *bisecting curve* γ between s and s'. Note that γ is either a straight line or a parabola according to the types of s and s'.

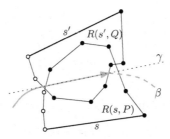

Fig. 6. Tracing a merge curve β in $R(s, P) \cap R(s', Q)$. To find the point on the boundary hit by β, test vertices of $R(s, P)$ and $R(s', Q)$ in the order along β which side of γ it lies in. Vertices marked with black dots will get tested in order and those marked with white dots will not because they are passed already by β.

In order to trace β from z, we walk along locally γ in the direction that $s \in P$ lies to the right and $s' \in Q$ lies to the left of γ. Note that γ coincides with β in $R(s, P) \cap R(s', Q)$. We thus trace γ until it hits the boundary $\partial R(s, P)$ or $\partial R(s', Q)$. Assume that γ hits $\partial R(s, P)$ before $\partial R(s', Q)$. Then, we proceed to the next region $R(s'', P)$ in $\mathsf{VD}^-(P)$, replacing s by s'' and repeating the above procedure. Since β is a closed curve, we are done when getting back to z.

In order to find the point on the boundary $\partial R(s, P)$ or $\partial R(s', Q)$ hit by γ, we test each vertex on $\partial R(s, P)$ and $\partial R(s', Q)$ in a certain order whether it lies to the left or to the right of the bisecting curve γ between s and s'. This test can be performed in $O(1)$ time per tested vertex by examining a point and a line or a parabola γ. The order of vertices is induced by the tracing direction of β. Since $\beta \cap R(s, P) \cap R(s', Q)$ is monotone with respect to s and s', simultaneously, by Lemma 6, one can find the next vertex from a vertex in $O(1)$ time along the boundary of each region $R(s, P)$ and $R(s', Q)$. See Fig. 6. We thus test the vertices in this order. If the tested vertex v is of $R(s, P)$ and it is the first that lies to the right of γ, then γ crosses $R(s, P)$ between v and its preceding vertex on $\partial R(s, P)$; if v is of $R(s', Q)$ and it is the first that lies to the left of γ, then γ crosses $R(s', Q)$ between v and its preceding vertex on $\partial R(s', Q)$. Hence, as soon as we find a vertex of $R(s, P)$ lying to the right of γ or one of $R(s', Q)$ lying to the left of γ, γ hits the boundary $\partial R(s, P)$ or $\partial R(s', Q)$, respectively.

Assume without loss of generality that γ hits $\partial R(s, P)$ before $\partial R(s', Q)$. Then, β is trace up to the hitting point, and is about to enter the next region $R(s'', P)$ from $R(s, P)$ for some $s'' \in P$. Since we know the entering point $y \in \gamma$ at $\partial R(s'', P)$, we know the next vertex along $\partial R(s'', P)$ from y. Note that we do not have to test vertices on $\partial R(s', Q)$ that have been tested before by Lemma 6. Hence, we substitute s by s'' and proceed on testing next vertices on $\partial R(s'', P)$ and $\partial R(s', Q)$ as above without repetition.

The Time Complexity. We now analyze the cost of finding a merge curve β, which corresponds to a single while loop in the algorithm described in Fig. 5.

Lemma 9. *Specifying a merge curve $\beta \subseteq B$ can be done in time $O(\log t + M_\beta + T_\beta + |\beta|)$, where M_β denotes the total complexity of Voronoi regions $R(s', Q)$ of $\mathsf{VD}(Q)$ that are intersected by β^-, T_β denotes the number of Voronoi vertices of $\mathsf{VD}^-(P)$ lying inside β^-, and $|\beta|$ denotes the number of Voronoi edges of $\mathsf{VD}_D(P \cup Q)$ along β.*

We then bound the total running time of our merging algorithm.

Lemma 10. *Merging two Voronoi diagrams $\mathsf{VD}^-(P)$ and $\mathsf{VD}(Q)$ into $\mathsf{VD}_D(P \cup Q)$ can be done in $O(h \log t + m + t)$ time.*

Proof. The total time bound is obtained by summing $O(\log t + M_\beta + T_\beta + |\beta|)$ over all merge curves $\beta \subseteq B$ by Lemma 9. We show that (i) $\sum_\beta M_\beta = O(m)$, (ii) $\sum_\beta T_\beta = O(t)$, and (iii) $\sum_\beta |\beta| = O(m + t)$.

(i) The value M_β counts the total complexity of $R(s', Q)$ for all $s' \in Q_j$ and $Q_j \subset \beta^-$, and each hole Q_j is contained in a unique merge curve by Lemma 5. This implies that $\sum_\beta M_\beta$ does not exceed the complexity of the Voronoi diagram $\mathsf{VD}(Q)$, bounded by $O(m)$.

(ii) T_β counts the number of vertices of $\mathsf{VD}^-(P)$ lying in β^-. Since the merge curves are not nested by Lemma 5, $\sum_\beta T_\beta$ does not exceed the total number of vertices of $\mathsf{VD}^-(P)$, bounded by $O(t)$.

(iii) The merge curves $\beta \subseteq B$ are the union of Voronoi edges of $\mathsf{VD}_D(P \cup Q)$. Thus, its total complexity $\sum_\beta |\beta|$ does not exceed the complexity of the merged diagram $\mathsf{VD}_D(P \cup Q)$, bounded by $O(m + t)$.

Therefore, we have

$$\sum_\beta O(\log t + M_\beta + T_\beta + |\beta|) = \sum_\beta O(\log t) + O(m + t) = O(h \log t + m + t),$$

since the number of merge curves is at most h by Lemma 5.   □

We are finally ready to conclude the main theorem of this section.

Theorem 1. *Let D be a polygonal domain whose outer boundary and holes may be weakly simple polygons. The Voronoi diagram $\mathsf{VD}^-(D)$ or the medial axis $\mathsf{MA}(D)$ can be computed in $O(m \log(m + t) + t)$ time, where m is the total complexity of the holes of D and t is the complexity of the outer boundary of D.*

4   Voronoi Diagram of Line Segments

Let S be a set of n line segments in the plane $\mathbb{R}^2$ with k pairwise intersections. In this section, we describe our algorithm that computes the Voronoi diagram $\mathsf{VD}(S)$ of line segments.

Our algorithm runs based on the decomposition of the plane $\mathbb{R}^2$ into faces of the arrangement of S. The *arrangement* $\mathcal{A}(S)$ of S is a decomposition of $\mathbb{R}^2$ into vertices, edges, and faces induced by the line segments in S. We regard edges and faces of $\mathcal{A}(S)$ relatively open sets: each edge as a set does not contain its end vertices and each face does not contain its incident edges and vertices.

For any face σ of $\mathcal{A}(S)$, we consider its boundary $\partial\sigma$, defined to be the union of vertices and edges that are incident to σ. Since a face in the line segment arrangement can have holes, the boundary $\partial\sigma$ may consist of several connected components, and each of them indeed forms a weakly simple polygon. This means that σ forms a polygonal domain whose outer boundary and holes can be weakly simple polygons, while the unbounded face of $\mathcal{A}(S)$ is regarded as a polygonal domain with outer boundary at infinity. Note that the complexity of the boundary $\partial\sigma$ of a face is bounded by $O(n\alpha(n))$, where $\alpha(n)$ denotes the functional inverse of the Ackermann function, and this bound can be realized [15].

We observe the following.

Lemma 11. *Let σ be any face of the arrangement $\mathcal{A}(S)$ and $x \in \sigma$ be any point. The Voronoi region $R(s, S)$ of site $s \in S$ intersects σ if and only if the site s appears on the boundary $\partial\sigma$ of σ.*

Lemma 11 implies that the Voronoi diagram $\mathsf{VD}(S)$ cropped by a face σ coincides with the Voronoi diagram of a polygonal domain σ, that is, $\mathsf{VD}_\sigma(S) = \mathsf{VD}^-(\sigma)$. Hence, we compute the diagram $\mathsf{VD}(S)$ in a face-by-face fashion. For example, the polygonal domain D in Fig. 4 is identical to a face σ of the arrangement of the line segments S in Fig. 2, and it holds that $\mathsf{VD}_\sigma(S) = \mathsf{VD}^-(D)$.

Our algorithm is described in three steps: (1) Compute all the intersection points in S, (2) compute the arrangement $\mathcal{A}(S)$ and the description of its faces, and then (3) for every face σ of $\mathcal{A}(S)$, compute the Voronoi diagram $\mathsf{VD}^-(\sigma)$ of σ. Then, the diagram $\mathsf{VD}(S)$ of the original n line segments can be obtained by taking the union of the face diagrams $\mathsf{VD}^-(\sigma)$ over all faces σ of $\mathcal{A}(S)$.

The first and second steps can be finished in $O(n \log n + h)$ time by Chazelle and Edelsbrunner [5]. The third step is handled by applying Theorem 1 to every face σ of $\mathcal{A}(S)$. Let $m = m_\sigma$ be the total complexity of the holes of σ, and $t = t_\sigma$ be the complexity of the outer boundary of σ. Also, let n_σ be the number of segments in S that appear in the boundary of a hole of σ. Since the boundary of σ forms a polygonal domain whose outer boundary and holes are weakly simple polygons, we apply Theorem 1 to compute $\mathsf{VD}^-(\sigma)$ in time $O(m \log(m + t) + t)$.

We are then ready to conclude our main result.

Theorem 2. *Let S be a set of n line segment with k pairwise intersections. Then, the Voronoi diagram $\mathsf{VD}(S)$ of S can be computed in $O(n\alpha(n) \log n + k)$ time, where $\alpha(\cdot)$ denotes the functional inverse of the Ackermann function.*

Proof. As described above, once the arrangement $\mathcal{A}(S)$ is computed, we apply Theorem 1 to every face σ of the arrangement $\mathcal{A}(S)$. The correctness of our algorithm simply follows from Lemma 11.

We now analyze the time complexity of our algorithm. We spend $O(n \log n + k)$ time for Steps (1) and (2). The total time spent in Step (3) of our algorithm is the sum of $O(m_\sigma \log(m_\sigma + t_\sigma) + t_\sigma)$ over all faces σ of $\mathcal{A}(S)$.

We claim that (i) $\sum_\sigma t_\sigma \leq 2n + 4k$ and (ii) $\sum_\sigma n_\sigma \leq n$. Since $m_\sigma = O(n_\sigma \alpha(n_\sigma)) = O(n_\sigma \alpha(n))$ [15], if our claim is true, then we have

$$\sum_\sigma O(m_\sigma \log(m_\sigma + t_\sigma) + t_\sigma) = \sum_\sigma O(n_\sigma \alpha(n) \log(n_\sigma \alpha(n) + t_\sigma) + t_\sigma)$$

$$= O(n\alpha(n)\log(n\alpha(n) + k) + k)$$
$$= O(n\alpha(n)\log n + k).$$

Therefore, we are done by showing that our claim is true. Consider any sub-segment s' after cutting segments in S at every intersection point. Observe that s' can appear in the outer boundary of a face σ of $\mathcal{A}(S)$ at most twice, since s' is incident to at most two faces of the arrangement $\mathcal{A}(S)$. Thus, the sum of t_σ over all faces σ is at most twice the number of such sub-segments, that is, $2n + 4k$, so claim (i) is shown. Claim (ii) easily follows from the fact that each segment $s \in S$ cannot participate the boundary of the holes of two distinct faces of $\mathcal{A}(S)$; otherwise, s must be placed inside two different faces, which is impossible. ⊡

References

1. Alt, H., Cheong, O., Vigneron, A.: The Voronoi diagram of curved objects. Discrete Comput. Geom. 34(3), 439–453 (2005)
2. Aurenhammer, F., Klein, R.: Voronoi diagrams. In: Sack, J.R., Urrutia, J. (eds.) Handbook of Computational Geometry. Elsevier (2000)
3. Chang, H., Erickson, J., Xu, C.: Detecting weakly simple polygons. In: Proc. 26th ACM-SIAM Sympos. Discrete Algo (SODA 2015) (2015)
4. Chazelle, B.: Triangulating a simple polygon in linear time. Discrete Comput. Geom. 6, 485–524 (1991)
5. Chazelle, B., Edelsbrunner, H.: An optimal algorithm for intersecting line segments in the plane. J. ACM 39, 1–54 (1992)
6. Chin, F., Snoeyink, J., Wang, C.A.: Finding the medial axis of a simple polygon in linear time. Discrete Comput. Geom. 21(3), 405–420 (1999)
7. Fortune, S.J.: A sweepline algorithm for Voronoi diagrams. Algorithmica 2, 153–174 (1987)
8. Held, M. (ed.): On the Computational Geometry of Pocket Machining. LNCS, vol. 500. Springer, Heidelberg (1991)
9. Kirkpatrick, D.: Optimal search in planar subdivisions. SIAM J. Comput. 12(1), 28–35 (1983)
10. Kirkpatrick, D.G.: Efficient computation of continuous skeleton. In: Proc. 20th Annu. IEEE Sympos. Found. Comput. Sci., pp. 18–27 (1979)
11. Klein, R.: Concrete and Abstract Voronoi Diagrams. LNCS, vol. 400. Springer, Heidelberg (1989)
12. Klein, R., Mehlhorn, K., Meiser, S.: Randomized incremental construction of abstract Voronoi diagrams. Comput. Geom.: Theory Appl. 3(3), 157–184 (1993)
13. Lee, D.T.: Medial axis transformation of a planar shape. IEEE Trans. Pattern Anal. Mach. Intell. PAMI 4(4), 363–369 (1982)
14. Okabe, A., Boots, B., Sugihara, K., Chiu, S.N.: Spatial Tessellations: Concepts and Applications of Voronoi Diagrams, 2nd edn. John Wiley and Sons, New York (2000)
15. Sharir, M., Agarwal, P.K.: Davenport-Schinzel Sequences and Their Geometric Applications. Cambridge University Press, New York (1995)
16. Srinivasan, V., Nackman, L.R.: Voronoi diagram for multiply-connected polygonal domains I: Algorithm. IBM J. Research Development 31(3), 361–372 (1987)
17. Yap, C.K.: An $O(n \log n)$ algorithm for the Voronoi diagram of a set of simple curve segments. Discrete Comput. Geom. 2(1), 365–393 (1987)

PTAS's for Some Metric p-source Communication Spanning Tree Problems*

Santiago V. Ravelo and Carlos E. Ferreira

Instituto de Matemática e Estatística
Universidade de São Paulo, Brasil
{ravelo,cef}@ime.usp.br

Abstract. In this work we consider some NP-hard cases of the metric p-source communication spanning tree problem (metric p-OCT). Given an undirected complete graph $G = (V, E)$ with non-negative length $\omega(e)$ associated to each edge $e \in E$ satisfying the triangular inequality, a set $S \subseteq V$ of p vertices and non-negative routing requirements $\psi(u, v)$ between all pairs of nodes $u \in S$ and $v \in V$, the metric p-OCT's objective is to find a spanning tree T of G, that minimizes: $\sum_{u \in S} \sum_{v \in V} \psi(u, v) d(T, u, v)$, where $d(H, x, y)$ is the minimum distance between nodes x and y in a graph $H \subseteq G$. This problem is a particular case of the optimum communication spanning tree problem (OCT). We prove a general result which allows us to derive polynomial approximation schemes for some NP-hard cases of the metric p-OCT improving the existing ratios for these problems.

1 Introduction

In this work we consider NP-hard particular cases of the metric p-source optimum communication spanning tree problem (metric p-OCT) which is a particular case of the optimum communication spanning tree problem (OCT). In the OCT, introduced by Hu in 1974 ([Hu, 1974, Wu and Chao, 2004]), the input is an undirected graph $G = (V, E)$ with non-negative lengths $\omega(e)$ associated to each edge $e \in E$ and non-negative requirements $\psi(u, v)$ between each pair of nodes $u, v \in V$. The problem is to find a spanning tree T of G which minimizes the total communication cost given by $C(T) = \sum_{u \in V} \sum_{v \in V} \psi(u, v) d(T, u, v)$, where $d(H, x, y)$ denotes the minimum distance between the nodes x and y in the subgraph H of G. In the p-OCT it is additionally given a set of p nodes (sources) $S \subseteq V$, that are considered in the objective function: $C(T) = \sum_{u \in S} \sum_{v \in V} \psi(u, v) d(T, u, v)$.

In [Johnson et al., 1978] it was proven, by a reduction from the 3-exact cover problem (3-EC), that the minimum routing cost spanning tree problem (MRCT) is NP-hard. MRCT is a particular case of OCT where for each pair of nodes the communication requirement between them is equal to one, i.e., $\psi(u, v) = 1$ for all $u, v \in V$. In [Wu et al., 2000c] a PTAS was given for the MRCT. Also in this

---

* This research is supported by the following projects: FAPESP 2013/03447−6, CNPq 477203/2012 − 4 and CNPq 302736/2010 − 7.

M.S. Rahman and E. Tomita (Eds.): WALCOM 2015, LNCS 8973, pp. 137–148, 2015.

work a reduction from the general case to the metric one is presented and it was proven that MRCT with edge-lengths that satisfy the triangular inequality is also NP-hard. Also, in [Wu et al., 2000c] a $O(log^2(n))$-approximation for OCT was presented applying a result from [Bartal, 1996] which was improved to a $O(log(n))$-approximation by [Talwar et al., 2003].

A particular case of p-OCT is weighted p-MRCT introduced in [Wu, 2002]. In this case a non-negative sending requirement $\sigma(u)$ is given for each source $u \in S$, and the requirement between a source $u \in S$ and a node $v \in V$ is $\psi(u, v) = \sigma(u)$. When the sending requirements σ for the p sources are equal to one, the problem is called p-MRCT. In [Wu, 2002] it was proven that 2-MRCT is NP-hard, moreover PTASs were shown for 2-MRCT and for the metric case of weighted 2-MRCT.

In [Wu et al., 2000a] was introduced the minimum sum-requirement communication spanning tree problem (SROCT), in which the sending requirement $\sigma(u)$ is given for each $u \in V$. Observe that weighted p-MRCT is a particular case of SROCT, where $\sigma(u) = u$ for all $u \in V - S$. Also, the authors presented a 2-approximation algorithm for this problem.

In [Wu, 2004] a 2-approximation algorithm was given for the metric p-OCT and a 3-approximation for the general 2-OCT.

In this work we present a general result that allows us to derive PTASs for natural special cases of p-OCT. We introduce three particular NP-hard cases of p-OCTand we derive PTASs for the metric cases of these problems. Also, we give new PTASs for the metric p-MRCT and for the fixed parameter weighted p-MRCT.

This work is organized as follows. In section 2, we introduce some definitions. In section 3 a polynomial time algorithm to find an optimum k-star for a given k is presented. In section 4 we prove the main result of this work, which allows us to obtain approximation algorithms for different metric p-OCT problems. In section 5 we introduce particular cases of p-OCT giving a PTAS for the metric case of that problem. We finish the paper in section 6 with some conclusions.

2   Definitions

Unless specified all graphs in this work are undirected. Given a graph G we denote the set of its nodes as V_G and the set of its edges as E_G (when G is implicit by the context we use V or E instead of V_G or E_G and $n = |V|$). Also, when G has non-negative lengths associated to its edges, the **length of a path** in G is defined as the sum of the lengths of its edges (a path with no edges has length zero). The **distance** between node x and node y in a sub-graph H of G is the length of a path with minimum length between x and y in H and is denoted by $d(H, x, y)$. Now, for each positive integer p we can define the p-OCT as:

Problem 1. p-OCT - p-source Optimum Communication spanning Tree problem.

Input: A graph G, a non-negative length function over the edges of G, ω : $E \to \mathbb{Q}_+$, a set of p sources $S \subseteq V$ and a non-negative routing requirement function from each source node to each node of V, $\psi : S \times V \to \mathbb{Q}_+$.

Output: A spanning tree T of G which minimizes the total requirement routing cost: $C(T) = \sum_{u \in S} \sum_{v \in V} \psi(u,v) d(T,u,v)$.

This paper considers the metric p-OCT, which is the particular case of p-OCT where the graph G is complete and the length function over the edges satisfies the triangular inequality. To find a feasible solution of the metric p-OCT we use a valid k-star[1]:

Definition 1. *Given a graph G, a set S of nodes of G and an integer $k \geq |S|$, a* **k-star** *of G is a spanning tree of G with no more than k internal nodes (that is, at least $n - k$ leaves). A* **core** *of a k-star T of G is a tree resulting by eliminating $n - k$ leaves from T. A core τ is* **valid** *for the set S of nodes (or just valid) if τ contains all the nodes of S. Then, a* **valid k-star** *of G is a k-star of G with at least a valid core.*

The problem of finding an optimal valid k-star for the metric p-OCT can be defined as follows. Also, in section 3, we show how to solve it efficiently.

Problem 2. Optimum valid k-star for the metric p-OCT.

Input: A positive integer k and an instance of metric p-OCT: a complete graph G, a non-negative length function over the edges of G which satisfies the triangular inequality, $\omega : E \to \mathbb{Q}_+$, a set of p sources $S \subseteq V$ and a non-negative routing requirement function between each node of S and each node of V, $\psi : S \times V \to \mathbb{Q}_+$. (Notice that $k \geq p = |S|$)

Output: A valid k-star T of G which minimizes the total requirement routing cost: $C(T) = \sum_{u \in S} \sum_{v \in V} \psi(u,v) d(T,u,v)$.

3   Optimal Valid k-star for Metric p-OCT

First note that if k and p are constants, the number of possible valid cores of k-stars of G is polynomial. Indeed, since the core of a valid k-star must contain the p vertices in S and $k - p$ other vertices, one can enumerate the $\binom{n}{k-p} = O(n^{k-p})$ possibilities. For each different choice one has to enumerate all possible trees with k vertices and there are $O(k^k)$ possible trees. Then, the number of all possible valid cores is limited by $O(k^k n^{k-p})$. Our approach is to find an optimal valid k-star with core τ for each valid core τ, selecting the minimum k-star among them.

---

[1] The definition of k-star used in this paper is similar to the one used by [Wu et al., 2000c, Wu et al., 2000a, Wu et al., 2000b], which is different from the usual definition of k-star in graph theory (a tree with k leaves linked to a single vertex of degree k).

Given a valid core τ, to obtain a valid k-star T with core τ each node of $V_G - V_\tau$ must be adjacent to some node of τ (i.e., these nodes will be leaves of T).

Let u_v be the node of τ adjacent to $v \in V_G - V_\tau$ in T. Since all nodes in S belong to τ then: $C(T) = C(\tau) + \sum_{v \in V_G - V_\tau} \sum_{w \in S} \psi(w, v)(d(\tau, w, u_v) + \omega(u_v, v))$.

Thus, in order to find the best vertex of the core τ to link each vertex $v \in V - V_\tau$ it suffices to consider the node $u_v^* \in V_\tau$ that minimizes:

$$\sum_{w \in S} \psi(w, v)(d(\tau, w, u_v) + \omega(u_v, v)).$$

To compute it efficiently, first we pre-calculate for each $w \in S$ and $u \in V_\tau$ all the distances $d(\tau, w, u)$ (it can be done in $O(|S|k) = O(pk)$). After that, we calculate for each pair of nodes $v \in V_G - V_\tau$ and $u \in V_\tau$ ($k(n-k)$ pairs) the cost of linking v to vertex u in τ which can be computed in $O(|S|) = O(p)$ using the pre-calculated distances. Therefore, we can obtain an optimal valid k-star with core τ in $O(k(n-k)p + pk) = O(npk)$ time.

From the ideas above, we conclude that it is possible to find and optimal valid k-star in $O(k^{k+1} n^{k-p+1} p)$ time.

Lemma 1. *An optimum valid k-star for metric p-OCT with fixed $k \geq p$ can be found in $O(n^{k-p+1})$ time.*

4   Approximation Lemma (for Metric p-OCT Problems)

In this section we present the main result of the paper. First we introduce the notion of δ-balanced-path, $0 < \delta \leq \frac{1}{2}$. This definition is based on similar concepts for related problems introduced in [Wu et al., 2000c], [Wu et al., 2000a] and [Wu et al., 2000b]. Using it we derive a general lemma that applies for different special cases of metric p-OCT.

Definition 2. *Given a spanning tree T of G, we denote by S_T the minimal subtree of T which contains all the nodes in S. It is easy to see that every leaf of S_T must be a node of S.*

Definition 3. *We define $\psi(S', U) = \sum_{u \in S'} \sum_{v \in U} \psi(u, v)$ for every $S' \subseteq S$ and $U \subseteq V$.*

Definition 4. *Given a spanning tree T of G and a path $P = w_1, ..., w_h$ of T, we denote:*

- *$f_P = w_1$ and $l_P = w_h$ the endpoints of P;*
- *V_P^f: the set of nodes in T connected to P through vertex f_P (including f_P itself);*
- *V_P^m: the set of nodes in T connected to P through an internal node of P (including these nodes);*
- *V_P^l: the set of nodes in T connected to P through vertex l_P (including l_P itself).*

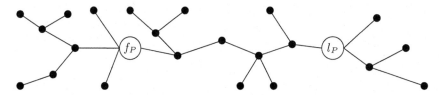

Fig. 1. Example of V_P^f, V_P^m and V_P^l for a path P of a tree T. Observe that V_P^f is the set of nodes to the left of f_P (including f_P), V_P^l is the set of nodes to the right of l_P (including l_P), V_P^m is the set containing the rest of the nodes and P is the path connecting f_P to l_P in T.

Notice that $V_P^f \cup V_P^m \cup V_P^l = V$ and these sets are disjoint. We also denote by S_P^i the set of vertices in $V_P^i \cap S$, where $i \in \{f, m, l\}$. We say that P is m-**source-free** if $S_P^m = \emptyset$. If P is m-source-free, $S_P^f \neq \emptyset$ and $S_P^l \neq \emptyset$ we say that P is a **connecting-source** path.

Now we introduce the definition for δ-balanced-path, that is an m-source-free path (P) for which the routing requirement delivered to its interior (V_P^m) is small, i.e., at most a portion (δ) of the routing requirement that passes through P. Formally:

Definition 5. *Given $0 < \delta \leq \frac{1}{2}$ and a spanning tree T of G, an m-source-free path P of T is a δ-**balanced-path** if $\psi(S_P^f \cup S_P^l, V_P^m) = \psi(S, V_P^m) \leq \delta\left(\psi(S_P^f, V_P^l) + \psi(S_P^l, V_P^f)\right).$*

The following proposition gives a basis for the main result in this work, since it provides a valid star whose cost is bounded by the cost of the given tree.

Proposition 1. *Consider $0 < \delta \leq \frac{1}{2}$, a spanning tree T of G and a set Y of internally disjoint δ-balanced-paths of T whose union results in S_T. Then there exists a valid $(|Y| + 1)$-star X such that: $C(X) \leq (1 + 2\delta) C(T)$. Where two paths are internally disjoint if the intersection of their sets of nodes is empty or contains nodes that are endpoints of both paths.*

Proof. First we note that there exists a spanning tree $\overline{T}$ such that $S_{\overline{T}} = S_T$, all vertices of $\overline{T}$ that do not belong to S_T are leaves and $C(\overline{T}) \leq C(T)$. If all vertices of T that do not belong to S_T are leaves, then $\overline{T} = T$, otherwise there exists a leaf u in T that is not adjacent to any node of S_T. Let v be the nearest node to u in S_T. Since all the nodes of S are in S_T and the graph is metric, the spanning tree T' of G, resulting from removing the edge adjacent to u in T and adding the edge (v, u) satisfies $C(T') \leq C(T)$. It is easy to see that repeating this process to all the leaves that are not adjacent to a node in S_T we obtain tree $\overline{T}$. With this property we can suppose that in tree T all vertices in $V - S_T$ are leaves. Now we construct a valid $(|Y| + 1)$-star X of G as follows:

- The core τ of X has the set of nodes that are endpoints of the paths in Y. Two nodes $u, v \in \tau$ are adjacent in τ if in Y there exists a path with endpoints u and v. Since the paths in Y are internally disjoint and their union results in the tree S_T, we conclude that τ is a tree over the endpoints of the paths in Y.

- For every node $u \in \tau$ and for every leaf $v \in V - \tau$ adjacent to u in T we also include an edge (u, v) in X.
- Observe that each node $u \in T$ not included in X by the previous steps belongs to V_P^m for some path $P \in Y$. Then, we include edge (u, f_P) in X if $\omega(u, f_P) \leq \omega(u, l_P)$, otherwise we include edge (u, l_P) in X.

Our construction guarantees X to be a $(|Y| + 1)$-star of G with core τ. Then, we only need to analyze its associated communication cost, which can be calculated by adding over each edge e the communication amount passing over e times the length of e. Since every edge is a path with exactly two vertices we also use the notation given by definition 4 on the edges. Then:

$$
\begin{aligned}
C(X) &= \sum_{e \in E_X} \left(\psi(S_e^f, V_e^l) + \psi(S_e^l, V_e^f) \right) \omega(e) \\
&= \sum_{e \in E_\tau} \left(\psi(S_e^f, V_e^l) + \psi(S_e^l, V_e^f) \right) \omega(e) \\
&\quad + \sum_{e \in E_{X-\tau}} \left(\psi(S_e^f, V_e^l) + \psi(S_e^l, V_e^f) \right) \omega(e).
\end{aligned}
$$

Observe that, by construction, each edge $e \in E_\tau$ corresponds to a δ-balanced-path $P \in Y$, such that: $\psi(S_e^f, V_e^l) + \psi(S_e^l, V_e^f) \leq \psi(S_P^f, V_P^l) + \psi(S_P^l, V_P^f) + \psi(S, V_P^m)$. Also, by the triangular inequality: $\omega(e) \leq \omega(P)$, then:

$$
\begin{aligned}
C(X) \leq &\sum_{P \in Y} \left(\psi(S_P^f, V_P^l) + \psi(S_P^l, V_P^f) + \psi(S, V_P^m) \right) \omega(P) \\
&+ \sum_{e \in E_{X-\tau}} \left(\psi(S_e^f, V_e^l) + \psi(S_e^l, V_e^f) \right) \omega(e).
\end{aligned}
$$

Notice that for every edge $e \in E_{X-\tau}$ one of its endpoints is a leaf outside of τ and the other one a node of τ. Let $p(u)$ be the node in τ adjacent in X to a leaf $u \in V - \tau$, then:

$$
\begin{aligned}
C(X) \leq &\sum_{P \in Y} \left(\psi(S_P^f, V_P^l) + \psi(S_P^l, V_P^f) + \psi(S, V_P^m) \right) \omega(P) \\
&+ \sum_{u \in V_{X-\tau}} \psi(S, u) \omega(u, p(u)) \\
\leq &\sum_{P \in Y} \left(\psi(S_P^f, V_P^l) + \psi(S_P^l, V_P^f) + \psi(S, V_P^m) \right) \omega(P) \\
&+ \sum_{P \in Y} \sum_{u \in V_P^m} \psi(S, u) \omega(u, p(u)) + \sum_{u \in V_I} \psi(S, u) \omega(u, p(u)),
\end{aligned}
$$

where V_I is the set of leaves in X that are not in τ and have the same adjacent node in both trees T and X. Notice that all leaves in T outside of τ adjacent (in

T) to some node of τ belong to V_I (that is, only nodes in V_P^m for some $P \in Y$ may be out of $V_I \cup V_\tau$).

For every $P \in Y$ and every node $u \in V_P^m$, $p(u)$ is one of the endpoints of the path P. Thus $w(u, p(u)) = \min\{w(u, f_P), w(u, l_P)\}$:

$$C(X) \leq \sum_{P \in Y} \left(\psi(S_P^f, V_P^l) + \psi(S_P^l, V_P^f) + \psi(S, V_P^m) \right) w(P)$$

$$+ \sum_{P \in Y} \sum_{u \in V_P^m} \psi(S, u) \min\{w(u, f_P), w(u, l_P)\} + \sum_{u \in V_I} \psi(S, u) w(u, p(u)).$$

Let $q(u)$ be the node in S_T adjacent to $u \in T - S_T$ and for $u \in S_T$ consider $q(u) = u$, then:

$$C(X) \leq \sum_{P \in Y} \left(\psi(S_P^f, V_P^l) + \psi(S_P^l, V_P^f) + \psi(S, V_P^m) \right) w(P)$$

$$+ \sum_{P \in Y} \sum_{u \in V_P^m} \psi(S, u) \min\{w(u, f_P), w(u, l_P)\} + \sum_{u \in V_I} \psi(S, u) w(u, q(u)).$$

By the triangular inequality $w(u, f_P) \leq w(q(u), f_P) + w(u, q(u))$ for $P \in Y$ and $u \in V_P^m$ (the same applies for l_P). Since $q(u) \in P$, $w(q(u), f_P) \leq w(P)$ and $w(q(u), l_P) \leq w(P)$. So, $\min\{w(u, f_P), w(u, l_P)\} \leq w(P) + w(u, q(u))$. Then:

$$C(X) \leq \sum_{P \in Y} \left(\psi(S_P^f, V_P^l) + \psi(S_P^l, V_P^f) + \psi(S, V_P^m) \right) w(P)$$

$$+ \sum_{P \in Y} \sum_{u \in V_P^m} \psi(S, u) (w(P) + w(u, q(u))) + \sum_{u \in V_I} \psi(S, u) w(u, q(u))$$

$$= \sum_{P \in Y} \left(\psi(S_P^f, V_P^l) + \psi(S_P^l, V_P^f) + \psi(S, V_P^m) \right) w(P)$$

$$+ \sum_{P \in Y} w(P) \psi(S, V_P^m) + \sum_{u \in V_{T-\tau}} \psi(S, u) w(u, q(u))$$

$$\leq \sum_{P \in Y} \left(\psi(S_P^f, V_P^l) + \psi(S_P^l, V_P^f) + 2\psi(S, V_P^m) \right) w(P)$$

$$+ \sum_{u \in V_{X-\tau}} \psi(S, u) w(u, q(u)).$$

Since each $P \in Y$ is δ-balanced, $\psi(S, V_P^m) \leq \delta \left(\psi(S_P^f, V_P^l) + \psi(S_P^l, V_P^f) \right)$. Also, $w(u, q(u)) = 0$ for each node u in S_T, then:

$$C(X) \leq \sum_{P \in Y} (1 + 2\delta) \left(\psi(S_P^f, V_P^l) + \psi(S_P^l, V_P^f) \right) w(P)$$

$$+ \sum_{u \in V_{T-S_T}} \psi(S, u) w(u, q(u))$$

$$\leq (1 + 2\delta) C(T).$$

$\square$

The above lemma gives us a general result: given $0 < \delta \leq \frac{1}{2}$, if a metric p-OCT problem Π satisfies that for any spanning tree T of G there exists a set of k internally disjoint δ-balanced-paths whose union results in S_T, then for each T there exists a $(k+1)$-star X_T such that $C(X_T) \leq (1+2\delta)C(T)$. Thus, for an optimal spanning tree T^* there exists a valid $(k+1)$-star of G which is a $(1+2\delta)$-approximation for Π. Then, an optimal valid $(k+1)$-star of G is a $(1+2\delta)$-approximation for Π. Remember that, by lemma 1, we can find an optimal valid $(k+1)$-star in $O(n^{k-p+2})$ (and therefore an approximation for the optimum value).

Lemma 2. *Consider a metric p-OCT problem for which we can guarantee for every spanning tree T the existence of k internally disjoint δ-balanced-paths whose union results in S_T. Thus, there exists a $(1+2\delta)$-approximation algorithm with time complexity $O(n^{k-p+2})$ where $0 < \delta \leq \frac{1}{2}$.*

The following proposition provides a sufficient condition for the existence of an internally disjoint δ-balanced-path set whose union results in S_T for any spanning tree T of a metric p-OCT problem.

Proposition 2. *Consider a metric p-OCT problem. Given $0 < \delta \leq \frac{1}{2}$, if it satisfies that every connecting-source path of any spanning tree T of G can be divided in, at most k internally disjoint δ-balanced-paths, then there exists a set of internally disjoint δ-balanced-paths of T with at most $2k(p-1)$ elements whose union results in S_T.*

Proof. Let T be a spanning tree of G, and consider the corresponding sub-tree S_T, whose leaves are nodes of S (S_T has at most p leaves), and S_T is also a tree. Then, the number of nodes with degree greater than two in S_T is at most $p-1$.

Consider now the collection of paths Y constructed from S_T such that each path has as endpoints nodes with degree different of two in S_T or nodes of S, and all the internal nodes of these paths are not in S and has degree two in S_T. It is easy to see that the number of paths in Y is at most $2p-2$ (that is, the number of nodes in S which contains the leaves of S_T plus the number of nodes with degree greater than two in S_T minus one). Also, observe that each path in Y is a connecting-source path, so it can be divided in at most k internally disjoint δ-balanced-paths. Then, by applying that division we obtain a set of δ-balanced-paths with at most $2k(p-1)$ elements whose union results in S_T. $\square$

Finally, from the results given by lemma 2 and proposition 2 we conclude:

Lemma 3. *Consider a metric p-OCT problem. If, for this problem it holds that for every connecting-source path P of any spanning tree of G there exists at most k internally disjoint δ-balanced-paths whose union results in P, there exists a $(1+2\delta)$-approximation algorithm with time complexity $O(n^{(2k-1)(p-1)+1})$ where $0 < \delta \leq \frac{1}{2}$.*

In next section we use this result to prove PTASs for three different metric p-OCT problems, for which the condition given above holds.

5   PTAS for Three p-OCT Metric Problems

In the previous section we provide a general result to obtain good approximations for p-OCT metric problems. In this section we show some special cases for which lemma 3 applies and consequently we are able to provide PTASs.

The first problem we introduce is the p-source Weighted Source Destination Optimum Communication spanning Tree problem (p-WSDOCT) which is a particular case of the p-OCT, where there is a sending requirement for each source node and also all the nodes have a receiving requirement associated:

Problem 3. p-WSDOCT - p-source Weighted Source Destination Optimum Communication spanning Tree problem.

Input: A graph G, a non-negative length function over the edges of G, $\omega :$ $E \to \mathbb{Q}_+$, a set of p sources $S \subseteq V$, a positive sending requirement $\sigma : S \to \mathbb{Q}_+$ and a non-negative receiving requirement $\lambda : V \to \mathbb{Q}_+$. The requirement function between $u \in S$ and $v \in V$ is given by $\psi(u, v) = \sigma(u)\lambda(v)$.

Output: A spanning tree T of G which minimizes the total requirement routing cost: $C(T) = \sum_{u \in S} \sum_{v \in V} \sigma(u)\lambda(v)d(T, u, v)$.

Now, using the result of lemma 3 we prove the following theorem:

Theorem 1. *There exists a PTAS for metric p-WSDOCT with fixed parameter. Given $0 < \delta \leq \frac{1}{2}$, the algorithm can find a $(1 + 2\xi\delta)$-approximation in*
$$O\left(n^{\left(2\frac{1}{\delta - \delta^2} - 1\right)(p-1)+1}\right) \text{ time complexity where } \xi = \frac{\sum_{u \in S} \sigma(u)}{\min\{\sigma(u)\}_{u \in S}}.$$

Proof. Consider a connecting-source path P for any spanning tree T of G, where:
$$\sum_{u \in V_P^m} \lambda(u) \leq \delta(1 - \delta) \sum_{u \in V} \lambda(u).$$

Observe that:

$$\psi(S, V_P^m) = \left(\sum_{u \in S} \sigma(u)\right)\left(\sum_{u \in V_P^m} \lambda(u)\right) \leq \left(\sum_{u \in S} \sigma(u)\right)\delta(1 - \delta) \sum_{u \in V} \lambda(u)$$

$$\leq \left(\sum_{u \in S} \sigma(u)\right)\delta\left(1 - (\delta - \delta^2)\right) \sum_{u \in V} \lambda(u)$$

$$= \left(\sum_{u \in S} \sigma(u)\right)\delta\left(\sum_{u \in V} \lambda(u) - (\delta - \delta^2)\sum_{u \in V} \lambda(u)\right)$$

$$\leq \left(\sum_{u \in S} \sigma(u)\right)\delta\left(\sum_{u \in V} \lambda(u) - \sum_{u \in V_P^m} \lambda(u)\right).$$

Since $V = V_P^l \cup V_P^f \cup V_P^m$:

$$\psi(S, V_P^m) \leq \left(\sum_{u \in S} \sigma(u)\right)\delta\left(\sum_{u \in V_P^l} \lambda(u) + \sum_{u \in V_P^f} \lambda(u)\right)$$

$$\leq \left(\sum_{u \in S} \sigma(u)\right) \delta \frac{\sum_{u \in S_P^f} \sigma(u)}{\min\{\sigma(u)\}_{u \in S}} \sum_{u \in V_P^l} \lambda(u)$$

$$+ \left(\sum_{u \in S} \sigma(u)\right) \delta \frac{\sum_{u \in S_P^l} \sigma(u)}{\min\{\sigma(u)\}_{u \in S}} \sum_{u \in V_P^f} \lambda(u)$$

$$= \frac{\sum_{u \in S} \sigma(u)}{\min\{\sigma(u)\}_{u \in S}} \delta \left(\psi(S_P^f, V_P^l) + \psi(S_P^l, V_P^f)\right).$$

So, P is a $\frac{\sum_{u \in S} \sigma(u)}{\min\{\sigma(u)\}_{u \in S}} \delta$-balanced-path and every connecting-source path of T can be divided in at most $\frac{1}{\delta - \delta^2}$ paths like P, which is a connecting-source path that satisfies $\sum_{u \in V_P^m} \lambda(u) \leq \delta (1 - \delta) \sum_{u \in V} \lambda(u)$.

Then, by applying lemma 3 we conclude that for metric p-WSDOCT there exists a $\left(1 + 2\frac{\sum_{u \in S} \sigma(u)}{\min\{\sigma(u)\}_{u \in S}} \delta\right)$-approximation algorithm with time complexity $O\left(n^{\left(2\frac{1}{\delta - \delta^2} - 1\right)(p-1)+1}\right)$. □

Another particular case of p-WSDOCT is the p-source Weighted Destination Optimum Communication spanning Tree problem (p-WDOCT), where all vertices of S have unitary sending requirement ($\sigma(S) = 1$).

Problem 4. p-WDOCT - p-source Weighted Destination Optimum Communication spanning Tree problem.

Input: A graph G, a non-negative length function over the edges of G, $\omega :$ $E \to \mathbb{Q}_+$, a set of p sources $S \subseteq V$ and a non-negative receiving requirement $\lambda : V \to \mathbb{Q}_+$. The requirement function between $u \in S$ and $v \in V$ is given by $\psi(u, v) = \lambda(v)$.

Output: A spanning tree T of G which minimizes the total requirement routing cost: $C(T) = \sum_{u \in S} \sum_{v \in V} \lambda(v) d(T, u, v)$.

Observe that p-WDOCT is a particular case of p-WSDOCT in which:

$$\xi = \frac{\sum_{u \in S} \sigma(u)}{\min\{\sigma(u)\}_{u \in S}} = \frac{p}{1} = p.$$

Then, using theorem 1 we conclude the following result for p-WDOCT:

Corollary 1. *There exists a PTAS for metric p-WDOCT, which can find a $(1 + 2p\delta)$-approximation in $O\left(n^{\left(2\frac{1}{\delta - \delta^2} - 1\right)(p-1)+1}\right)$ time complexity where $0 <$ $\delta \leq \frac{1}{2}$.*

Notice that the weighted p-MRCT is also a particular case of p-WSDOCT and the p-MRCT is a particular case of p-WDOCT so the results above imply also new PTASs for the fixed parameter metric weighted p-MRCT and for the metric p-MRCT.

Table 1. Comparison of our approach with current state of art in the literature. As it can be seen, for almost all the problems our approach improve the previous approximation ratios, only for the case of 2-MRCT it does not give a better solution (we obtain the same approximation scheme but with a greater complexity time). In the case of the PTASs it was considered the approximation $(1 + \epsilon)$ being $\epsilon \in (0, 1]$.

Problem	Best Approach In Literature			Our Approach	
	Ratio	Complexity	References	Ratio	Complexity
2-MRCT	PTAS	$O\left(n^{\frac{1}{\epsilon}+1}\right)$	[Wu, 2002]	PTAS	$O\left(n^{\frac{8}{\epsilon(2-\epsilon)}}\right)$
p-MRCT	2	$O\left(n^3\right)$	[Wu et al., 2000a]	**PTAS**	$O\left(n^{\left(\frac{8p^2}{\epsilon(2p-\epsilon)}-1\right)(p-1)+1}\right)$
weighted p-MRCT	2	$O\left(n^3\right)$	[Wu et al., 2000a]	**PTAS**	$O\left(n^{\left(\frac{8\xi^2}{\epsilon(2\xi-\epsilon)}-1\right)(p-1)+1}\right)$
p-WDOCT	2	$O\left(n^{p-1}\right)$	[Wu, 2004]	**PTAS**	$O\left(n^{\left(\frac{8p^2}{\epsilon(2p-\epsilon)}-1\right)(p-1)+1}\right)$
fixed parameter p-WSDOCT	2	$O\left(n^{p-1}\right)$	[Wu, 2004]	**PTAS**	$O\left(n^{\left(\frac{8\xi^2}{\epsilon(2\xi-\epsilon)}-1\right)(p-1)+1}\right)$
p-USCOCT	2	$O\left(n^{p-1}\right)$	[Wu, 2004]	**PTAS**	$O\left(n^{\left(4\log_{\frac{2}{2+\epsilon}}(r)-1\right)(p-1)+1}\right)$

Another particular case of p-OCT in which we can apply our results is the p-Uniform Source Connecting Optimum Communication spanning Tree problem (p-USCOCT). In this problem the minimum routing requirement between two sources must be at least a fixed ratio of the sum of requirements between all the pairs of nodes[2]:

Problem 5. p-USCOCT - p-Uniform Source Connecting Optimum Communication spanning Tree problem.

Input: A graph G, a non-negative length function over the edges of G, $\omega : E \to \mathbb{Q}_+$, a set of p sources $S \subseteq V$, a fixed ratio $r > 0$ and a non-negative routing requirement function between each node of S and each node of V, $\psi : S \times V \to \mathbb{Q}_+$, where each pair of sources $u, v \in S$ satisfies $\psi(u, v) \geq r \sum_{u \in S} \sum_{v \in V} \psi(u, v)$.

Output: A spanning tree T of G that minimizes the total requirement routing cost: $C(T) = \sum_{u \in S} \sum_{v \in V} \psi(u, v) d(T, u, v)$.

Similar ideas allow us to prove that for $0 < \delta \leq \frac{1}{2}$ we can divide any connecting-source path P of a spanning tree of G in an instance of metric p-USCOCT with no more than $2 \log_{\frac{1}{1+\delta}}(r)$ internally disjoint δ-balanced-paths. Then by applying lemma 3 we obtain:

---

[2] Metric p-USCOCT is NP-hard, a reduction from SAT is given in the full version of the paper.

Theorem 2. *There exists a* PTAS *for the metric p-*USCOCT. *Given* $0 < \delta \leq \frac{1}{2}$, *the algorithm guarantees a* $(1 + 2\delta)$-*approximation of the optimum solution in time complexity* $O\left(n^{\left(4log_{\frac{1}{1+\delta}}(r)-1\right)(p-1)+1}\right)$.[3]

6 Conclusions

In this work we consider different NP-hard variants of p-OCT: metric case of p-MRCT, p-WDOCT fixed parameter WSDOCTand p-USCOCT. We prove a lemma that allows us to present PTAS's for these problems, being possible to use that result in order to obtain other approximations for some other particular cases of p-OCT. Also we prove that metric USCOCT is NP-hard. In table 1 we summarize our approaches and compare with previous results in literature. Many questions remain open regarding p-OCT and related problems. For example, no PTAS for metric p-OCT is known. Also, when we do not consider metric problems much is still to be researched.

References

[Bartal, 1996] Bartal, Y.: Probabilistic approximation of metric spaces and its algorithmic applications. In: Proceedings of the 37th Annual IEEE Symposium on Foundations of Computer Science, pp. 184–1963 (1996)

[Hu, 1974] Hu, T.C.: Optimum communication spanning trees. SIAM J. Comput. 3(3), 188–195 (1974)

[Johnson et al., 1978] Johnson, D.S., Lenstra, J.K., Kan, A.H.G.R.: The complexity of the network design problem. Networks 8, 279–285 (1978)

[Talwar et al., 2003] Talwar, K., Fakcharoenphol, J., Rao, S.: A tight bound on approximating arbitrary metrics by tree metrics. In: Proceedings of the 35th Annual ACM Symposium on Theory of Computing, pp. 448–455 (2003)

[Wu, 2002] Wu, B.Y.: A polynomial time approximation scheme for the two-source minimum routing cost spanning trees. J. Algorithms 44, 359–378 (2002)

[Wu, 2004] Wu, B.Y.: Approximation algorithms for the optimal p-source communication spanning tree. Discrete and Applied Mathematics 143, 31–42 (2004)

[Wu and Chao, 2004] Wu, B.Y., Chao, K.M.: Spanning Trees and Optimization Problems. Chapman & Hall / CRC (2004) ISBN: 1584884363

[Wu et al., 2000a] Wu, B.Y., Chao, K.M., Tang, C.Y.: Approximation algorithms for some optimum communication spanning tree problems. Discrete and Applied Mathematics 102, 245–266 (2000)

[Wu et al., 2000b] Wu, B.Y., Chao, K.M., Tang, C.Y.: A polynomial time approximation scheme for optimal product-requirement communication spanning trees. J. Algorithms 36, 182–204 (2000)

[Wu et al., 2000c] Wu, B.Y., Lancia, G., Bafna, V., Chao, K.M., Ravi, R., Tang, C.Y.: A polynomial time approximation scheme for minimum routing cost spanning trees. SIAM J. on Computing 29(3), 761–778 (2000)

---

[3] The complete proof of this theorem can be found in the full version of the paper.

Fault-Tolerant Gathering of Asynchronous Oblivious Mobile Robots under One-Axis Agreement

Subhash Bhagat[1], Sruti Gan Chaudhuri[2], and Krishnendu Mukhopadhyaya[1]

[1] Indian Statistical Institute, Kolkata, India
[2] Jadavpur University, Kolkata, India
{sbhagat_r,krishnendu}@isical.ac.in, srutiganc@it.jusl.ac.in

Abstract. In this paper, we have studied one of the fundamental coordination problems for multi robot system, namely *gathering*, for $n \geq 2$ asynchronous, oblivious mobile robots in the presence of $f < n$ faulty robots. Earlier works have reported that, in general, to solve gathering problem for asynchronous robots, many assumptions are required, like multiplicity detection or total agreement in coordinate axis or constant amount of memory bits. However, in this paper we have proved that gathering of asynchronous robots is possible with less number of such assumptions and even in the presence of any number of faulty robots. In our case, the robots only agree on the direction and orientation of any one axis.

Keywords: Gathering, Crash fault, Asynchronous, Oblivious, Swarm robots.

1 Introduction

A system of multiple autonomous mobile robots working in collaboration, is a very emerging topic of research in the field of swarm robots. A group of small, inexpensive robots is easy to deploy even in hazardous situations, like space, deep sea or after some environmental disaster. They are also applicable for many other tasks which are supposed to be performed by a group of objects, e.g., moving a big object, cleaning a big surface, etc. Since, the system has multiple robots with similar capabilities, if some of the robots fail i.e., they can not perform their tasks, the remaining robots can complete the job. *Gathering*, (i.e., collecting the robots to a point not defined in advance) of such robots is a fundamental coordination problem for a group of mobile robots. In this paper, we have addressed this problem and presented a distributed algorithm for gathering a set of robots. The robots are distributed in nature, i.e., they have their own computational unit and they act independently. They do not communicate by sending or receiving messages. The robots are indistinguishable by their appearances; they perform a given task cooperatively. They are represented by points on a 2D plane. Each robot treats its own position as its origin in its local coordinate system. They agree on the direction and orientation of any one axis (e.g.

M.S. Rahman and E. Tomita (Eds.): WALCOM 2015, LNCS 8973, pp. 149–160, 2015.
© Springer International Publishing Switzerland 2015

Y axis/direction of north and south, for this paper). However, they do not have any agreement on the orientation of other axis (X axis, for this paper). At any point of time, a robot may be active or inactive. When the robots are active, they operate by executing a cycle repeatedly. Under this cycle a robot has following three states:

- *Look*: in this state, the robots sense or observe the positions of the other robots in their surroundings with the help of some sensing devices. The robots plot the positions of other robots in their local coordinate systems.
- *Compute*: in this state, depending upon what they have observed and the requirement of the given task, the robots compute destinations to move to.
- *Move*: in this state the robots move to their destinations. The movement is nonrigid i.e., the robots may stop before reaching their destinations.

The operation cycle is scheduled asynchronously, i.e., at any point of time a set of robots may be in look state while some other sets of robots are computing or moving. A robot can not differentiate between a static robot and a moving robot. The robots are oblivious, i.e., at the end of a cycle, the robots remove all computed data of that cycle. Some of the robots may be faulty i.e, they stop their activity for ever but remain in the system. However, the robots can not decide whether a robot they see is functional or defective. The robots can not detect if more than one robots lie at a single point. We propose an distributed algorithm that will gather all non-faulty robots (under the above model), starting from any arbitrary set of initial positions, to a point not fixed in advance, in a finite number of cycles.

1.1   Earlier Works

Gathering problem is the most visited research topic [2,6,4,5,9,10,12,15,16,17,19] from the birth of the multi robot systems. The primary aim of these investigations has always been to find out the sets of minimum capabilities which the robot should have to be gathered at a point in finite time[1]. Depending on the activation scheduling the system of robots have been viewed in following two models; (i) CORDA ([18]): under this model the robots are asynchronous, i.e., they independently execute the phases of the cycles; (ii) SYm (Suzuki, Yamashita model [20]): the robots are semi-synchronous, i.e., a set of robots are active at some time and perform the phases of the cycles simultaneously. There is a third type called *fully synchronous* where all the robots are active and perform the phases of the cycle at the same time. Prencipe [19] showed that, deterministic gathering of $n > 2$ robots is impossible in CORDA and SYm model without the assumption on multiplicity detection, i.e., the robots can detect a point consists of multiple robots. If there is no agreement about the coordinate system Gathering of even two robots is not possible without remembering the past. The results

---

[1] There is some variation of gathering as, convergence [7]: where the robots come as close as possible but do not gather at a single point. However, in this paper we only discuss about the gathering at a single point.

continues to hold even with multiplicity detection. Flocchini et al. [14], have reported an algorithm for gathering two robots using constant number of memory bits. Flocchini et al. [12], have showed that gathering is possible if the robots have an agreement in direction and orientation of both the axes, even when the robots can observe limited regions of certain radius, around themselves. All these studies are based on a fact that the robots are fault free.

Two types of fault model have been reported in the literature; (i) crash fault: the robots stop working for ever. (ii) byzantine fault: the robots exhibit arbitrary behaviors or movements. In this paper we will focus only on crash fault. Agmon and Peleg [1] proved that gathering of non faulty robots is possible under the SYm model even in presence of a single faulty robot under crash fault model. Défago et. al. [8], have studied fault tolerant gathering algorithm under SYm model. Recently, Bouzid et. al.[3], reported that gathering is possible with arbitrary number of faults by the robots in ATOM model with chirality (agreement in clockwise orientation) and multiplicity detection.

1.2 Our Contribution

To the best of our knowledge, there are no results reported on the possibility of gathering, in presence of faulty robots, under one axis agreement. Flocchini et. al. [13], have characterized all form-able patterns by a set of robots from arbitrary initial configurations. However, point formation or gathering does not lie in their classifications. In this paper, we have proposed an algorithm for gathering of $n \geq 2$ asynchronous oblivious robots under the agreement of only one axis. Gathering of two oblivious robots has been proved to be impossible even with multiplicity detection. However, in this paper we prove that gathering of two oblivious robots is possible if they agree on the direction of any one axis. We also prove that our algorithm can tolerate $f < n$ crash faults i.e., our algorithm guarantees gathering of all non-faulty robots in presence of $f < n$ faulty robots under crash faults. Earlier works have reported the possibility of gathering for robot under the assumption on multiplicity detection and chirality. We propose an algorithm for fault tolerant gathering of asynchronous robots without having chirality and multiplicity detection but only having agreement in any one axis. Our algorithm will gather all non-faulty robots in finite time.

2   Terminology

We consider a set, $\mathcal{R} = \{r_1, r_2, \ldots, r_n\}$, of oblivious, indistinguishable n point robots on a 2D plane. The robots follow the basic characteristics of the traditional model [11] with some additional features. A robot can be either active or inactive(idle). The active robots execute the cycle *look-compute-move* repeatedly in asynchrony following CORDA model.

Under this model the robots can stop (and start a new cycle) before reaching their respective destinations. However, they must move, on each movement, at least a distance $\delta > 0$ if their destination is more than δ distance apart from

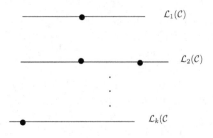

Fig. 1. An example of $\mathcal{L}(C)$

it[2]. This assumption in the description of the model [11] is necessary to assure that a robot will not take infinite time to reach its destination. An inactive robot will not remain idle for an infinite time [11]. There is no global coordinate system (defined by origin and axes) for the robots. Each robot has its own local coordinate system which may differ from the others. However, the robots only agree on Y axis which is conventionally north-south direction[3]. The robots have unlimited visibility range, i.e., they can observe all robots in R and plot them in their own coordinate systems. The robots may become faulty at any stage of execution cycle. In this paper, we only consider *crash faults* model where a faulty robot stops executing cycles forever. If there is any faulty robot in the system, then it is not removed from the system i.e., it physically remains in the system without executing the cycles. The robots can not identify the faulty robots. A *crash fault* model is denoted by (n, f), in which at most $f < n$ robots may become faulty at any stage of execution.

By a configuration $\mathcal{C}$, we mean the set of points on the plane occupied by the robots in $\mathcal{R}$ i.e. $\mathcal{C} = \{p_1, p_2, \ldots, p_k\}$, $k \le n$, where p_i is the position of one or multiple robots in $\mathcal{R}$ (a point in $\mathcal{C}$ may be occupied by more than one robot, however the robots can not differentiate between a point having single robot and a point having multiple robots). By $\mathcal{C}_t$, we denote the geometric configuration of robots at time t. $\tilde{\mathcal{C}}$ is the set of all such possible configurations of the robots by $\mathcal{R}$. Following notations are used in our proposed algorithm.

- The agreement on north-south directions enables us to classify the robots into different groups according to their positions and to obtain an ordering among these groups as follows: Horizontal line is drawn through each point (Fig. 1) in $\mathcal{C}$. Note that these lines may contain multiple robots. These horizontal lines are sorted according to there positions along north to south

---

[2] If the destination is less than δ distance apart then the robot reach it without stopping in between.

[3] Note that we can also take the agreement on X axis. In that case, all computations that require the notion of north-south direction, have to be modified for east-west direction.

direction. Let $\mathcal{L}_i(\mathcal{C})$ be the ith horizontal line and $\mathcal{L}(\mathcal{C})$ be the collection of all such lines in the sorted order. Let $HL_i(\mathcal{C})$ be the set of distinct robot positions on $\mathcal{L}_i(\mathcal{C})$ (if two or more robots occupy same position, we consider it once) in $\mathcal{C}$. If there are k number of such horizontal lines in $\mathcal{C}$, we define $\mathcal{H}(\mathcal{C}) = \{HL_1(\mathcal{C}), \ldots, HL_k(\mathcal{C})\}$.

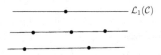

$\mathcal{L}_1(\mathcal{C})$

Fig. 2. An example of $\widetilde{\mathcal{C}}_1$

$\mathcal{L}_1(\mathcal{C})$

Fig. 3. An example of $\widetilde{\mathcal{C}}_{>1}$

- We define an equivalence relation $\prec$ on $\widetilde{\mathcal{C}}$ as follows: $\forall\ \mathcal{C}, \mathcal{C}' \in \widetilde{\mathcal{C}}, \mathcal{C} \prec \mathcal{C}'$ iff $|HL_1(\mathcal{C})| = |HL_1(\mathcal{C}')| = 1$ or both $|HL_1(\mathcal{C})|$ and $|HL_1(\mathcal{C}')|$ are greater than 1. Note that this relation yields only two equivalence classes. In one class, each member $\mathcal{C}$ has $|HL_1(\mathcal{C})| = 1$ and we denote this class by $\widetilde{\mathcal{C}}_1$ (Fig. 2). In the other class each member $\mathcal{C}'$ has $|HL_1(\mathcal{C}')| > 1$ and we denote this class by $\widetilde{\mathcal{C}}_{>1}$ (Fig. 3).
- **ComputeLevels**() is a function which takes the current configuration $\mathcal{C}$ as argument and returns $\mathcal{L}(\mathcal{C})$ in sorted order according to the positions of the lines from north to south direction.
- **ComputeGroups**() is a function which takes a configuration $\mathcal{C}$ as argument and returns $\mathcal{H}()$. The robots compute $\mathcal{H}(\mathcal{C})$ as follows:
  - Sort the points in $\mathcal{C}$ in descending order according to y-coordinates.
  - Group the points having same y-coordinate in the above sorted list and denote the group by $HL_i(\mathcal{C})$ if the y-coordinate of the group is the ith largest element in the list.
  - return $\mathcal{H}(\mathcal{C})$, the set of all such groups $HL_i(\mathcal{C})$, in sorted order.
- **CheckLevel**() is a function which takes $\mathcal{L}(\mathcal{C})$ and a robot r_i as arguments and returns k such that r_i lies on $\mathcal{L}_k(\mathcal{C})$.
- **ComputeCorners**() is a function which takes $HL_i(\mathcal{C})$ as argument and returns the two corner positions occupied by robots of $|HL_i(\mathcal{C})|$ on $\mathcal{L}_i(\mathcal{C})$. By $\mathcal{M}_t$, we denote the two robot positions computed by $ComputeCorners()$ at time t.

- Sort the points in $HL_i(\mathcal{C})$ according to their own x-coordinates.
- Take the two points having the smallest and the largest x-coordinates and put them in a set $\mathcal{M}$.
- return $\mathcal{M}$.

Note that here the robots compute the corner robots using the ordering information w.r.t. their own coordinate system, however, the agreement on the corner robots is unique for all robots.

- **ComputeTriangle**() is a function which takes two robot positions, p_i and p_j lying on the same horizontal line $\mathcal{L}_k(\mathcal{C})$ for some k, as arguments and computes the equilateral triangle $\triangle p_i T p_j$ where length of each side is equal to the distance between p_i and p_j i.e., $|\overline{p_i p_j}|$ and the point T lies in the north to the line $\mathcal{L}_k(\mathcal{C})$. It returns the point T. By $\triangle_t$, we denote the triangle computed by *ComputeTriangle*() at time t. Note that in lemma 3, we shall prove that T remains invariant whenever it is possible to define it. So through out this paper, we use same notation T to represent this position.
- **Closest**() is a function which takes a robot r_i and a set of positions $\mathcal{M}$ as arguments and returns the closest position in $\mathcal{M}$ from r_i (break the tie arbitrarily).

Using above functions we describe an algorithm **ComputeDestination**() which computes a destination point for a robot on the plane. Finally **Gathering Algorithm**() combines all mentioned functions and builds the final gathering configuration (a point). These two algorithms are presented in section 3. Each robot in $\mathcal{R}$ executes **GatheringAlgorithm**() independently in its "compute" stage. Our proposed algorithm is *wait-free*, i.e., all robots are allowed to move simultaneously, if their movement strategies support.

3   Algorithm

In this section we describe a distributed gathering algorithm which can tolerate f ($< n$) crash faults for a set of robots $\mathcal{R}$. The objective of our algorithm is to find a unique point so that all non-faulty robots can agree on it and gather at that point within finite time. As the robots are oblivious and asynchronous, after some steps the gathering point should remain invariant even as the robots change their positions during the execution of the algorithm. If the initial configuration of the robots provides us such an invariant point then we are done. In that case the only responsibility of the robots is to move to that point in such a way that this point remains intact till all robots reach there. On the other hand, if such a point is not available, some of the robots change their positions so that an invariant gathering point becomes computable after a finite number of movements of these robots.

Our algorithm is based on the ordering information of $\mathcal{R}$ as reflected in $\mathcal{H}(\mathcal{C}) = \{HL_1(\mathcal{C}), \ldots, HL_k(\mathcal{C})\}$ for some k. We also exploit the partition information of $\widetilde{\mathcal{C}}$ provided by the equivalence relation $\prec$. For an arbitrary robot configuration $\mathcal{C}$, the algorithm first checks whether $\mathcal{C} \in \widetilde{\mathcal{C}}_1$ or $\mathcal{C} \in \widetilde{\mathcal{C}}_{>1}$ and accordingly decides the strategy for the movements of the robots. If $\mathcal{C} \in \widetilde{\mathcal{C}}_1$, the robots which are not in $HL_1(\mathcal{C})$ move towards the robot in $HL_1(\mathcal{C})$. If $\mathcal{C} \in \widetilde{\mathcal{C}}_{>1}$, then the present configuration is converted into one which belongs to $\widetilde{\mathcal{C}}_1$ and then the strategy for the previous case is followed. The different scenarios and the corresponding solution strategies are:

– **Case 1:** $\mathcal{C} \in \widetilde{\mathcal{C}}_1$
  The robot $r_i \in HL_1(\mathcal{C})$ retains its current position and all other robots not in $HL_1(\mathcal{C})$ move towards r_i along a path not crossing $\mathcal{L}_1(\mathcal{C})$ (Fig. 4).

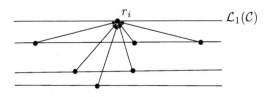

Fig. 4. An example of Case 1

– **Case 2:** $\mathcal{C} \in \widetilde{\mathcal{C}}_{>1}$
  Consider the two corner robots, say r_j and r_k, on $\mathcal{L}_1(\mathcal{C})$. Compute the equilateral triangle $\triangle r_j T r_k$. The robots r_j and r_k move towards T along the respective sides of the triangle. The rest non-faulty robots compute the nearest robot among r_j and r_k and move towards it (break the tie arbitrarily) along the line segment joining them to their respective destinations (Fig. 5).

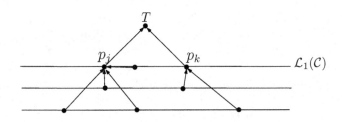

Fig. 5. An example of Case 2

Next we present the formal description of the algorithms. The algorithms are executed in all robot sites in their *compute* state independently and asynchronously. Using *ComputeDestination*() the robots determine the destination points to move to. Then they reach their respective destinations using

GatheringAlgorithm(). In the next section, we discuss the correctness of these algorithms.

Algorithm 1. ComputeDestination()

    Input: $r_i, \mathcal{L}(\mathcal{C}), \mathcal{H}(\mathcal{C})$.
    Output: A destination point of r_i.
  1 $l \leftarrow CheckLevels(r_i, \mathcal{L}(\mathcal{C}))$;
  2 $h \leftarrow |HL_1(\mathcal{C})|$;
  3   **if** $l == 1$ *and* $h == 1$ **then**
  4 | $r \leftarrow r_j \in HL_1(\mathcal{C})$;
  5   **else**
  6 | $M \leftarrow ComputeCorners(HL_1(\mathcal{C}))$;
  7 | **if** *position of* $r_i \in M$ **then**
  8 | | $T \leftarrow ComputeTriangle(M)$;
  9 | | $r \leftarrow T$;
10 | **else**
11 | | $r \leftarrow Closest(r_i, M)$;

12 return r;

Algorithm 2. GatheringAlgorithm()

    Input: $r_i \in R$
    Output: r_i moves towards its destination.
  1 $\mathcal{L}(\mathcal{C}) \leftarrow ComputeLevels(\mathcal{C})$;
  2 $\mathcal{H}(\mathcal{C}) \leftarrow ComputeGroups(\mathcal{C})$;
  3 $r \leftarrow ComputeDestination(r_i, L(\mathcal{C}), \mathcal{H}(\mathcal{C}))$;
  4   Move towards r along the line segment $\overline{r_i r}$;

---

4   Correctness

In this section, we prove that all non-faulty robots will gather at a point in finite time. The robots can not detect the presence of multiple robots (multiplicity detection) at one point. They interpret multiple robots, occupying the same location, as a single robot. If $HL_1(t)$ has a single location, it is the destination point for all the robots who might compute at time t. If there are robots at different locations in $HL_1(t)$, the two farthest locations define an equilateral triangle. The tip of that triangle would be the point of gathering if at least one of the extreme robots completes journey to its destination. If not and both of them stop on the same horizontal line similar situation continues with the tip remaining invariant. If they stop on different horizontal lines, we have $HL_1()$ having a single point and gathering takes place there. In any case no robot moves to a point outside the equilateral triangle initially defined.

Observation 1. *Suppose* $\triangle ABC$ *is an equilateral triangle. If* D *and* E *are two points on side* AB *and* AC *such that* $|BD| = |CE|$, *then* $\triangle ADE$ *is also equilateral.*

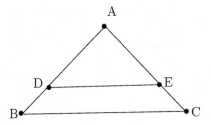

Fig. 6. An example of Observation 1

Lemma 1. *Suppose for the initial configuration $\mathcal{C}_{t_0}$, $|HL_1(\mathcal{C}_{t_0})| \geq 2$. Then there exists some $t' \geq t_0$ such that all the robots on $\mathcal{L}_1(\mathcal{C}_t)$ lie on the non-horizontal sides of $\triangle_{t_0}$, $\forall\, t \geq t'$.*

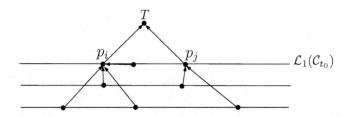

Fig. 7. An example of Lemma-1

Proof. The active robots in $\mathcal{C}_{t_0}$, move in one of the following ways (Fig. 7): (i) the robots occupying the positions in $\mathcal{M}_{t_0}$ move towards T along the respective non-horizontal sides of $\triangle_{t_0}$ and (ii) rest of the robots move to the points in $\mathcal{M}_{t_0}$ along the line segment joining them to their respective destinations. Until a robot in $\mathcal{M}_{t_0}$ moves or a robot reaches its destination corner point, no robot crosses $\mathcal{L}_1(\mathcal{C}_{t_0})$. Suppose at time t', at least one active robot in $\mathcal{M}_{t_0}$ starts moving towards T along the corresponding side of $\triangle_{t_0}$. At time $t \geq t'$, any active robot will find itself either on $\mathcal{L}_1(\mathcal{C}_t)$ or below it. Using observation 1, any robot on $\mathcal{L}_1(\mathcal{C}_t)$ has reached its current position by either moving along a non-horizontal side of $\triangle_{t_0}$ or moving towards its current position occupied by the robot(s) on the respective non-horizontal side of $\triangle_{t_0}$. Thus all the robots on $\mathcal{L}_1(\mathcal{C}_t)$ lie on the non-horizontal sides of $\triangle_{t_0}$. □

Lemma 2. *There exits some t' such that $|HL_1(\mathcal{C}_t)|$ does not exceed two for all $t \geq t'$.*

Proof. If $|HL_1(\mathcal{C}_{t_0})| = 1$, all the non-faulty robots move towards the robot in $HL_1(\mathcal{C}_{t_0})$ without crossing $\mathcal{L}_1(\mathcal{C}_{t_0})$. In this case the result holds with $t' = t_0$.

If $|HL_1(\mathcal{C}_{t_0})| \geq 2$, using lemma 1, let t' be the time from which $\mathcal{L}_1()$ contains only the robot positions which lie on the two non-horizontal sides of $\triangle_{t_0}$. For $t \geq t'$, $|HL_1(\mathcal{C}_t)|$ can not exceed two.   □

Lemma 3. *Whenever ComputeCorners() and ComputeTriangle() are called, they always return positions of the robots on the non-horizontal sides of the triangle $\triangle_{t_0}$ and T respectively.*

Proof. *ComputeCorners()* and *ComputeTriangle()* are called by the active robots only when they find $|HL_1()| \geq 2$. If a robot finds itself as one of the corner points detected by *ComputeCorners()*, it calls *ComputeTriangle()*. Until at least one robot occupying a position in $\mathcal{M}_{t_0}$ moves, *ComputeCorners()* and *ComputeTriangle()* compute the same points on every call and these points are the three vertices of $\triangle_{t_0}$. When at least one robot in $\mathcal{M}_{t_0}$ starts moving, combining observation 1, lemma 1 and lemma 2, we can say the following: *ComputeCorners()* and *ComputeTriangle()* are called by the active robots after finding that (i) $|HL_1()| = 2$ and (ii) the robots in $\mathcal{L}_1()$ lie on the non-horizontal sides of $\triangle_{t_0}$. Thus, the lemma follows.   □

Theorem 1. *GatheringAlgorithm() solves the gathering problem for $(n, n-1)$ crash fault model in finite time.*

Proof. Our approach is to find a unique point so that all the non-faulty robots can agree upon this point for gathering and the point remains intact even when the robots move towards it. If $|HL_1(\mathcal{C}_{t_0})| = 1$ for the initial configuration $\mathcal{C}_{t_0}$, we have a unique point for gathering, namely the point in $HL_1(\mathcal{C}_{t_0})$. All the non-faulty robots compute this point as their destination point and move towards it. Since robots move towards this destination point along the line segments joining them to this destination point, they never touch or cross $\mathcal{L}_1(\mathcal{C}_{t_0})$ except at the destination point. This implies that the destination point remains intact under the motion of the robots. No robot waits for the completion of any action by any other robot. Thus *GatheringAlgorithm()* can solve gathering problem for $(n, n-1)$ fault model.

Next consider the case when $|HL_1(\mathcal{C}_{t_0})| \geq 2$. Robots try to make $|HL_1()| = 1$ whenever $|HL_1()|$ is more than 1. By lemma 2, there exists a moment after which $|HL_1()| = 1$ or 2. During the execution of our algorithm, if $|HL_1()|$ becomes 1 and all other non-faulty active robots become aware of it, then the point in $HL_1()$ becomes the desired gathering point. Otherwise by lemma 3, some active robots reach T and make $|HL_1()| = 1$. Note that if any robot reaches the point T, then T becomes a static point on which all non-faulty robots can agree for gathering. Once $|HL_1()|$ is fixed to 1, the correctness proof is same as above. We can see that till there is single non-faulty robot in the system, our algorithm guarantees that at some time $|HL_1()|$ would become 1. Thus, *GatheringAlgorithm()* solves gathering problem for $(n, n-1)$ fault model. It is worthwhile to note that our approach is wait-free.

In each computational cycle, the functions used by the robots run in polynomial time. Since the number of robots and the number of functions used by the robots are finite, each computational cycle would take polynomial time. On each non-null movement, any active robot moves at least $\delta > 0$ distance. As the distances between the robots are finite, to reach their destinations, robots have to spend a finite number of computational cycles (as our algorithm is wait-free, no dead-lock occurs in the system). So we conclude that $GatheringAlgorithm()$ is capable of gathering all non-faulty robots in finite time. $\square$

5   Conclusion

In this paper, we have proposed a distributed algorithm that gathers $n \geq 2$ asynchronous oblivious robots in finite time, under the agreement of only one axis even in the presence of $f(< n)$ faulty robots. To the best of our knowledge, this is the first reported result on the possibility of gathering, in the presence of faulty robots, under one axis agreement. In this paper, the visibility range is assumed to be unlimited and not obstructed by the presence of other robots (i.e., the robots are assumed to be see through), fault tolerant gathering under limited or obstructed visibility is still an open problem.

References

1. Agmon, N., Peleg, D.: Fault-tolerant gathering algorithms for autonomous mobile robots. In: Proceedings of the Fifteenth Annual ACM-SIAM Symposium on Discrete Algorithms, SODA, pp. 1070–1078. Society for Industrial and Applied Mathematics, Philadelphia (2004)
2. Bouzid, Z., Das, S., Tixeuil, S.: Wait-free gathering of mobile robots. CoRR, abs/1207.0226 (2012)
3. Bouzid, Z., Das, S., Tixeuil, S.: Gathering of mobile robots tolerating multiple crash faults. In: 2013 IEEE 33rd International Conference on Distributed Computing Systems (ICDCS), pp. 337–346 (July 2013)
4. Cieliebak, M., Flocchini, P., Prencipe, G., Santoro, N.: Solving the robots gathering problem. In: Baeten, J.C.M., Lenstra, J.K., Parrow, J., Woeginger, G.J. (eds.) ICALP 2003. LNCS, vol. 2719, pp. 1181–1196. Springer, Heidelberg (2003)
5. Cieliebak, M., Flocchini, P., Prencipe, G., Santoro, N.: Distributed computing by mobile robots: Gathering. SIAM Journal on Computing 41(4), 829–879 (2012)
6. Cieliebak, M., Prencipe, G.: Gathering autonomous mobile robots. In: In Proc. SIROCCO, pp. 57–72 (2002)
7. Cohen, R., Peleg, D.: Convergence properties of the gravitational algorithm in asynchronous robot systems. SIAM Journal on Computing 34(6), 1516–1528 (2005)
8. Défago, X., Gradinariu, M., Messika, S., Raipin-Parvédy, P.: Fault-tolerant and self-stabilizing mobile robots gathering. In: Dolev, S. (ed.) DISC 2006. LNCS, vol. 4167, pp. 46–60. Springer, Heidelberg (2006)
9. Degener, B., Kempkes, B., Langner, T., Meyer auf der Heide, F., Pietrzyk, P., Wattenhofer, R.: A tight runtime bound for synchronous gathering of autonomous robots with limited visibility. In: Proceedings of the 23rd ACM Symposium on Parallelism in Algorithms and Architectures, SPAA 2011, pp. 139–148. ACM Press, New York (2011)

10. Dieudonné, Y., Petit, F.: Self-stabilizing gathering with strong multiplicity detection. Theoretical Computer Science 428(0), 47–57 (2012)
11. Flocchini, P., Prencipe, G., Santoro, N.: Distributed Computing by Oblivious Mobile Robots. Synthesis Lectures on Distributed Computing Theory. Morgan & Claypool Publishers (2012)
12. Flocchini, P., Prencipe, G., Santoro, N., Widmayer, P.: Gathering of asynchronous robots with limited visibility. Theoretical Computer Science 337(1-3), 147–168 (2005)
13. Flocchini, P., Prencipe, G., Santoro, N., Widmayer, P.: Arbitrary pattern formation by asynchronous, anonymous, oblivious robots. Theor. Comput. Sci. 407(1-3), 412–447 (2008)
14. Flocchini, P., Santoro, N., Viglietta, G., Yamashita, M.: Rendezvous of two robots with constant memory. In: Moscibroda, T., Rescigno, A.A. (eds.) SIROCCO 2013. LNCS, vol. 8179, pp. 189–200. Springer, Heidelberg (2013)
15. Gordon, N., Elor, Y., Bruckstein, A.: Gathering multiple robotic agents with crude distance sensing capabilities. In: Dorigo, M., Birattari, M., Blum, C., Clerc, M., Stützle, T., Winfield, A.F.T. (eds.) ANTS 2008. LNCS, vol. 5217, pp. 72–83. Springer, Heidelberg (2008)
16. Izumi, T., Katayama, Y., Inuzuka, N., Wada, K.: Gathering autonomous mobile robots with dynamic compasses: An optimal result. In: Pelc, A. (ed.) DISC 2007. LNCS, vol. 4731, pp. 298–312. Springer, Heidelberg (2007)
17. Katayama, Y., Tomida, Y., Imazu, H., Inuzuka, N., Wada, K.: Dynamic compass models and gathering algorithms for autonomous mobile robots. In: Prencipe, G., Zaks, S. (eds.) SIROCCO 2007. LNCS, vol. 4474, pp. 274–288. Springer, Heidelberg (2007)
18. Prencipe, G.: Instantaneous actions vs. full asynchronicity: Controlling and coordinating a set of autonomous mobile robots. In: Restivo, A., Ronchi Della Rocca, S., Roversi, L. (eds.) ICTCS 2001. LNCS, vol. 2202, pp. 154–171. Springer, Heidelberg (2001)
19. Prencipe, G.: Impossibility of gathering by a set of autonomous mobile robots. Theoretical Computer Science 384(2-3), 222–231 (2007); Structural Information and Communication Complexity (SIROCCO 2005)
20. Suzuki, I., Yamashita, M.: Formation and agreement problems for anonymous mobile robots. In: Proc. 31st Annual Conference on Communication, Control and Computing, pp. 93–102 (1993)

Enumerating Eulerian Trails
via Hamiltonian Path Enumeration

Hiroyuki Hanada[1], Shuhei Denzumi[2], Yuma Inoue[2], Hiroshi Aoki[2],
Norihito Yasuda[1], Shogo Takeuchi[1], and Shin-ichi Minato[1,2]

[1] ERATO Minato Discrete Structure Manipulation System Project,
Japan Science and Technology Agency, Sapporo, Hokkaido, Japan
[2] Graduate School of Information Science and Technology,
Hokkaido University, Sapporo, Hokkaido, Japan
hana-hiro@live.jp

Abstract. Given an undirected graph G, we consider enumerating all
Eulerian trails, that is, walks containing each of the edges in G just once.
We consider achieving it with the enumeration of Hamiltonian paths
with the *zero-suppressed decision diagram* (ZDD), a data structure that
can efficiently store a family of sets satisfying given conditions. First we
compute the *line graph* $L(G)$, the graph representing adjacency of the
edges in G. We also formulated the condition when a Hamiltonian path
in $L(G)$ corresponds to an Eulerian trail in G because every trail in G
corresponds to a path in $L(G)$ but the converse is not true. Then we
enumerate all Hamiltonian paths in $L(G)$ satisfying the condition with
ZDD by representing them as their sets of edges.

Keywords: Eulerian trail, Hamiltonian path, path enumeration, line
graph, zero-suppressed binary decision diagram.

1 Introduction

In the graph theory, an *Eulerian trail* of an undirected graph G is a walk that
contains each of the edges in G just once. We can easily judge whether a connected undirected graph G has an Eulerian trail: G has an Eulerian trail if and
only if it has no or just two vertices of odd degree [1, 2]. In addition, it is also
known that we can obtain an Eulerian trail of G in a simple manner called
Fleury's algorithm [1]. However, it is considered difficult to enumerate *all* Eulerian trails: its time complexity is proved to be "#P-complete" (roughly speaking,
time complexity "P-complete" for each trail) [3–5].

To solve such a problem with feasible computational time and space, we consider enumerating Eulerian trails by way of enumerating *Hamiltonian paths*. A
Hamiltonian path of an undirected graph G is a walk that contains each of
the vertices in G just once. Although enumerating all Hamiltonian paths is not
so easy in general, either, many approaches have been proposed to enumerate
them [6–10], and we especially focus on the algorithm using *zero-suppressed
binary decision diagram* (ZDD) [11] with the advantage described next.

M.S. Rahman and E. Tomita (Eds.): WALCOM 2015, LNCS 8973, pp. 161–174, 2015.

Here we consider enumerating all Eulerian trails in a *simple graph G*, that is, there exists at most one edge between every pair of vertices and there does not exist any loop (an edge whose two ends are the same vertex) [1]. The algorithm is explained as the following three parts:

- First we compute the *line graph* $L(G)$ so that every Eulerian trail in a graph G correspond to a Hamiltonian path in $L(G)$ (Sect. 3.1). Note that the converse does not hold, that is, not all paths in $L(G)$ correspond to trails in G.
- Then we derive the condition when a path in $L(G)$ corresponds to a trail in G (Sect. 3.2). We formulate the condition by the "labels" defined for the edges in $L(G)$.
- Finally we enumerate Hamiltonian paths in $L(G)$ satisfying the condition above with the algorithm based on ZDD [11]. ZDD is a data structure that can store a family of sets with small memory. We store such paths as sets of edges in a ZDD (Sect. 4.1). We also use the operation on ZDD of excluding sets satisfying a given condition, in order to exclude Hamiltonian paths that do not satisfy the condition above.

If the graph G is not simple, in order to obtain a simple one, we insert some vertices to G (Sect. 4.2).

2   Definitions

We denote by (V, E) the graph whose set of vertices is V and whose set of edges is E, respectively.

For an undirected graph G, a pair of a sequence of vertices $(v_1, v_2, \ldots, v_m)$ and a sequence of edges $(e_1, e_2, \ldots, e_{m-1})$ is called a *walk* if the two ends of e_i are v_i and v_{i+1} for $i \in \{1, 2, \ldots, m-1\}$. We call a sequence of either vertices or edges also a walk if there exists the other sequence satisfying the condition above.

A walk is called to be *closed* if its sequence of vertices $(v_1, v_2, \ldots, v_m)$ satisfies $v_1 = v_m$. A *trail* is a walk whose edges in the sequence are distinct. A *path* is a walk whose vertices in the sequence are distinct (except for the precondition $v_1 = v_m$ if the walk is closed). Note that any path is also a trail. A closed path is called a *cycle*. [1]

For a connected undirected graph G, it is called a *semi-Eulerian graph* (or an *Eulerian graph*) if there exists a trail (or a closed trail) containing all edges in G. Such a trail is called an *Eulerian trail*. Similarly, for a connected undirected graph G, it is called a *semi-Hamiltonian graph* (or a *Hamiltonian graph*) if there exists a path (or a cycle) containing all vertices in G. Such a path is called a *Hamiltonian path*. [1]

3   Representing Eulerian Trails as Hamiltonian Paths in the Line Graph

3.1   Line Graph

Given a connected undirected graph G, we use its *line graph* $L(G)$ so that every Eulerian trail in G gives a Hamiltonian path in $L(G)$. The line graph $L(G)$ is a

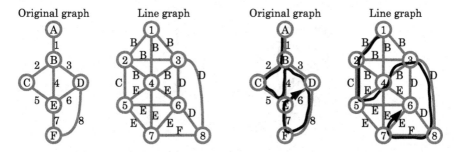

Fig. 1. An example of a line graph. In the line graph, characters on edges represent the vertices in the original graph where the edges in the original graph are adjacent (see the "label" defined in Definition 2).

Fig. 2. An example of a trail in a graph and the corresponding path in the line graph

graph characterizing the adjacency of the edges in G as follows: $L(G)$ has vertices corresponding one-to-one to the edges in G, and there exists an edge in $L(G)$ between the two vertices u and v if and only if two edges in G corresponding to u and v are adjacent [2,12]. An example is shown in Fig. 1. In this paper we define it formally as follows:

Definition 1. *Given two undirected graphs $G = (V, E)$ and $G' = (V', E')$, G' is called the* line graph *of G, denoted by $L(G)$, if*

- *V' corresponds to E one-to-one, that is, there exists a bijection $l : E \to V'$, and*
- *For any $v_1', v_2' \in V'$, there exists an edge between v_1' and v_2' if $l^{-1}(v_1')$, $l^{-1}(v_2') \in E$ are adjacent in G, or no edge otherwise.*

It is known in 1960s that $L(G)$ is a (semi-)Hamiltonian graph if G is a (semi-)Eulerian graph [13,14]. Moreover, a sufficient and necessary condition for G is known when $L(G)$ is (semi-)Hamiltonian.

Property 1. [13] For an undirected graph $G = (V, E)$, $L(G)$ is either Hamiltonian or semi-Hamiltonian[1] if and only if G is *sequential*, where G is called sequential if there exists a permutation of E: $(e_1, e_2, \ldots, e_m)$ $(e_i \in E$, $m = |E|)$ such that e_i and e_{i+1} are adjacent for all $i \in \{1, 2, \ldots, m-1\}$.

If the sequence of edges $(e_1, e_2, \ldots, e_m)$ is an Eulerian trail of G then it is also sequential, but the converse does not always hold. Therefore, every Eulerian trail in G corresponds to a Hamiltonian path in $L(G)$ but the converse does not always hold. For example of Fig. 1, an Eulerian trail in G "1→2→5→4→3→8→7→6" corresponds to a Hamiltonian path in $L(G)$ (Fig. 2); however, "1→2→4→3→8→6→7 →5" is a Hamiltonian path in $L(G)$ but not an Eulerian trail in G.

---

[1] The original work [13] treats only Hamiltonian case, however, it is easy to prove semi-Hamiltonian case with the similar way.

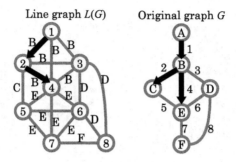

Fig. 3. An example of a path in the line graph $L(G)$ of a simple graph G that does not have a corresponding trail in the original graph

Thus, to enumerate Eulerian trails in G as Hamiltonian paths in $L(G)$, we consider excluding such excessive paths. However, the condition of exclusion has not been derived as far as the authors know. (For directed graphs, it is known in 1963 at latest that there is a one-to-one correspondence between the trails in G and the paths in $L(G)$, that is, no exclusion is needed[2] [6].) In the next section we derive the condition when G is a simple graph.

3.2 The Condition When a Path in a Line Graph Represents a Trail in the Original Graph

Let us assume G is a connected undirected simple graph and consider when a path p in $L(G)$ does not correspond to a trail in G (not limited to Hamiltonian or Eulerian). An example is shown in Fig. 3. In this case, the path in $L(G)$ with three vertices does not correspond to a trail in the original graph because three successive edges in G share a vertex.

In this section we prove that a path in $L(G)$ does not correspond to a trail in G only if the case above occurs, that is, three edges in G corresponding to three successive vertices in the path in $L(G)$ shares a vertex.

First, to state the fact formally, we define *labels* of the edges in $L(G)$ as follows:

Definition 2. *For an undirected simple graph G and an edge $e' = (u', v')$ ($e' \in E'$, $u', v' \in V'$) in the line graph $L(G) = (V', E')$, we define the label of e', denoted by $\lambda(e')$, by the only vertex in G where the two edges in G: $l^{-1}(u')$ and $l^{-1}(v')$ are adjacent.*

Note that the label must be unique for any e' because no two edges in a simple graph G exist between the same pair of vertices. See Fig. 1 in Sect. 3.1 for an example.

From the definition of the label, in case three edges in G share a vertex like in Fig. 3, the labels in the corresponding two edges in $L(G)$ must be the same. The fact can be formulated as follows:

---

[2] In this paper we omit the definition of the line graph of a directed graph. See the reference.

Theorem 1. *Let $G = (V, E)$ be a connected undirected simple graph and p be a path in $L(G) = (V', E')$ whose sequence of vertices is $(v'_1, v'_2, \ldots, v'_m)$ $(v'_i \in V')$. Then the followings are equivalent: (A) there exists a trail in G whose sequence of edges is $(l^{-1}(v'_1), l^{-1}(v'_2), \ldots, l^{-1}(v'_m))$ $(l^{-1}(v'_i) \in E)$, and (B) the same edge label does not appear successively in the sequence of edges for p.*

Proof. Let $v'_a \xrightarrow{e'_a} v'_b \xrightarrow{e'_b} v'_c$ $(v'_i \in V', e'_i \in E')$ be a subpath of p. Then we prove the corresponding sequence of edges $(l^{-1}(v'_a), l^{-1}(v'_b), l^{-1}(v'_c))$ is a subtrail in G if and only if the condition (B) is satisfied.

We focus on how $l^{-1}(v'_a), l^{-1}(v'_b), l^{-1}(v'_c) \in E$ are connected in G, which is classified to the following three cases:

(X) In case the labels of e'_a and e'_b are the same, the three edges $l^{-1}(v'_a), l^{-1}(v'_b)$, $l^{-1}(v'_c) \in E$ shares the vertex of the label. Thus these three edges are connected at a vertex in G (Case 1 in Fig. 4).

(Y) In case the labels of e'_a and e'_b are different,

  (Y1) If the two edges $l^{-1}(v'_a)$ and $l^{-1}(v'_c)$ are adjacent in G, then there exists an edge between v'_a and v'_c in $L(G)$, where its label is different from the other two. Thus the three edges yield a cycle (Case 2 in Fig. 4).

  (Y2) If the two edges $l^{-1}(v'_a)$ and $l^{-1}(v'_c)$ are not adjacent in G, then they yield a non-cycle path (Case 3 in Fig. 4).

From the consideration, we prove (A) and (B) are equivalent.

Proof of (A) $\Rightarrow$ (B): Suppose p, a path in $L(G)$, has two successive edges with the same label, that is, there exist a subpath $v'_a \xrightarrow{e'_a} v'_b \xrightarrow{e'_b} v'_c$ in $L(G)$ with $\lambda(e'_a) = \lambda(e'_b)$. In this case $l^{-1}(v'_a), l^{-1}(v'_b)$ and $l^{-1}(v'_c)$, three edges in G, must be adjacent with the form of (X) among (X), (Y1) and (Y2) above. This contradicts the precondition that $(l^{-1}(v'_a), l^{-1}(v'_b), l^{-1}(v'_c))$ is a subtrail in G.

Proof of (B) $\Rightarrow$ (A): Let p be a path in $L(G)$ without any two successive edges with the same label. Then, for any three successive vertices in p, corresponding three edges in G must take the form of (Y1) or (Y2). This implies no branching edges exist in the sequence of edges and thus the whole p corresponds to a trail in G. □

4 Enumerating Hamiltonian Paths in the Line Graph Corresponding to Eulerian Trails

4.1 Representing Hamiltonian Paths by Zero-Suppressed Binary Decision Diagram

As an algorithm of enumerating Hamiltonian paths satisfying given conditions, we use an enumeration algorithm based on the *zero-suppressed binary decision diagram* (ZDD) [11], a data structure originally for representing binary functions and also for storing families of sets. A famous algorithm for the enumeration with ZDD is proposed by Knuth [15], called *SIMPATH* in his website [16]. First we show the outline of ZDD.

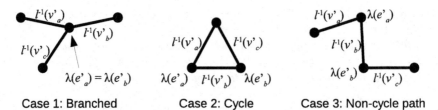

Case 1: Branched   Case 2: Cycle   Case 3: Non-cycle path

Fig. 4. All possible relationships of connections of three edges $l^{-1}(v'_a)$, $l^{-1}(v'_b)$, $l^{-1}(v'_c)$ given as a subpath in the line graph (v'_a, v'_b, v'_c) of a simple graph

Definition 3. *[11,15] Given a sequence of boolean variables $A = (a_1, a_2, \ldots, a_n)$: $\{0,1\}^n$ and a boolean function $f(a_1, a_2, \ldots, a_n)$: $\{0,1\}^n \to \{0,1\}$, zero-suppressed binary decision diagram (ZDD) for f is a minimal directed acyclic graph (DAG) such that:*

- *There are two vertices "0-terminal" and "1-terminal". These vertices are sinks, that is, they do not have any outgoing edges.*
- *All other vertices are named by elements in A. (Two or more vertices with the same name may exist.) Each of them has two outgoing edges named "0-edge" and "1-edge". For any vertex named a_i, the edges are connected to a vertex named a_j ($j > i$), "0-terminal" or "1-terminal".*
- *$f(a_1, a_2, \ldots, a_n)$ takes 1 for the arguments defined by paths from the root vertex to "1-terminal" in the diagram as follows: for every path above, a_i ($i = 1, 2, \ldots, n$) takes 1 if there exists a vertex named a_i which is a source of "1-edge" in the path, otherwise a_i takes 0. For the other arguments $f(a_1, a_2, \ldots, a_n)$ takes 0.*

ZDD can represent a family of sets by regarding A as the universal set, the assignments for variables $a_1, a_2, \ldots, a_n$ as the existence of the elements in a set, and the function value $f(a_1, a_2, \ldots, a_n)$ as taking 1 if the set is contained in the family or 0 otherwise. ZDD is invented as a variant of BDD (binary decision diagram) [17] so that the diagram becomes smaller when f takes zero for most of the elements in A, that is, the number of sets stored in the family is much fewer than 2^n (the number of all possible sets). An example is shown in Fig. 5.

As stated in the definition, ZDD must be minimal, that is, the vertices in ZDD must be removed or merged as long as the resulted binary function (or family of sets) is not changed. Concretely, we apply the operations in Fig. 6 to make the ZDD minimal [11].

Not only expressing a family of sets by a ZDD, we can conduct set operations like "excluding sets containing certain elements" on it [11,15]. We use the operations to implement the condition when a Hamiltonian path in $L(G)$ corresponds to an Eulerian trail in G (Theorem 1(B)).

4.2 Algorithm for Enumerating Eulerian Trails

To represent paths in a graph with a ZDD, we represent every path as a set of edges, with the universal set for the ZDD being the set of all edges in the

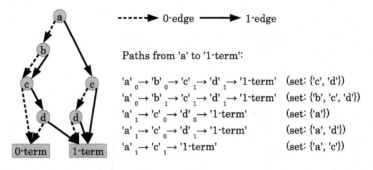

Fig. 5. An example of ZDD for a family of sets. It represents the family of five sets {'c', 'd'}, {'b', 'c', 'd'}, {'a'}, {'a', 'd'} and {'a', 'c'} over the universal set {'a', 'b', 'c', 'd'}.

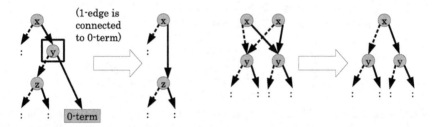

Fig. 6. The reduction rules of ZDD [11]. The first one is to remove an excessive vertex: in case there is a vertex whose 1-edge is connected to 0-term, remove it and connect its parent to its destination of 0-edge. The second one is to merge two vertices contributing to the same binary function.

graph [15] (Fig. 7(B)(C)). Note that different paths have different sets of edges, which is not the case for trails.

To enumerate all Hamiltonian paths in $L(G)$ satisfying the condition of Theorem 1(B), however, the condition cannot be directly applied because the orders of the edges the paths traverse are not stored in the ZDD. Thus, for each Hamiltonian path p in $L(G)$ given as a set of edges, we instead examine the condition of Theorem 1(B) by "for every pair of adjacent edges in $L(G)$ with the same label, they does not appear simultaneously in a path" rather than examining every pair of adjacent edges only in the path. We can apply the condition in the following two manners:

1. A straightforward manner is that we first store all Hamiltonian paths in $L(G)$ to a ZDD with SIMPATH algorithm, and then remove all paths not satisfying the condition. Concretely, we repeat the following for every pair of adjacent edges X, Y in $L(G)$ with the same label: remove all paths (set of edges) in the ZDD containing both X and Y.
2. The other manner is based on the behavior of SIMPATH algorithm: for each edge in $L(G)$ (sorted in a certain order), it adds edges to a ZDD one by one with excluding sets of edges that cannot be paths. Thus we simultaneously

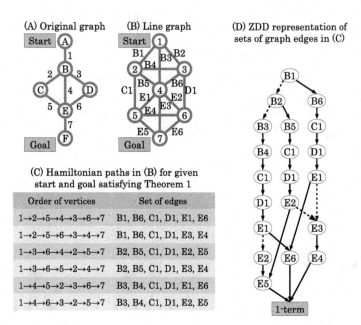

(A) Original graph (B) Line graph (D) ZDD representation of sets of graph edges in (C)

(C) Hamiltonian paths in (B) for given start and goal satisfying Theorem 1

Order of vertices	Set of edges
1→2→5→4→3→6→7	B1, B6, C1, D1, E1, E6
1→2→5→6→3→4→7	B1, B6, C1, D1, E3, E4
1→3→6→4→2→5→7	B2, B5, C1, D1, E2, E5
1→3→6→5→2→4→7	B2, B5, C1, D1, E3, E4
1→4→5→2→3→6→7	B3, B4, C1, D1, E1, E6
1→4→6→3→2→5→7	B3, B4, C1, D1, E2, E5

Fig. 7. An example of ZDD representation of Hamiltonian paths satisfying Theorem 1 in a line graph. In the figure of (D), unspecified ZDD edges are regarded as being connected to 0-terminal. (For example, the destination of 0-edge for 'B6' is 0-terminal.

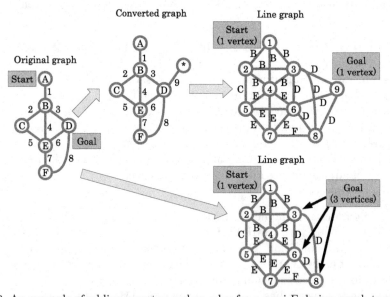

Fig. 8. An example of adding a vertex and an edge for a semi-Eulerian graph to assure unique start/goal vertices in $L(G)$. In this case, because the degree of the vertex 'D' is 3, an odd number larger than 1, we add a dummy vertex '*' and an edge '9'.

exclude sets of edges that do not satisfy the condition above. (We adopted this way in the experiment.)

Lastly we show the whole algorithm of enumerating Eulerian trails including for non-simple graphs.

1. Given a connected undirected graph G, make G a simple graph without changing the number of the Eulerian trails in it so that Theorem 1 can be applied. Precisely,
    - In case there exists a pair of vertices with two or more edges between them, split each of the edges into two by inserting a vertex except for arbitrary one edge.
    - In case there exists a loop edge (Sect. 2), split it into three by inserting two vertices[3].
2. Add some vertices and edges to G so that the start and goal vertices of Hamiltonian paths in $L(G)$ become unique, without changing the number of the Eulerian trails in it. Precisely,
    - In case G is semi-Eulerian, for each of the two vertices with odd degree (Sect. 1), create a dummy vertex and an edge to connect to the vertex of odd degree unless the degree is 1. This assures the start and goal edges of the Eulerian trails in G being unique, that is, the start and goal vertices of the Hamiltonian paths in $L(G)$ being unique (See Fig. 8).
    - In case G is Eulerian, (1) we add two new vertices u_1 and u_2 to G, (2) remove arbitrary edge (v_1, v_2) from G and (3) create two edges (u_1, v_1) and (u_2, v_2). (Namely, we "split" an edge in G into two.) As a result, G becomes semi-Eulerian.
3. Create $L(G)$ from G. Simultaneously, classify all edges in $L(G)$ by their labels.
4. Enumerate all Hamiltonian paths in $L(G)$ satisfying the condition of Theorem 1(B) stated before.

5 Experiment

5.1 Setting

We implemented the algorithm of Sect. 4.2 with Graphillion [18], a Python library for graphs and their paths based on ZDD in the manner in Sect. 4.1. We used the implementation of the Hamiltonian path enumeration in Graphillion with the default parameter.

We enumerated the Eulerian trails in the four types of graphs shown in Figs. 9 to 12. Their numbers of vertices, edges and degrees are shown in Tables 1 and 2. Because the line graph $L(G)$ has $d(d-1)/2$ edges for each vertex of degree

---

[3] In this case we treat two ends of the loop edge are distinguished: for example, we treat there are two Eulerian trails (not one) in the graph with three vertices $\{v_1, v_2, v_3\}$ and three edges $\{(v_1, v_2), (v_2, v_2), (v_2, v_3)\}$ starting at v_1 and ending at v_3. The algorithm for treating them not distinguished is not developed yet.

d in G, the time and space for the computation are expected to grow much for increasing vertex degree even if the number of edges in G is not so increased. Thus we experimented graphs with constant maximum degree (Ring, Diamond) and increasing degree (Bunch, Complete). As seen in Table 2, the number of edges in Bunch and Complete are multiplied by $\Theta(k)$ after the conversions to the line graphs.

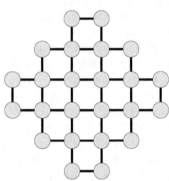

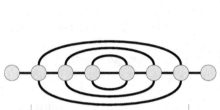

k cycles ($k = 3$ in this figure)

Fig. 9. The graph Ring(k) ($k = 3$)

k large cycles ($k = 3$ in this figure)

Fig. 10. The graph Diamond(k) ($k = 3$). This graph is a variant of *Aztec diamond* [19].

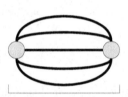

k edges ($k = 5$ in this figure)

k vertices ($k = 5$ in this figure)

Fig. 11. The graph Bunch(k) ($k = 5$). The start and the goal vertices are the two points if k is odd (i.e. the graph is semi-Eulerian), otherwise one of the edges are divided into two to set the start and the goal (Operation 2 of the whole algorithm in Sect. 4.2). As a result, number of Eulerian trails are the same for Bunch($2m-1$) and Bunch($2m$) for any m.

Fig. 12. The graph Complete(k) ($k = 5$). Such a graph is called the *complete graph* [1,2,12]. It has Eulerian trails if and only if k is odd.

We measured the computation times of the enumeration; times for setting up graph structures (converting given graphs to their line graphs, adding dummy vertices for making them simple and unique the start and goal vertices) and obtaining paths which are Hamiltonian and satisfying the condition of Theorem 1. The experiment was conducted on a Linux (Xubuntu 14.04) computer with the CPU "AMD A4-5000 APU" (clock: 1.5GHz) and 4GB RAM. The running time for each graph is limited to one hour.

Table 1. Graphs examined in the experiment

Name	Structure	#Vertices	#Edges	Maximum degree
Ring(k)	Fig. 9	$2k+2$	$4k+1$	4
Diamond(k)	Fig. 10	$2k(k+1)$	$4k^2$	4
Bunch(k)	Fig. 11	2	k	k
Complete(k)	Fig. 12	k	$\dfrac{k(k-1)}{2}$	$k-1$

Table 2. The properties of the graphs ·after making the graph simple and adding dummy vertices (Sect. 4.2).

Name	#Vertices	#Edges	#LineGraphEdges	Maximum degree
Ring(k)	$3k+3$	$5k+2$	$13k+1$	4
Diamond(k)	$2k(k+1)+2$	$4k^2+1$	$4k(3k-2)$	4
Bunch(k)	$k+3$	$\begin{cases} 2k+1 \ (k\text{: odd}) \\ 2k \ (k\text{: even}) \end{cases}$	$\begin{cases} k^2-1 \ (k\text{: odd}) \\ k^2+2k-1 \ (k\text{: even}) \end{cases}$	$\begin{cases} k+1 \ (k\text{: odd}) \\ k \ (k\text{: even}) \end{cases}$
Complete(k)	$k+2$	$\dfrac{k(k-1)}{2}+1$	$\dfrac{k^2(k-1)}{2}$	$k-1$

Table 3. Number of Eulerian trails and computation times (sec) of four types of graphs

k	Ring(k) #trails	Ring(k) time	Diamond(k) #trails	Diamond(k) time	Bunch(k) #trails	Bunch(k) time	Complete(k) #trails	Complete(k) time
1	6	0.0093	1	0.0066	1	0.0073	—	
2	36	0.0139	40	0.0152	1	0.0065	—	
3	216	0.0184	132,160	0.0487	6	0.0099	1	0.0063
4	1,296	0.0222	33,565,612,800	2.6198	6	0.0101	—	
5	7,776	0.0266	Memory out		120	0.0224	132	0.0169
6	46,656	0.0309			120	0.0238	—	
7	279,936	0.0356			5,040	0.2487	64,988,160	49.6530
8	1,679,616	0.0402			5,040	0.2678	—	
9	10,077,696	0.0450			362,880	10.5978	Time out	
10	60,466,176	0.0493			362,880	10.5284		
11	362,797,056	0.0555			Memory out			
12	2,176,782,336	0.0597						
13	13,060,694,016	0.0647						
14	78,364,164,096	0.0697						
15	470,184,984,576	0.0747						
20	3.6×10^{15}	0.1066						
30	2.2×10^{23}	0.1887						

Table 4. Numbers of vertices and edges in the line graph. Note that the numbers of vertices and edges are equivalent to #Edges and #LineGraphEdges in Table 2, respectively. Underlined numbers denote the cases of failed computations (memory-out or time-out).

k	$L(\text{Ring}(k))$ vertices	edges	$L(\text{Diamond}(k))$ vertices	edges	$L(\text{Bunch}(k))$ vertices	edges	$L(\text{Complete}(k))$ vertices	edges
1	7	14	5	4	3	2	—	
2	12	27	17	32	4	3	—	
3	17	40	37	84	7	14	4	3
4	22	53	65	160	8	15	—	
5	27	66	<u>102</u>	<u>269</u>	11	34	11	30
6	32	79			12	35	—	
7	37	92			15	62	22	105
8	42	105			16	63	—	
9	47	118			19	98	<u>37</u>	<u>252</u>
10	52	131			20	99		
11	57	144			<u>23</u>	<u>142</u>		
12	62	157						
13	67	170						
14	72	183						
15	77	196						
⋮								
20	102	261						
30	152	391						

5.2 Result

We show the results of computation times and numbers of trails in Table 3. Table 4 describes the numbers of vertices and edges in the line graph.

From Table 4 with the three graphs Diamond(k), Bunch(k) and Complete(k), it seems to be possible to enumerate Eulerian trails if there are about 120 edges or less in the line graph. However, as shown by Ring(k), more edges would be acceptable according to the shape of graphs. It is easily assumed, but yet to be examined, that the Hamiltonian paths in $L(\text{Ring}(k))$ satisfying Theorem 1 are well compressed by ZDD.

As for the computation times, they grow rapidly for increasing k except for Ring(k) in almost linear against k, which is natural since the number of edges in the line graphs grow in $O(k^2)$ or $O(k^3)$ (see also Table 2). However, the linear time for Ring(k) is unexpectedly fast because, in general, we need $O(2^n)$ time and space to compute ZDD for a universal set of size n.

There remains the problem of what parameter is essential for fast computation. From the property of ZDD, it is clear that the number of edges affects much. However, we should examine other parameters from the result of Ring(k): in fact, Eulerian trails in Ring(20) (102 vertices and 261 edges in the line graph) is much easier to be computed than that in Diamond(5) (102 vertices and 269 edges in the line graph).

6   Conclusion

In this research we considered enumerating all Eulerian trails in an undirected graph G, which in general requires high computational cost. We focus on ZDD-based Hamiltonian path enumeration, which can enumerate not only all Hamiltonian paths but also Hamiltonian paths satisfying certain conditions efficiently. We consider converting G into the line graph $L(G)$, where every Eulerian trail in G corresponds to a Hamiltonian path in $L(G)$ (Sect. 3.1). In addition, because not all Hamiltonian paths in $L(G)$ correspond to Eulerian trails in $L(G)$, we formulated the condition by defining "labels" of the edges in $L(G)$ (Theorem 1 in Sect. 3.2). As a result of the experiment, we could enumerate Eulerian trails in G if $L(G)$ has 120 or less edges, although more edges can be accepted for certain type of graphs.

We consider the following problems as future works: finding parameters of graphs determining the computational time other than the number of vertices and edges, and developing more memory-efficient data structure.

References

1. Wilson, R.J.: Introduction to Graph Theory, 4th edn. Pearson Education (1996)
2. Harary, F.: Graph Theory, 1st edn. Addison-Wesley (1969)
3. Mihail, M., Winkler, P.: On the number of Eulerian orientations of a graph. In: Proceedings of the Third Annual ACM-SIAM Symposium on Discrete Algorithms, SODA 1992, pp. 138–145 (1992)
4. Creed, P.: Sampling Eulerian orientations of triangular lattice graphs. Journal of Discrete Algorithms 7(2), 168–180 (2009)
5. Ge, Q., Štefankovič, D.: The complexity of counting Eulerian tours in 4-regular graphs. Algorithmica 63(3), 588–601 (2012)
6. Kasteleyn, P.W.: A soluble self-avoiding walk problem. Physica 29(12), 1329–1337 (1963)
7. Rubin, F.: A search procedure for Hamilton paths and circuits. Journal of the ACM 21(4), 576–580 (1974)
8. Mateti, P., Deo, N.: On algorithms for enumerating all circuits of a graph. SIAM Journal on Computing 5(1), 90–99 (1976)
9. van der Zijpp, N.J., Catalano, S.F.: Path enumeration by finding the constrained k-shortest paths. Transportation Research Part B: Methodological 39(6), 545–563 (2005)
10. Liu, H., Wang, J.: A new way to enumerate cycles in graph. In: International Conference on Internet and Web Applications and Services/Advanced International Conference on Telecommunications, p. 57 (2006)
11. Minato, S.: Zero-suppressed BDDs and their applications. International Journal on Software Tools for Technology Transfer 3(2), 156–170 (2001)
12. Diestel, R.: Graph Theory, 4th edn. Springer (2010)
13. Chartrand, G.: On Hamiltonian line-graphs. Transactions of the American Mathematical Society 134, 559–566 (1968)
14. Harary, F., Nash-Williams, C.S.J.A.: On Eulerian and Hamiltonian graphs and line graphs. Canadian Mathematical Bulletin 8, 701–709 (1965)

15. Knuth, D.E.: 7.1.4 Binary Decision Diagrams. In: Combinatorial Algorithms, vol. 4A. The Art of Computer Programming, vol. 4A. Pearson Education (2011)
16. Knuth, D.E.: Don Knuth's home page, http://www-cs-staff.stanford.edu/~uno/
17. Bryant, R.E.: Graph-based algorithms for boolean function manipulation. IEEE Transactions on Computers C-35(8), 677–691 (1986)
18. Inoue, T., Iwashita, H., Kawahara, J., Minato, S.: Graphillion: Software library designed for very large sets of graphs in python. Technical Report TCS-TR-A-13-65, Division of Computer Science, Hokkaido University (2013)
19. Elkies, N., Kuperberg, G., Larsen, M., Propp, J.: Alternating-sign matrices and domino tilings (part I). Journal of Algebraic Combinatorics 1(2), 111–132 (1992)

The Impact of Communication Patterns on Distributed Self-Adjusting Binary Search Trees*

Thim Strothmann

Computer Science Department, University of Paderborn, Germany
thim@mail.upb.de

Abstract. This paper introduces the problem of communication pattern adaption for a distributed self-adjusting binary search tree. We propose a simple local algorithm that is closely related to the nearly thirty-year-old idea of splay trees and evaluate its adaption performance in the distributed scenario if different communication patterns are provided. To do so, the process of self-adjustment is modeled similarly to a basic network creation game in which the nodes want to communicate with only a certain subset of all nodes. We show that, in general, the game (i.e., the process of local adjustments) does not converge, and convergence is related to certain structures of the communication interests, which we call conflicts. We classify conflicts and show that for two communication scenarios in which convergence is guaranteed, the self-adjusting tree performs well. Furthermore, we investigate the different classes of conflicts separately and show that, for a certain class of conflicts, the performance of the tree network is asymptotically as good as the performance for converging instances. However, for the other conflict classes, a distributed self-adjusting binary search tree adapts poorly.

Keywords: Binary Search Tree, Self Optimization, Basic Network Creation Game, Sink Equilibrium, Distributed Data Structure.

1 Introduction

Over 30 years ago, Sleator and Tarjan [15] introduced an interesting paradigm to design efficient data structures. Instead of optimizing general metrics, like tree depth, they proposed a *self-adjusting* data structure. To be more precise, the authors introduced *splay trees*, self-adjusting binary search trees in which frequently accessed elements are closer to the root. This therefore improves the average access times weighted by the popularity of the elements. Avin et al. [4] recently proposed *SplayNet*, a *distributed generalization* of splay trees, which is heavily inspired by [15]. In contrast to classical splay trees where requests (i.e., lookups) always originate from the root of the tree, communication in SplayNets happens between arbitrary node pairs in the network. As such, SplayNets can be

---

* This work was partially supported by the German Research Foundation (DFG) within the Collaborative Research Center "On-The-Fly Computing" (SFB 901).

M.S. Rahman and E. Tomita (Eds.): WALCOM 2015, LNCS 8973, pp. 175–186, 2015.
© Springer International Publishing Switzerland 2015

interpreted as a distributed data structure, e.g., a structured peer-to-peer (p2p) system or distributed hash table (DHT). Following the ideas of Avin et al., we further investigate the dynamics of a distributed locally self-adjusting tree.

An intuitive requirement to a distributed data structure is that nodes that communicate more frequently with each other become topologically closer to each other. An important factor that influences the performance of a distributed data structure is the peculiarity of the underlying communication interest pattern. Likewise to the original concept of splay trees, each node in the distributed splay tree should only have access to local information to decide whether it needs to change its position in the tree. In our specific scenario, the only kinds of information that each node has access to are its parent, its children and information about the distances to nodes it wants to communicate with. With only little knowledge about the structure of the tree and only limited possibilities to change the structure (called *rotations*), a distributed self-adjusting tree can be seen as a local algorithm whose performance is affected by the communication interests. We want to focus on this specific aspect and try to answer the question of how the performance of a distributed self-adjusting tree is influenced by different communication patterns. However, instead of using *empirical entropies* as a building block for the analysis (as done in [4]), the analytical method we use is heavily inspired by the concept of *Basic Network Creation Games (BNCG)* [2]. By doing so we can extend the analysis of [4] in convergent scenarios to a wider variety of instances. Furthermore, we contrast the previous positive results of [4] by giving concrete examples in which a distributed self-adjusting tree performs poorly, compared to an optimal static network.

We focus on a binary search tree network structure, since trees are one of the most elemental networks. They allow a simple and local routing strategy and are a fundamental constituent of more complex networks. Additionally, many network protocols rely on spanning trees or cycle-free backbones. Taking the same line as [4], we do not see our work as an introduction for a new network structure, but as a step towards a better understanding of the inherent dynamics of self-adjusting networks and their limitations.

1.1 Model and Notions

We model the dynamic process of a distributed self-adjusting tree whose structure is changed as a game in which the nodes of a binary search tree are the players. An instance of the *Self-Adjusting Binary Search Tree Game (SABST-game)* $\Gamma = (G_C, G_I)$ is given by an initial *connection graph* $G_C = (V, E_C)$ with $V = \{1, \ldots, n\}$ being the set of players, which is required to be a binary search tree (BST), and a *(communication) interest graph* $G_I = (V, E_I)$. G_C is undirected, whereas G_I is directed. The connection graph represents the distributed self-adjusting tree network and can be altered during the game. We use $IS(v) := \{u \in V : (v, u) \in E_I\}$ to refer to the neighborhood of player v in G_I and denote it as the *interest set* of player v. Since the connection graph is a binary search tree, we can compare two nodes by comparing their identifiers. The depth of a node v is the length of a path from the root to v. If v has a smaller depth than some node u, we say that v is above u, otherwise v is below u. We say that two edges $(u, v), (x, y)$ from G_I *intersect* if x is in the interval $[u, v]$ for $u < v$ or $[v, u]$ for $u > v$ and y is not, or vice versa.

Given a connection graph, we formalize the *private cost* of a player v as the sum over all distances to the nodes in its interest set: $c(v) := \sum_{u \in IS(v)} d(v, u)$. Here $d(v, u)$ denotes the shortest path distance between u and v in G_C. Note that by using the sum, each player tries to minimize the average distance. To improve its private cost, a player may perform *rotations* in the connection graph. These rotations are closely related to the *splay* operation of *splay trees* [15]; a single right rotation of a node (abbreviated with $RR(x)$) is visualized in Figure 1 (node x *rotates over* the node y). For a *response*, a player u is not only allowed to perform a single rotation on itself, but also multiple rotations on itself. Additionally, u can tell nodes from $IS(u)$ to perform rotations. This is due to the fact that by performing rotations on only itself, a node can only move upwards in the tree. Thus, u can only move closer to a node $v \in IS(u)$ that is in its subtrees in G_C, if it can tell v to perform rotations. Consequently, players have the opportunity to decrease their private cost as much as possible, instead of being restricted by the current connection graph. If a player u decreases its private cost by a series of rotations, we refer to this as a *better response*. If the decrease is maximal compared to all other possible better responses, we refer to this as a *best response*. To provide an easy way of computing best responses, we will stick close to the idea of the *double splay* algorithm of [4]. A node u first rotates itself upward such that it is the lowest common ancestor of all $v \in IS(u)$ (i.e., it becomes the root of this particular subtree), then all nodes v are rotated as close as possible to u. Note that according to [4] a general optimal solution as well as best responses can be computed in polynomial time. We denote the connection graph to be in a *rotation equilibrium*, if no node can perform a better response. We say that a game *converges* if every sequence of best responses converges, irrespective of the initial connection graph. Otherwise, we say that the game is *non-convergent*.

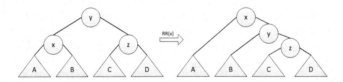

Fig. 1. A single right rotation of node x. The triangles represent (possibly empty) subtrees that are not changed by the rotation.

The dynamic process of changing the connection graph (i.e., the game) proceeds in rounds. A round is finished when all players with non-empty interest sets have played a better response at least once. However, we do not enforce an order in a single round, but consider an arbitrary order. The overall quality of a connection graph G_C is measured by the *social cost* $c(G_C) = \sum_{v \in V} c(v)$.

Our goal is to analyze the social cost of worst-case rotation equilibria and compare them with a general optimal solution. We use the ratio of the two measures, the *Price of Anarchy (PoA)*, to do so.

1.2 Related Work

Self-adjusting networks have many possible application scenarios, varying from self-optimizing peer-to-peer topologies (e.g., [11]) over green computing [10] (because of reduced energy consumption) to adaptive virtual machine migrations [3,13]. Self-adjusting routing schemes were examined to deal with congestion, e.g., in scale-free networks [16].

Our work combines ideas from two interesting and very different research areas: self-adjusting binary search trees and basic network creation games. Self-adjusting binary search trees have a long history [1,5,15]. The focus of this paper is on splay trees [15]. Introduced in 1985, they have an amortized time bound of $\mathcal{O}(\log n)$ for the standard tree operations of searching, insertion and deletion. Additionally, splay trees are as efficient as static, optimal search trees for a sufficiently long sequence of node accesses. Splay trees achieve this by applying a restructuring operation for each access in the tree. This splay operation moves the recently accessed node to the root of the tree by performing rotations on the nodes. Since their establishment, splay trees have been extensively analyzed and many variants have been proposed (see [8,14,17] which all use the dynamics of splay trees). Closest to our work is the aforementioned paper of [4], in which a fully decentralized generalization of splay trees called *SplayNet* is presented. SplayNets adapt to a communication pattern σ. The upper bound for the amortized communication cost is based on the empirical entropies of σ. Furthermore, SplayNets have a provable online optimality under special requests scenarios.

Basic Network Creation Games (BNCG) were introduced by Alon et al. in 2010 [2]. They are a variant of the original *Network Creation Game (NCG)* by Fabrikant et al. [7]. In the BNCG model, an initial connection graph is given and players are allowed to change the graph by performing what are called improving *edge swaps*. For an edge swap, a node is able to exchange a single incident edge with a new edge to an arbitrary other node. In contrast to the original NCG, best responses are polynomially computable. The cost for a single node is either induced by the sum of the distances to all other nodes (SUM-version) or by the maximal distance (MAX-version). The authors showed that for the SUM-version of the game all trees in an equilibrium have a diameter of 2, and that the diameter of all swap equilibria is $2^{\mathcal{O}(\sqrt{\log n})}$. For the MAX-version they showed that all trees in an equilibrium have a diameter of at most 3, and that the diameter of general swap equilibria is $\Omega(\sqrt{n})$. Lenzner [12] proved that if the game is played on a tree, it admits an ordinal potential function, which implies guaranteed convergence to a pure nash equilibrium. However, when played on general graphs, this game allows best response cycles. For computing a best response, they show a similar contrast: a linear-time algorithm for computing a best response on trees is provided, which works even if players are allowed to swap multiple edges at a time. On the other hand, they proved that this task is NP-hard even on simple general graphs, in case more than one edge can be swapped. [6] extended the BNCG model by introducing what are called interests to the game. Thus, the players are now no longer interested in communicating with all other nodes, but only with a specific subset. For the MAX-version they give a tight upper bound of $\Theta(\sqrt{n})$ for the Price of Anarchy, if the connection graph is a tree, and $\Theta(n)$ for general connection graphs.

1.3   Our Contribution

To the best of our knowledge, this is the first work that evaluates dynamics of self-adjusting topologies by using (basic) network creation games. We introduce a new BNCG that is closely related to the model of [2] but incorporates the dynamics inherent to self-adjusting binary search trees. We show that the game does not converge in general, and the distributed self-adjusting binary search tree will never stop changing its structure. However, for certain interest graphs which guarantee convergence, we prove a tight upper bound on the Price of Anarchy of $\Theta(1)$. For non-convergent game instances, we use an altered variant of the concept sink equilibria (introduced in [9]). We define the corresponding measure *worst-case Price of Sinking* to evaluate the worst-case performance of the distributed self-adjusting tree, in contrast to an optimal solution. We prove that there exists an interest graph class such that the worst-case Price of Sinking is constant. However, we also show that, for other interest graph classes, the worst-case Price of Sinking is $\Omega(\frac{n}{\log n})$.

2   Analysis

In general, the *SABST*-game does not converge and the dynamic process never settles on stable binary search tree. In fact, it is possible to construct a simple *SABST*-game with four nodes that can never converge (see Figure 2). Consequently, the Price of Anarchy cannot be computed for general instances of the game. In Section 2.1 we identify two classes of interest graphs that do converge and have a constant Price of Anarchy.

Fig. 2. An example *SABST*-game instance that does not converge. Interest graph edges are dashed, connection graph edges are continuous.

However, we can relate non-convergent behavior to properties of G_I, called *conflicts*. Once an interest graph contains a conflict, it is easy to show that the game can never converge to an equilibrium. We can observe three classes of conflicts: *cyclic conflicts, BST conflicts* and *focal point conflicts* (see Figure 3 for examples). Cyclic conflicts are cycles in G_I. A BST conflict occurs, if nodes have more than two outgoing edges in G_I (with one small exception, see Section 2.1) or if either two edges of G_I intersect in case the nodes are ordered according to their identifier. Focal point conflicts are nodes in G_I with an indegree greater than one. In Section 2.2 we analyze the conflict classes individually.

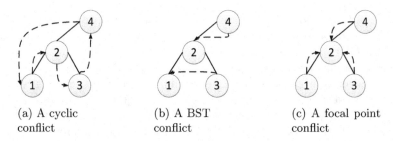

(a) A cyclic conflict

(b) A BST conflict

(c) A focal point conflict

Fig. 3. Small examples for the three conflict classes

2.1 Convergence and Rotation Equilibria

Two classes of interest graphs imply convergence: interest graphs that are binary search trees, and interest graphs that are star graphs (a central node v has interest in all other nodes).

Theorem 1. *Let $\Gamma = (G_C, G_I)$ be a SABST-game with G_I either forming a binary search tree or a star graph. Then, any sequence of best responses converges independent of the initial connection graph. The Price of Anarchy is at most 2.*

Theorem 1 implies that, for the two mentioned communication interest patterns, a distributed self-adjusting binary search tree converges to a steady BST and has almost optimal cost for communication: i.e., it has an approximation factor of at most 2 compared to the optimal BST. Theorem 1 follows from the following two lemmas.

Lemma 1. *Let $\Gamma = (G_C, G_I)$ be a SABST-game with G_I forming a binary search tree. Γ converges to a social optimum.*

Proof. We call a node/player *happy* if it cannot perform a rotation to improve its private cost. Let $\mathcal{H}$ denote the set of all nodes $v \in V$, with the property that the complete subtree of G_I rooted at v is happy. To prove convergence we show that the size of $\mathcal{H}$ is monotonically increasing.

We first show that once a node has entered $\mathcal{H}$, it will never leave $\mathcal{H}$. Let v be a node from $\mathcal{H}$ whose parent in G_I is not happy. Consequently, v and all nodes in the subtree rooted at v in G_I are happy and they cannot decrease their private cost and form a connected component in G_C. Let CC_v be this connected component and v' be a node that is unhappy and performs a rotation. If v' and $IS(v')$ are both above or below CC_v in G_C, then the rotations performed by v' do not affect v and its subtrees. If v' is below CC_v and $IS(v')$ is above CC_v (or vice-versa), v' has to rotate over CC_v. To do so, it performs only right or only left rotations, because v' is either smaller or greater than all nodes in CC_v. But from the definition of a rotation (see Figure 1), we can deduce that performing only left or only right rotations does not affect the structure of the subgraph that v' rotates over. Thus, all nodes in CC_v remain happy. The last case is if v' is interested in v, above v in G_C and v' rotates v upwards. This implies that there exists at least an unhappy node v^- that is on the path from v' to v in G_C. Consequently, v^- is either above v' in G_I, a sibling of v' in G_I, or in the other subtree of v' than v in G_I. But in none of these cases can v^- be in between v' and v in G_C, since G_C is a binary search tree. Thus, v does not leave $\mathcal{H}$ and the size of $\mathcal{H}$ does not decrease. $\mathcal{H}$ is monotonically increasing, because in each round the parents of the nodes already in $\mathcal{H}$ will enter $\mathcal{H}$ and initially all leaves from G_I are in $\mathcal{H}$, since their interest set is empty.

Now assume that Γ does not converge to a social optimum. Let T' be the connection graph in a rotation equilibrium and $T' \neq G_I$: i.e., $\exists u \in V$ with $pos_{G_I}(u) \neq pos_{T'}(u)$, where $pos_{G_I}(u)$ and $pos_{T'}(u)$ denote the position of u in G_I and T' depending on v's depth. Let v' be the node with minimal depth in T' which has a child u with $pos_{G_I}(u) \neq pos_{T'}(u)$. Consequently, v' is unhappy and can perform a rotation to decrease its private cost, which contradicts the fact that T' is in a rotation equilibrium. Consequently, the connection graph in the rotation equilibrium is the same as G_I and the PoA is 1. □

Note that this result only holds for binary trees, since for general trees the size of $\mathcal{H}$ is not monotonically increasing: i.e., if a unhappy node performs a better response, happy nodes can become unhappy again.

Lemma 2. *Let $\Gamma = (G_C, G_I)$ be a SABST-game with G_I forming a star graph: i.e., all edges point from one single center node to all other nodes. Γ converges and has a PoA of at most 2.*

Note, that the star graph, is an exception to the conflict class of BST conflicts. However, this is the only exception, because by observation one can show that the game does not converge anymore if there is an edge $(u, v) \in E_I$ with u being not the center node. The proof of Lemma 2 can be found in the full version. Lemma 1 and 2 prove Theorem 1. The rest of this section justifies the approach of focusing on a single connected component of edges from G_I. We say a node w *affects* the private cost of a node v in a rotation equilibrium if w lies on the the shortest path from v to a node u with $u \in IS(v)$.

Lemma 3. *Consider a connected component E_I' of edges without conflicts from the interest graph $G_I = (V, E_I)$, the corresponding node set $V' = \{v \in V | \exists u \in V \wedge (u, v) \in E_I' \vee (v, u) \in E_I'\}$ and a single interest edge $e_I = (u', v')$. If e_I is* *neither a part of E'_I nor induces a conflict with E'_I, u and v do not affect the private cost of the nodes from V' in a rotation equilibrium and vice-versa.*

Again the proof can be found in the full version. We can easily extend Lemma 3 such that the single edge e_I can be replaced by a set of edges. Therefore, we can analyze multiple connected components from G_I separately. Furthermore, the proof can be extended in such a way that G'_I contains conflicts, instead of being connected. The game will not converge anymore, but has the property that a single edge (or even a set of edges) will no longer affect the private cost of G'_I eventually.

2.2   Non-convergence and Sink Equilibria

As mentioned before, the three identified classes of conflicts imply non-convergent behavior. Therefore, rotation equilibria do not necessarily exist and the Price of Anarchy is no longer well defined. To overcome this obstacle, we use the solution concept *sink equilibrium*, which was introduced by Goemans et al. [9]. A sink equilibrium is not defined for a single connection graph G_C of a game instance, but for the *configuration graph* of an instance. The configuration graph $G_S = (V^*, E^*)$ of an instance $\Gamma = ((V, E_C), (V, E_I))$ has a vertex which is equal to the set of *valid connection graphs* (i.e., all possible BSTs) for the given node set V. The edge set E^* corresponds to better responses of the players: i.e., an edge (u, v) is in E^* if a response of a single player in the connection graph represented by u leads the connection graph in v. A *sink equilibrium* is a strongly connected component without outgoing edges in the configuration graph. Analogical to the Price of Anarchy we define a new measurement of how well selfish players perform compared to a social optimum. [9] uses the expected social cost of a sink equilibrium to compute what is called *Price of Sinking (PoS)*. However, we want to focus on the worst-case behavior of nodes. Therefore, instead of looking at the expected social cost of sink equilibria, we choose a state with worst-case social cost of all sink equilibria and compare it to the social cost of a social optimum. We call this measure the *worst-case Price of Sinking (wcPoS)*. If the wcPoS is low, then every state in a sink equilibrium has social cost close to the optimal social cost and therefore the self-adjusting binary search tree still performs well, even though it does not converge to a fixed tree.

Before analyzing the different classes of conflicts separately and giving results on their worst-case Price of Sinking, we first prove a general result about sink equilibria in the *SABST*-game. Due to the definition of the wcPoS, we are faced with the problem of finding a state in a sink equilibrium with maximal social cost. Lemma 4 simplifies this task. A *response order* τ is a permutation of the players V. We say a response order is *applied* to connection graph G_C (respectively, a state from the configuration graph), when the players of the game play their responses according to τ starting from G_C.

Lemma 4. *Given an instance of the SABST-game Γ, a response order τ and a state s from the configuration graph $G_S = (V^*, E^*)$ of Γ. If $\forall s' \in V^*$ it holds that τ applied on s' results in s, then s lies in a unique sink equilibrium of G_S.*

Proof. Assume that there is another sink equilibrium $\mathcal{SE}'$ and let v' be a state from $\mathcal{SE}'$. We know that v^* can be reached from v' by τ. But by the definition of a sink equilibrium this implies that v^* and v' are in the same sink equilibrium, which is a contradiction to the original assumption.

Therefore, we can deduce a worst-case sink equilibrium state s, if we can give a response order τ that constructs the connection graph represented in s.

Cyclic Conflicts. We first take a closer look on interest graphs with only cyclic conflicts. We only need to consider interest graphs that are simple cycles (i.e., cycles that do not intersect and are not contained in each other) because these cases imply a BST conflict or a focal point conflict. W.l.o.g. we focus on the cyclic conflict over all nodes $G_I^{c.c.} = (V, E_I)$ with $V = \{1, \ldots, n\}$ and $E_I = \{(n, 1) \cup (i, i+1) : i = 1, \ldots, n - 1\}$.

Theorem 2. *Let $\Gamma^{c.c.} = (G_C, G_I^{c.c.})$ be a SABST-game, the wcPoS is $\mathcal{O}(1)$.*

Consequently, as long as the communication interests contain only cyclic conflicts, the performance of the self-adjusting tree is asymptotically as good as the performance without conflicts. To prove Theorem 2, we need to show the following to lemmas.

Lemma 5. *For the SABST-game $\Gamma^{c.c.}$, every state in the unique sink equilibrium has social cost of $2(n - 1)$.*

Proof. Let $\tau' = (n, \ldots, 1)$ be a response order. If τ' is applied on G_C the resulting connection graph is the one visualized in Figure 4, which is in a unique sink equilibrium. The social cost is $2(n - 1)$. Now independent of a response order, there is only one unhappy node in the connection graph that can decrease its private cost. This leads to a connection graph with social cost $2(n - 1)$ and a single unhappy node again. Consequently, independent of a response order in each round there is a single unhappy node and social cost of $2(n - 1)$,   □

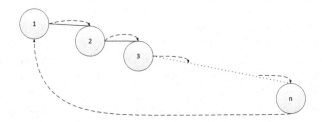

Fig. 4. The connection graph for $\Gamma^{c.c.}$ after response order τ' is applied

Lemma 6. *Every social optimum for $\Gamma^{c.c.}$ has social cost of $\Omega(2(n - 1))$.*

Proof. We call a connection graph edge e_C *traversed* by an interest graph edge $e_I = (u, v)$, if e_C is contained in the shortest path from u to v in the connection graph. We show that every connection graph edge of a social optimum is traversed by at least two interest graph edges. Let e'_C be an arbitrary connection graph edge from a socially optimal connection graph. If e'_C is removed, the connection graph is split in two connected components A and B. Since $G_I^{c.c.}$ is a simple cycle over all nodes, there exist interest graph edges $e'_I = (a', b')$ and $e''_I = (a'', b'')$ with $a', a'' \in A$, $a' \neq a''$ and $b', b'' \in B$, $b' \neq b''$. Consequently, e'_C is traversed twice. $\qquad\square$

Lemma 5 and Lemma 6 together conclude the proof.

BST Conflicts and Focal Point Conflicts. For BST conflicts and focal point conflicts we do not prove an upper bound for the wcPoS, but show that both conflict classes contain interest graphs such that the wcPoS is lower bounded by $\Omega(\frac{n}{\log(n)})$. Therefore, best responses of selfish players can lead to a state in a sink equilibrium, which has high social cost compared to a social optimum. This shows that the intuition of the *double splay* algorithm [4] performs poorly in these scenarios. We start with interest graphs with only BST conflicts. More specifically we focus on interest graphs with only *direct conflicts* in which two edges of G_I intersect if the nodes are ordered according to their identifier. Interest graphs with only direct conflicts have a node degree smaller than 2, since all other conflict types need a node degree of at least 2. We focus on interest graphs that maximize the number of direct conflicts. These are of the form $G_I^{d.c.} = (V, E_I)$ with $V = \{1, \ldots n\}$, n even and $E_I = \{(i, i + \frac{n}{2}) : i = 1, \ldots, \frac{n}{2}\}$, because every interest edge intersects with every other interest edge.

Theorem 3. *Let* $\Gamma^{d.c.} = (G_C, G_I^{d.c.})$ *be a SABST-game, the wcPoS is* $\Omega(\frac{n}{\log(n)})$.

To prove Theorem 3, we first prove that the configuration graph of Γ contains a unique sink equilibrium with a state that has social cost of $\Theta(n^2)$.

Lemma 7. *The configuration graph of* $\Gamma^{d.c.}$ *contains a state in the unique sink equilibrium with social cost of* $\Theta(n^2)$.

Proof. We pick the response order $\tau' = (1, \ldots, n)$ If we now apply τ' to any initial connection graph, we end up with the connection graph presented in Figure 5. The exact proof of this fact is skipped, but relies mainly on the idea that each player performs rotations such that the nodes from its interest set are in one of its subtrees. The social cost is $\sum_{i=0}^{\frac{n}{2}-1}(2i+1) = \frac{n^2}{4} = \Theta(n^2)$. Since this connection graph can be reached from any initial connection graph by τ', we know that it is a state in the unique sink equilibrium of $\Gamma^{d.c.}$. $\qquad\square$

Contrasting the last lemma, we now give a general upper bound for the social cost of a social optimum for Γ.

Lemma 8. *A social optimum for* $\Gamma^{d.c.}$ *has social cost of at most* $\mathcal{O}(n \log n)$.

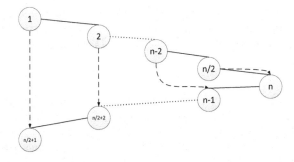

Fig. 5. The connection graph for a $\Gamma^{d.c.}$ if response order $\tau' = (1, \ldots, n)$ is applied

Proof. We arrange the connection graph nodes such that they form a balanced binary search tree. Since every node is only interested in at most a single other node we know that the private cost for a single node can be upper bound by $2 \log(n)$. Therefore, the social cost are at most $\mathcal{O}(n \log n)$. □

For interest graphs with only focal point conflicts we can state a similar result. We use the interest graph $G_I^{f.c.} = (V, E_I)$ with $V = \{1, \ldots n\}$ and $E_I = \{(1, n), (2, n), \ldots, (n - 1, n)\}$, which has the maximal possible focal point conflict.

Theorem 4. *Let $\Gamma^{f.c.} = (G_C, G_I^{f.c.})$ be a SABST-game the wcPoS is $\Omega(\frac{n}{\log(n)})$.*

The proof technique for Theorem 4 is analogous to Theorem 3. Moreover, Theorems 3 and 4 imply that a *SABST*-game $\Gamma = (G_C, G_I)$ in which G_I contains a subgraph G_I' of size k that is either $G_I^{f.c.}$ or $G_I^{d.c.}$, the wcPoS is $\Omega(\frac{k}{\log(k)})$. Therefore, we can conclude that the performance of a distributed self-adjusting binary search tree gets worse with increasing size of the communication patterns given by $G_I^{f.c.}$ or $G_I^{d.c.}$. Notice that $\mathcal{O}(n^2)$ is an upper bound for the social cost of a *SABST*-game with an interest graph with n many edges. Therefore, the upper bound for the wcPoS is $\mathcal{O}(n)$.

3 Conclusion and Open Problems

We analyzed the performance of a distributed self-adjusting binary search tree for different communication patterns. We have shown that, if the communication interests contain no conflicts or only cyclic conflicts, the performance of a self-adjusting tree is almost optimal (PoA of $\Theta(1)$ and wcPos of $\Theta(1)$). However, if the communication interests contain BST conflicts or focal point conflicts, a distributed generalization of splay trees performs poorly (wcPoS of $\Omega(\frac{n}{\log n})$).

There are a lot of different possibilities to extend our work. For example, it would be interesting to analyze the *SABST*-game with an arbitrary combination conflicts and give upper or lower bounds for the worst-case Price of Sinking. Moreover, it is interesting to compute the Price of Sinking as defined in [9] and thereby get statements about the average performance.

Acknowledgements. The author thanks Alexander Skopalik for stimulating discussions that improved the quality of the paper, and Christian Scheideler for his scientific guidance.

References

1. Adelson-Velsky, G.M., Landis, Y.M.: An algorithm for the organization of information. Deklady Akademii Nauk USSR 16 16(2), 263–266 (1962)
2. Alon, N., Demaine, E.D., Hajiaghayi, M., Leighton, T.: Basic network creation games. In: SPAA, pp. 106–113 (2010)
3. Arora, D., Bienkowski, M., Feldmann, A., Schaffrath, G., Schmid, S.: Online strategies for intra and inter provider service migration in virtual networks. In: IPTcomm (2011)
4. Avin, C., Haeupler, B., Lotker, Z., Scheideler, C., Schmid, S.: Locally self-adjusting tree networks. In: IPDPS, pp. 395–406 (2013)
5. Bayer, R.: Symmetric binary b-trees: Data structure and maintenance algorithms. Acta Inf. 1, 290–306 (1972)
6. Cord-Landwehr, A., Hüllmann, M., Kling, P., Setzer, A.: Basic network creation games with communication interests. In: SAGT, pp. 72–83 (2012)
7. Fabrikant, A., Luthra, A., Maneva, E., Papadimitriou, C.H., Shenker, S.: On a network creation game. In: PODC, pp. 347–351 (2003)
8. Galperin, I., Rivest, R.L.: Scapegoat trees. In: SODA, pp. 165–174 (1993)
9. Goemans, M.X., Mirrokni, V.S., Vetta, A.: Sink equilibria and convergence. In: FOCS, pp. 142–154 (2005)
10. Heller, B., Seetharaman, S., Mahadevan, P., Yiakoumis, Y., Sharma, P., Banerjee, S., McKeown, N.: Elastictree: Saving energy in data center networks. In: NSDI, pp. 249–264 (2010)
11. Leitao, J.C.A., da Silva Ferreira Moura Marques, J.P., Pereira, J.O.R.N., Rodrigues, L.E.T.: X-BOT: A protocol for resilient optimization of unstructured overlays. In: SRDS, pp. 236–245 (2009)
12. Lenzner, P.: On dynamics in basic network creation games. In: Persiano, G. (ed.) SAGT 2011. LNCS, vol. 6982, pp. 254–265. Springer, Heidelberg (2011)
13. Shang, Y., Li, D., Xu, M.: Energy-aware routing in data center network. In: Green Networking, pp. 1–8 (2010)
14. Sherk, M.: Self-adjusting k-ary search trees. In: Dehne, F., Santoro, N., Sack, J.-R. (eds.) WADS 1989. LNCS, vol. 382, pp. 381–392. Springer, Heidelberg (1989)
15. Sleator, D.D., Tarjan, R.E.: Self-adjusting binary trees. In: STOC, pp. 235–245 (1983)
16. Tang, M., Liu, Z., Liang, X., Hui, P.M.: Self-adjusting routing schemes for time-varying traffic in scale-free networks. Phys. Rev. E 80, 026114 (2009)
17. Wang, C.C., Derryberry, J., Sleator, D.D.: O(log log n)-competitive dynamic binary search trees. In: SODA, pp. 374–383 (2006)

An Efficient Silent Self-Stabilizing Algorithm for 1-Maximal Matching in Anonymous Networks

Yuma Asada and Michiko Inoue

Nara Institute of Science and Technology, Ikoma, Nara 630-0192 Japan
{asada.yuma.ar4,kounoe}@is.naist.jp

Abstract. We propose a new self-stabilizing 1-maximal matching algorithm which is *silent* and works for any *anonymous* networks without a cycle of a length of a multiple of 3 under a central *unfair* daemon. Let n and e be the numbers of nodes and edges in a graph, respectively. The time complexity of the proposed algorithm is $O(e)$ moves. Therefore, the complexity is $O(n)$ moves for trees or rings whose length is not a multiple of 3. That is a significant improvement from the best existing results of $O(n^4)$ moves for the same problem setting.

Keywords: distributed algorithm, self-stabilization, graph theory, matching problem.

1 Introduction

Self-Stabilization [3] can tolerate several inconsistencies of computer networks caused by transient faults, erroneous initialization, or dynamic topology change. It can recover and stabilize to consistent system configuration without restarting program execution.

Maximum or *maximal matching* is a well-studied fundamental problem for distributed networks. A matching is a set of pairs of adjacent nodes in a network such that any node belongs to at most one pair. It can be used in distributed applications where pairs of nodes, such as a server and a client, are required. This paper proposes an efficient anonymous self-stabilizing algorithm for *1-maximal matching*. A 1-maximal matching is a $\frac{2}{3}$-approximation to the maximum matching, and expected to find more matching pairs than a *maximal matching* which is a $\frac{1}{2}$-approximation to the maximum matching.

Self-stabilizing algorithms for the maximum and maximal matching problems have been well studied[5]. Table 1 summarizes the results, where n and e denote the numbers of nodes and edges, respectively.

Blair and Manne[1] showed that a *maximum* matching can be solved with $O(n^2)$ moves for non-anonymous tree networks under a read/write daemon. For *anonymous* networks, Karaata et al.[8] proposed a maximum matching algorithm with $O(n^4)$ moves for trees under a central daemon, and Chattopadhyay et al.[2] proposed a maximum matching algorithm with $O(n^2)$ rounds for bipartite networks under a central daemon.

M.S. Rahman and E. Tomita (Eds.): WALCOM 2015, LNCS 8973, pp. 187–198, 2015.

Table 1. Self-stabilizing matching algorithms

Reference	Matching	Topology	Anonymous	Daemon	Complexity
[1]	maximum	tree	no	read/write	$O(n^2)$ moves
[8]	maximum	tree	yes	central	$O(n^4)$ moves
[2]	maximum	bipartite	yes	central	$O(n^2)$ rounds
[7]	maximal	arbitrary	yes	central	$O(e)$ moves
[4]	1-maximal	tree, ring*	yes	central	$O(n^4)$ moves
[10]	1-maximal	arbitrary	no	distributed	$O(n^2)$ rounds
this paper	1-maximal	arbitrary*	yes	central	$O(e)$ moves

* without a cycle of length of a multiple of 3.

Hsu and Huang[7] proposed a *maximal* matching algorithm for anonymous networks with arbitrary topology under a central daemon. They showed the time complexity of $O(n^3)$ moves, and, it has been revealed that the time complexity of their algorithm is $O(n^2)$ moves by Tel[11] and Kimoto et al.[9] and $O(e)$ moves by Hedetniemi et al. [6].

Goddard et al.[4] proposed a *1-maximal* matching with $O(n^4)$ moves for anonymous trees and rings whose length is *not* a multiple of 3 under a central daemon. They also showed that there is no self-stabilizing 1-maximal matching algorithm for anonymous rings with length of a multiple of 3. Manne et al. [10] also proposed a 1-maximal matching algorithm for non-anonymous networks with any topology under a distributed unfair daemon. Their algorithm stabilizes in $O(n^2)$ rounds and $O(2^n \cdot \Delta \cdot n)$ moves, where Δ is the maximum degree of nodes.

Our Contribution. In this paper, we propose a new self-stabilizing 1-maximal matching algorithm. The proposed algorithm is *silent* and works for any *anonymous* networks without a cycle of a length of a multiple of 3 under a central *unfair* daemon. We will show that the time complexity of the proposed algorithm is $O(e)$ moves. Therefore, the complexity is $O(n)$ moves for trees or rings whose length is not a multiple of 3. That is a significant improvement of the best existing result of $O(n^4)$ for the same problem setting[4].

The remaining of the paper is organized as follows. In Section 2, we define distributed systems and the 1-maximal matching problem. A 1-maximal matching algorithm is proposed in Section 3, and proves for its correctness and performance are given in Section 4. Finally Section 5 concludes this paper.

2   Preliminaries

A distributed system consists of multiple asynchronous processes. Its topology is represented by an undirected connected graph $G = (V, E)$ where a node in V represents a process and an edge in E represents the interconnection between the processes. A node is a state machine which changes its states by actions. Each node has a set of actions, and a collection of actions of nodes is called a *distributed algorithm*. Let n and e denote the numbers of nodes and edges in a distributed system.

In this paper, we consider *state-reading model* as a communication model where each node can directly read the internal state of its neighbors. An action of a node is expressed $\langle label \rangle :: \langle guard \rangle \mapsto \langle statement \rangle$. A guard is a Boolean function of all the states of the node and its neighbors, and a statement updates its local state. We say a node is privileged if it has an action with a true guard. Only privileged node can *move* by selecting one action with a true guard and executing its statement.

Moves of nodes are scheduled by a *daemon*. Among several daemons considered for distributed systems, we consider an *unfair central daemon* in this paper. A central daemon chooses one privileged node at one time, and the selected node atomically moves. A daemon is unfair in a sense that it can choose any node among privileged nodes.

A problem $\mathcal{P}$ is specified by its legitimate configurations where configuration is a collection of states of all the nodes. We say a distributed algorithm $\mathcal{A}$ is *self-stabilizing* if $\mathcal{A}$ satisfies the following properties. 1) **convergence**: The system eventually reaches to a legitimate configuration from any initial state, and 2) **closure**: The system once reaches to a legitimate configuration, all the succeeding moves keep the system configuration legitimate. A self-stabilizing algorithm is *silent* if, from any arbitrary initial configuration, the system reaches a terminal configuration where no node can move. A self-stabilizing algorithm is *anonymous* if it does not use global IDs of nodes. We only assume that nodes have pointers and a node can determine whether its neighbor points to itself, some other nodes, or no node.

A *matching* in an undirected graph $G = (V, E)$ is a subset M of E such that each node in V is incident to at most one edge in M. We say a matching is *maximal* if no proper superset of M is a matching as well. A maximal matching M is *1-maximal* if, for any $e \in M$, any matching cannot be produced by removing e from M and adding two edges to $M - \{e\}$. A maximal matching is a $\frac{1}{2}$−approximation to the maximum matching. On the other hand, a 1-maximal matching is a $\frac{2}{3}$−approximation. In this paper, we propose a silent and anonymous self-stabilizing algorithm for the 1-maximal matching problem for graphs without a cycle of length of a multiple of 3.

3   Algorithm MM1

First, we will show an overview of a proposed self-stabilizing 1-maximal matching algorithm MM1. Each node i uses stages to construct 1-maximal matching. There are seven stages; *S1a*, *S1b*, *S2a*, *S2b*, *S3*, *S4*, and *S5*. Stages *S1a* and *S1b* mean that the node is not matching with any node. A stage *S2a* means the node is matching with a neighbor node, and, *S2b*, *S3*, *S4*, *S5* mean the node is trying to increase matches. A node i has three variables; $\texttt{level}_i$, $\texttt{m-ptr}_i$, $\texttt{i-ptr}_i$. We describe how to use the variables in our algorithm.

S1a, S1b, S2a. We say a node is *free* if the node is in *S1a* or *S1b*. A node in *S1a* does not invite any nodes, while a node in *S1b* invites its neighbor node. Fig.1 shows how free nodes make a match. When a free node i finds a free neighbor node j, i invites j by i-ptr$_i$ (i is in *S1b*). Then invited node j updates its level to 2 and points to i by m-ptr$_j$ to accept the invitation (j is in *S2a*). Finally i points to j by m-ptr$_i$ to make a match (i is in *S2a*). A node in *S2a* is at level 2 and does not invite any nodes. If two adjacent nodes i and j point to each other by m-ptr, we consider they are matching, that is $(i, j) \in M$.

S2b, S3, S4, S5. Matching nodes try to increase the number of matches if they have free neighbor nodes. Fig.2 shows how to increase matches, where matches are increased by breaking a match between i and j, and creating new matches between i and k, and j and l. In Fig.2(a), nodes i and j invite their free neighbors k and l if they do not invite i and j, respectively (i and j are in *S2b*). When both nodes notice that i and j invite free neighbor nodes, they change their level to 3 (i and j are in *S3*). That indicates that they are ready to be approved as in Fig.2(b). Then k and l point to the inviting nodes by i-ptr to approve their invitations (k and l are in *S1b*). Node i and j change their level to 4 if the neighbors approve the invitations (i and j are in *S4*) as in Fig.2(c), and change their level to 5 when they notice that both invitations are approved (i and j are in *S5*). This indicates that they are ready to break a match as in Fig.2(d). Then they create new matches with the free nodes, where k and l first move to *S2a* (Fig.2(e)) and then i and j move to *S2a* (Fig.2(f)), respectively. A node in *S1a* or *S1b* can make a match with the other node while an inviting node is in *S3*. However, once the inviting node moves to *S4*, it cannot change its i-ptr while the inviting node is in *S4*.

Reset. Each node always checks its validity, and resets to *S1a* if it finds its invalidity. We consider two kinds of validities, *one node validity* and *two nodes validity*. The one node validity means that a state represents some stage. For example, if a level is 1 and m-ptr points to some neighbor, the state is one node invalid. The two nodes validity means that a relation between states of two adjacent nodes is consistent. For example, if a node i is in *S2a*, a node pointed to by m-ptr should point to i by m-ptr at level 2 or by i-ptr at level 1 or 5. The full definition of the validity function is shown in Fig.3. A node does not move while some neighbor is one node invalid.

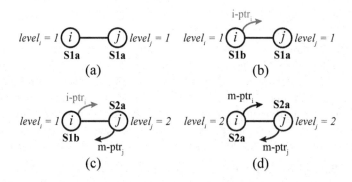

Fig. 1. Making a match between free nodes

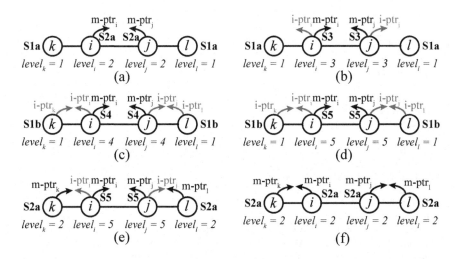

Fig. 2. Increasing matches

Cancel. A node cancels an invitation or progress to increase matches, if it detects that the invitation cannot be accepted or it cannot increase matches. When canceling, a node goes back to *S1a* if it is at level 1, and to *S2a* if it is at level 2 or higher.

The algorithm MM1 uses some statement macros and a guard function. The variables, validity functions, statement macros and a guard function are shown in Fig.3, and a code of MM1is shown in Fig.4. In the algorithm, each node i uses $N(i)$ to represent a set of its neighbors. That is a set of local IDs for each node and the algorithm does not use any global IDs. We only assume that each node can determine whether its neighbor point to itself, some other node, or no node by pointers i-ptr and m-ptr.

4   Correctness

Lemma 1. *There are no nodes at level 5 in any terminal configuration of MM1.*

Proof. By contradiction. Assume that a node i is in *S5* in a terminal configuration. In this case, $\text{i-ptr}_i = k$ holds for some k, and $\text{level}_k = 1 \wedge \text{i-ptr}_k = i$ or $\text{level}_k = 2 \wedge \text{m-ptr}_k = i$ holds since i is in *S5*. If it is $\text{level}_k = 1$, k can execute migrate1. If it is $\text{level}_k = 2$, i can execute migrate2. A contradiction.   □

Lemma 2. *A node that points to its neighbor node by m-ptr also pointed by the neighbor's m-ptr in any terminal configuration of MM1.*

Proof. By contradiction. There is no node at level 5 in any terminal configuration and all nodes are valid. Assume that there are adjacent nodes i and j such that $\text{m-ptr}_i = j \wedge \text{m-ptr}_j \neq i$. A node i is in *S2a* since validity $S2b(i)$, $S3(i)$ or $S4(i)$ do not hold. A node j is at level 1 and $\text{i-ptr}_j = i$ from $S2a(i)$. Since i is in *S2a* and j is $\text{level}_j = 1 \wedge \text{i-ptr}_j = i$, j can execute match3. A contradiction.   □

Variables
$\text{level}_i \in \{1,2,3,4,5\}$
$\text{m-ptr}_i \in N(i) \cup \{\bot\}$
$\text{i-ptr}_i \in N(i) \cup \{\bot\}$

Valid Predicates
$S1b\_valid(i,k)$: $\text{level}_i = 1 \wedge \text{m-ptr}_i = \bot \wedge \text{i-ptr}_i = k$
$S2a\_valid(i,j)$: $\text{level}_i = 2 \wedge \text{m-ptr}_i = j \wedge \text{m-ptr}_i = \bot$
$S2b\_valid(i,j,k)$: $\text{level}_i = 2 \wedge \text{m-ptr}_i = j \wedge \text{m-ptr}_i = k \wedge j \neq k$
$S3\_valid(i,j,k)$: $\text{level}_i = 3 \wedge \text{m-ptr}_i = j \wedge \text{m-ptr}_i = k \wedge j \neq k$
$S4\_valid(i,j,k)$: $\text{level}_i = 4 \wedge \text{m-ptr}_i = j \wedge \text{m-ptr}_i = k \wedge j \neq k$
$S5\_valid(i,j,k)$: $\text{level}_i = 4 \wedge \text{m-ptr}_i = j \wedge \text{m-ptr}_i = k \wedge j \neq k$

One Node Validity
$S1a\_valid1(i)$: $\text{level}_i = 1 \wedge \text{m-ptr}_i = \bot \wedge \text{i-ptr}_i = \bot$
$S1b\_valid1(i)$: $\exists k \in N(i)$ S1b_valid(i,k)
$S2a\_valid1(i)$: $\exists j,k \in N(i)$ S2a_valid(i,j)
$S2b\_valid1(i)$: $\exists j,k \in N(i)$ S2b_valid(i,j,k)
$S3\_valid1(i)$: $\exists j,k \in N(i)$ S3_valid(i,j,k)
$S4\_valid1(i)$: $\exists j,k \in N(i)$ S4_valid(i,j,k)
$S5\_valid1(i)$: $\exists j,k \in N(i)$ S4_valid(i,j,k)
$valid1(i)$: $S1a\_valid(i) \wedge S1b\_valid(i) \wedge S2a\_valid(i) \wedge S2b\_valid(i) \wedge S3\_valid(i) \wedge S4\_valid(i) \wedge S5\_valid(i)$
$invalid1(i)$: $\neg valid1(i)$

Valid Functions (One Node Validity and Two Node Validity)
$S1a(i)$: $S1a\_valid1(i)$
$S1b(i)$: $S1b\_valid1(i)$
$S2a(i)$: $\exists j \in N(i)(S2a\_valid(i,j) \wedge (\text{level}_j = 2 \wedge \text{m-ptr}_j = i) \vee (\text{level}_j = 1 \wedge \text{i-ptr}_j = i) \vee (\text{level}_j = 5 \wedge \text{i-ptr}_j = i))$
$S2b(i)$: $\exists j,k \in N(i)(S2b\_valid(i,j,k) \wedge (\text{level}_j = 2 \vee \text{level}_j = 3 \vee \text{level}_j = 4) \wedge \text{m-ptr}_j = i)$
$S3(i)$: $\exists j,k \in N(i)(S3\_valid(i,j,k) \wedge (\text{level}_j = 2 \vee \text{level}_j = 3 \vee \text{level}_j = 4) \wedge \text{m-ptr}_j = i)$
$S4(i)$: $\exists j,k \in N(i)(S4\_valid(i,j,k) \wedge (\text{level}_j = 2 \vee \text{level}_j = 3 \vee \text{level}_j = 4 \vee \text{level}_j = 5) \wedge \text{m-ptr}_j = i \wedge \text{i-ptr}_j \neq \bot \wedge \text{level}_k = 1 \wedge \text{i-ptr}_k = i)$
$S5(i)$: $\exists j,k \in N(i)(S5\_valid(i,j,k) \wedge (\text{level}_k = 1 \wedge \text{i-ptr}_k = i) \vee (\text{level}_k = 2 \wedge \text{m-ptr}_k = i))$
$valid(i)$: $S1a(i) \wedge S1b(i) \wedge S2a(i) \wedge S2b(i) \wedge S3(i) \wedge S4(i) \wedge S5(i)$
$invalid(i)$: $\neg valid(i)$

Statement Macros
make_match: $\text{i-ptr}_i = \bot, \text{m-ptr}_i = j, \text{level}_i = 2$
reset_state: $\text{i-ptr}_i = \bot, \text{m-ptr}_i = \bot, \text{level}_i = 1$
abort_exchange: $\text{i-ptr}_i = \bot, \text{level}_i = 2$

Guard Function
$no\_invalid1\_neighbor(i)$: $\forall x \in N(i)$ $valid1(x)$

Fig. 3. Variables, validity functions, statement macros and guard function

Reset
reset1 :: $invalid1(i) \mapsto$ reset_state
reset2 :: $invalid1(i) \wedge no\_invalid1\_neighbor(i) \mapsto$ reset_state

S1a
match1 :: $S1a(i) \wedge no\_invalid1\_neighbor(i) \wedge \exists x \in N(i)(\text{i-ptr}_x = i \wedge \text{level}_x = 1) \mapsto \text{i-ptr}_i =\perp, \text{m-ptr}_i = x, \text{level}_i = 2$
approve1 :: $S1a(i) \wedge no\_invalid1\_neighbor(i) \wedge \exists x \in N(i)(\text{i-ptr}_x = i \wedge \text{level}_x = 3) \mapsto \text{i-ptr}_i = x$
invite1 :: $S1a(i) \wedge no\_invalid1\_neighbor(i) \wedge \exists x \in N(i)\text{level}_x = 1 \mapsto \text{i-ptr}_i = x$

S1b
match2 :: $S1b(i) \wedge no\_invalid1\_neighbor(i) \wedge \exists x \in N(i)(\text{i-ptr}_x = i \wedge \text{level}_x = 1) \wedge \exists k \in N(i)(S1b\_valid(i,k) \wedge \text{level}_k < 4) \mapsto \text{i-ptr}_i =\perp, \text{m-ptr}_i = x, \text{level}_i = 2$
match3 :: $S1b(i) \wedge no\_invalid1\_neighbor(i) \wedge \exists k \in N(i)(S1b\_valid(i,k) \wedge \text{m-ptr}_k = i \wedge \text{level}_k = 2) \mapsto$ make_match
migrate1 :: $S1b(i) \wedge no\_invalid1\_neighbor(i) \wedge \exists k \in N(i)(S1b\_valid(i,k) \wedge \text{i-ptr}_k = i \wedge \text{level}_k = 5) \mapsto$ make_match
cancel1 :: $S1b(i) \wedge no\_invalid1\_neighbor(i) \wedge \exists k \in N(i)(S1b\_valid(i,k) \wedge (\text{level}_k = 2 \vee (\text{level}_k = 3 \wedge \text{i-ptr}_k \neq i) \vee (\text{level}_k = 4 \wedge \text{i-ptr}_k \neq i) \vee (\text{level}_k = 5 \wedge \text{i-ptr}_k \neq i))) \mapsto \text{i-ptr}_i =\perp$

S2a
invite2 :: $S2a(i) \wedge no\_invalid1\_neighbor(i) \wedge \exists x \in N(i)(\text{level}_x = 1 \wedge \text{i-ptr}_x \neq i) \wedge \exists j \in N(i)(S2a\_valid(i,j) \wedge \text{m-ptr}_j = i) \mapsto \text{i-ptr}_i = x$

S2b
cancel2 :: $S2b(i) \wedge no\_invalid1\_neighbor(i) \wedge \exists j,k \in N(i)(S2b\_valid(i,j,k) \wedge \text{level}_k \geq 2) \mapsto$ abort_exchange
proceed1 :: $S2b(i) \wedge no\_invalid1\_neighbor(i) \wedge \exists j,k \in N(i)(S2b\_valid(i,j,k) \wedge \text{i-ptr}_j \neq\perp) \mapsto \text{level}_i = 3$

S3
cancel3 :: $S3(i) \wedge no\_invalid1\_neighbor(i) \wedge \exists j,k \in N(i)(S3\_valid(i,j,k) \wedge ((\text{level}_j = 2 \wedge \text{i-ptr}_j =\perp) \vee \text{level}_k \geq 2)) \mapsto$ abort_exchange
proceed2 :: $S3(i) \wedge no\_invalid1\_neighbor(i) \wedge \exists j,k \in N(i)(S3\_valid(i,j,k) \wedge \text{i-ptr}_k = i \wedge \text{level}_k = 1) \mapsto \text{level}_i = 4$

S4
cancel4 :: $S4(i) \wedge no\_invalid1\_neighbor(i) \wedge \exists j,k \in N(i)(S4\_valid(i,j,k) \wedge \text{level}_j = 2 \wedge \text{i-ptr}_j =\perp) \mapsto$ abort_exchange
proceed3 :: $S4(i) \wedge no\_invalid1\_neighbor(i) \wedge \exists j,k \in N(i)(S4\_valid(i,j,k) \wedge (\text{level}_j = 4 \vee \text{level}_j = 5)) \mapsto \text{level}_i = 5$

S5
migrate2 :: $S5(i) \wedge no\_invalid1\_neighbor(i) \wedge \exists j,k \in N(i)(S5\_valid(i,j,k) \wedge \text{level}_k = 2 \wedge \text{m-ptr}_k = i \wedge \text{i-ptr}_k =\perp \wedge \text{level}_j = 5) \mapsto \text{i-ptr}_i =\perp, \text{m-ptr}_i = k, \text{level}_i = 2$

Fig. 4. Algorithm MM1

Lemma 3. *There are no two nodes i and j such that $\texttt{level}_i = 1$, $\texttt{level}_j = 3$ or 4, $\texttt{i-ptr}_i = j$ and $\texttt{i-ptr}_j = i$ in any termination configuration of MM1 for any graphs without a cycle of length of a multiple of 3.*

Proof. By contradiction. There is no node at level 5 in any terminal configuration and all nodes are valid. Assume that there are adjacent nodes i and j such that $\texttt{level}_i = 1$, $\texttt{level}_j = 3$ or 4, $\texttt{i-ptr}_i = j$, and $\texttt{i-ptr}_j = i$. If $\texttt{level}_j = 3$, j can execute $\texttt{proceed2}$ since j is in *S3*.

Consider the case of $\texttt{level}_j = 4$. There is a node $k \in N(j)$ such that $\texttt{level}_k = 2$ or 3 or 4, $\texttt{m-ptr}_j = k$, $\texttt{i-ptr}_k \neq \bot$. Node k can execute $\texttt{proceed1}$ if $\texttt{level}_k = 2$ and j can execute $\texttt{proceed3}$ if $\texttt{level}_k = 4$. Hence $\texttt{level}_k$ is limited to 3. Therefore, there is a node $l \in N(k)$ such that $\texttt{i-ptr}_k = l$ and $\texttt{level}_l = 1$. Node l satisfies $\texttt{i-ptr}_l \neq k$ because it is in a terminal configuration. Therefore, there is a node $m \in N(l)$ such that $\texttt{i-ptr}_l = m$ and $\texttt{level}_m = 4$. Repeating the above observation, we can show there is an infinite sequence of nodes at levels $1, 4, 3, 1, 4, 3, \cdots$. However, there is no such a sequence since there is no cycle of length of a multiple of 3. A contradiction. $\qquad\square$

Theorem 1. *A maximal matching is constructed in any terminal configuration of MM1 for any graphs without a cycle of length of a multiple of 3.*

Proof. By contradiction. There is no node at level 5 in any terminal configuration and all nodes are valid. Assume that a matching is not maximal in some terminal configuration. There are adjacent nodes i and j at level 1 by the assumption and Lemma 2.

If a node i or j is in *S1a*, it can execute $\texttt{invite1}$. Therefore, both nodes are in *S1b* (Observation 1). Let k be a node pointed by $\texttt{i-ptr}_i$. The level of k is not 5 by Lemma 1.

In case of $\texttt{level}_k = 1$, k is in *S1b* by Observation 1. Let x be a node pointed by $\texttt{i-ptr}_k$. A node k can execute $\texttt{match2}$ to make a match with i if $\texttt{level}_x \neq 4$. Therefore, $\texttt{level}_x = 4$ and this implies $\texttt{i-ptr}_x \neq k$ by Lemma 3, and k can execute $\texttt{cancel1}$. In case of $\texttt{level}_k = 2$, k can execute $\texttt{invite2}$ if k is in *S2a*. Node i can execute $\texttt{cancel1}$ if k is in *S2b* since $\texttt{m-ptr}_k \neq i$ by Lemma 2. If $\texttt{level}_k = 3$ or 4, i can execute $\texttt{cancel1}$ since $\texttt{i-ptr}_k \neq i$ by Lemma 3. A contradiction. $\qquad\square$

Theorem 2. *A 1-maximal matching is constructed in any terminal configuration of MM1 for any graphs without a cycle of length of a multiple of 3.*

Proof. By contradiction. Assume that a matching is not 1-maximal in some terminal configuration. Since it is terminal, a maximal matching is constructed by Theorem 1. Therefore, there are matching nodes i and j and both have neighbors at level 1 from Lemma 2.

Both i and j are at level 2 or higher since they are matching. They are not in *S2a* since they have level 1 neighbors and can execute $\texttt{invite1}$ if they are in *S2a*, or not at level 5 by Lemma 1. Since i and j are in *S2b*, *S3* or *S4*, both nodes point to some neighbor by $\texttt{i-ptr}$, and the neighbors are at level 1.

That is because, i or j can execute `cancel2` in $S2b$, `cancel3` in $S3$ and `reset2` in $S4$ if it points to a node at level 2 or higher.

Nodes i and j are not in $S2b$ since $\text{i-ptr}_i \neq \bot$ and $\text{i-ptr}_j \neq \bot$, and therefore, they can execute `proceed1` if they are in $S2b$.

Consider the case where i or j is in $S3$. Assume i is in $S3$ w.l.o.g., and let k be a level 1 node that i points to by `i-ptr`. A node k can execute `approve` if $\text{i-ptr}_k \neq \bot$, and node i can execute `proceed2` if $\text{i-ptr}_k = i$. Therefore, $\text{i-ptr}_k = x$ for some $x \neq i$. Since there is no adjacent level 1 nodes by Theorem 1, there is no level 5 node by Lemma 1, and `m-ptrs` point to each other between two matching nodes by Lemma 2, x is at level 2, 3, or 4, and $\text{m-ptr}_x \neq k$. A node x is not at level 2 since k can execute `cancel1` if x is at level 2. In case where x is at level 3 or 4, $\text{i-ptr}_x \neq k$ by Lemma 3, and therefore, k can also execute `cancel1`. Therefore, none of i and j is not in $S3$.

That is, both i and j are in $S4$, however, both can execute `proceed3` in this case. A contradiction.   □

Lemma 4. *If a node i at level 1 is valid, that is $S1a(i)$ or $S1b(i)$ holds, i is valid while it is at level 1 in MM1.*

Proof. Validity functions $S1a(i)$ and $S1b(i)$ check only the variables of a node i. That is the validity of a node at level 1 is independent of its neighbors' states. Any move for $S1a$ or $S1b$ keeps the state of node valid, a valid node at level 1 is valid while it is at level 1.   □

Lemma 5. *Once a node executes one of match1, match2, match3, migrate1 and migrate2, the node never executes reset1 or reset2 in MM1.*

Proof. By contradiction. Assume some nodes execute resets (`reset1` or `reset2`) after executing `match1`, `match2`, `match3`, `migrate1` or `migrate2`. Let i be a node that executes such a move r of a reset first. Let m be the last move of among `match1`, `match2`, `match3`, `migrate1` and `migrate2` before the reset. Since no move except `reset1` and `reset2` brings invalid states and i already executed m, when i executes r, i is two node invalid. Therefore, i detects some invalidity between i and some neighbor.

Let k be a node such that $\text{i-ptr}_i = k$ when i executes r. If k causes the reset r, i is at level 4 or 5 at that time. When i moves to $S4$ by `proceed2`, i confirms that k's validity, $\text{level}_k = 1$ and $\text{i-ptr}_k = i$. Node k never resets while it is at level 1 by Lemma 4 and the validity between i and k is preserved. Node k may move to $S2a$ by `migrate1` but never resets before r by the assumption, and therefore, the validity i and k is also preserved.

Therefore, i executes r by detecting invalidity between i and j such that $\text{m-ptr}_i = j$. Since m is the last chance to set `m-ptr` for i, i sets $\text{m-ptr}_i = j$ by m. When i executes m, j is in $S1b$, $S2a$, or $S5$.

In case of $S1b$, when i executes m, i confirms j's validity and $\text{i-ptr}_j = i$. Node j is valid while it is at level 1 by Lemma 4. Node i moves to $S2b$ after j sets $\text{m-ptr}_j = i$ and moves to $S2a$ by `match2` or `match3`. Therefore, while j is at level 1, $\text{i-ptr}_j = i$ always holds and therefore i cannot reset. After j moves to level 2 by `match2` or `match3`, j does not reset before r from the assumption. Therefore, the validity between i and j is preserved until r.

In case of $S5$, that is i migrates to j, when i executes m, i confirms `i-ptr`$_j = i$. Since the validity of a node in $S5$ only depends on its state and a state of a node pointing to by `i-ptr`, j is valid if the validity between i and j is preserved. Since i does not reset between m and r, the validity is preserved while j is in $S5$. After j moves to level 2 by `migrate2`, j does not reset before r from the assumption. Therefore, the validity between i and j is preserved until r.

In case of $S2a$, i confirms the validity between i and j and `m-ptr`$_j = i$ when i executes m. Since j is in $S2a$, `i-ptr`$_j$ does not point to any node. Therefore, even if j points to some node by `i-ptr` after m, the validity between j and the pointed node is preserved like between i and k. Therefore j is valid if the validity between i and j is preserved while `m-ptr`$_j = i$ and `level`$_j \leq 4$ (When j moves to $S5$, it does not take care of i). Since i does not reset between m and r, the validity is preserved. □

We say a move is a *progress move* if it is by `match1`, `match2`, `match3`, or `migrate1`. A level of node changes from 1 to 2 by a progress move.

Lemma 6. *Each node resets at most once in MM1.*

Proof. Once a node executes `reset1` or `reset2`, it moves to $S1a$. The node never resets while it is at level 1 from Lemma 4. The node executes a progress move to move to level 2, and never resets after that by Lemma 5. □

Lemma 7. *Each node execute a progress move at most once in MM1.*

Proof. A progress move changes levels of a node from 1 to 2, and a node never resets if it executes a progress move by Lemma 5. That is the node never goes back to level 1. Therefore, once a node executes a progress move it never executes a progress move again. □

Lemma 8. *In MM1, `cancel1`, `cancel2`, `cancel3` and `cancel4` are executed $O(e)$ times.*

Proof. In MM1, a node i executes a cancel (`cancel1`, `cancel2`, `cancel3` or `cancel4`) when it is initially possible, some neighbor node executed a cancel, or some neighbor node executed a progress move.

Consider that some node j executes a progress move that changes a stage of j to $S2a$. Nodes that point to j by `i-ptr` will execute a cancel as follows. If such a node k is in $S1b$, k will execute `cancel1`, and if such a node k is in $S2b$ or $S3$, k will execute `cancel2` or `cancel3`.

If some node executes `cancel2` or `cancel3`, it causes more cancels. If there is an adjacent node x and trying to increase matches, it will also cancels by `cancel3` or `cancel4`. That cancel may further causes one more cancel. If x already invited some node y to migrate to x, y will execute `cancel1`.

Now we classify cancels with *direct cancels* and *indirect cancels*. The direct cancel is a cancel caused by some progress move or its initial state. The indirect cancel is a cancel caused by a cancel of its neighbor.

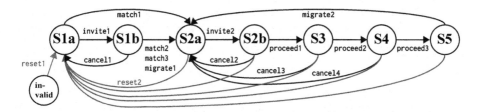

Fig. 5. Transitions of stages

From the above observation, any cancel causes at most two indirect cancels. Let deg_j be the degree of j. There are at most deg_j nodes that execute a cancel due to the progress move of j. From Lemma 7, j executes a progress move at most once, and therefore there are at most $\Sigma_{i \in V} deg_i = e$ direct cancels caused by progress moves. Moreover, there are at most n direct cancels caused by initial states. Therefore, the total number of moves by cancels are $O(e)$.   □

Lemma 9. *In MM1*, `migrate2` *is executed $O(n)$ times.*

Proof. Let m_1 and m_2 be two consecutive moves by `migrate2` of a node i. The node i moves to *S2a* by m_1 and then invites some neighbor node j at level 1 to migrate to i. Then, node j executes `migrate1` that points to i by `m-ptr`. That is, there is a move by `migrate1` that points to i between two consecutive moves by `migrate2` of node i. Therefore, the total number of moves by `migrate2` $\leq$ the total number of moves by `migrate1` $+n$. From Lemma 7, it is bounded by $O(n)$.   □

Theorem 3. *MM1 is silent and takes $O(e)$ moves to construct 1-maximal matching for any graphs without a cycle of length of a multiple of 3.*

Proof. Fig. 5 shows stage transition in MM1. In MM1, each node moves to a higher stage from the current stage in the order of *S1a, S1b, S2a, S2b, S3, S4* and *S5* except `reset1`, `reset2`, `cancel1`, `cancel2`, `cancel3`, `cancel4` and `migrate2`. Therefore, if a node does not execute these actions, the number of moves is at most 6.

Let R_i, C_i and M_i be the numbers of moves of a node i by reset (`reset1` or `reset2`), cancel (`cancel1`, `cancel2`, `cancel3` or `cancel4`), and `migrate2`. Let MOV_i denote the total number of moves of a node i. From the observation, it is bounded as follows.

$$MOV_i \leq 7(R_i + C_i + M_i + 1)$$

From Lemmas 6, 8 and 9, we have

$$\Sigma_{i \in V} R_i = O(n), \Sigma_{i \in V} C_i = O(e), \text{ and } \Sigma_{i \in V} M_i = O(n).$$

Therefore, the total number of moves in MM1 can be derived as follows.

$$\Sigma_{i \in V} MOV_i \leq 7(\Sigma_{i \in V} R_i + \Sigma_{i \in V} C_i + \Sigma_{i \in V} M_i + \Sigma_{i \in V} 1) = O(e)$$

Since each node always takes a finite number of moves, MM1 always reaches a terminal configuration where 1-maximal matching is constructed by Theorem 2. This also implies MM1 is silent.   □

5 Conclusion

We proposed a 1-maximal matching algorithm MM1 that is silent and works for any anonymous networks without a cycle of a length of a multiple of 3 under a central unfair daemon. The time complexity of MM1 is $O(e)$ moves. Therefore, it is $O(n)$ moves for trees or rings whose length is not a multiple of 3. We had a significant improvement from Goddard et al.[4] that is also an anonymous 1-maximal matching algorithm but works for only trees or rings which length is not a multiple of 3 and the time complexity is $O(n^4)$.

References

1. Blair, J.R., Manne, F.: Efficient self-stabilizing algorithms for tree networks. In: Proceedings of 23rd International Conference on Distributed Computing Systems, pp. 20–26. IEEE (2003)
2. Chattopadhyay, S., Higham, L., Seyffarth, K.: Dynamic and self-stabilizing distributed matching. In: Proceedings of the Twenty-first Annual Symposium on Principles of Distributed Computing, pp. 290–297. ACM (2002)
3. Dijkstra, E.W.: Self-stabilizing systems in spite of distributed control. Commun. ACM 17(11), 643–644 (1974), http://doi.acm.org/10.1145/361179.361202
4. Goddard, W., Hedetniemi, S.T., Shi, Z., et al.: An anonymous self-stabilizing algorithm for 1-maximal matching in trees. In: Proc. International Conference on Parallel and Distributed Processing Techniques and Applications, pp. 797–803 (2006)
5. Guellati, N., Kheddouci, H.: A survey on self-stabilizing algorithms for independence, domination, coloring, and matching in graphs. Journal of Parallel and Distributed Computing 70(4), 406–415 (2010)
6. Hedetniemi, S.T., Jacobs, D.P., Srimani, P.K.: Maximal matching stabilizes in time O(m). Information Processing Letters 80(5), 221–223 (2001)
7. Hsu, S.C., Huang, S.T.: A self-stabilizing algorithm for maximal matching. Information Processing Letters 43(2), 77–81 (1992)
8. Karaata, M.H., Saleh, K.A.: Distributed self-stabilizing algorithm for finding maximum matching. Comput. Syst. Sci. Eng. 15(3), 175–180 (2000)
9. Kimoto, M., Tsuchiya, T., Kikuno, T.: The time complexity of Hsu and Huang's self-stabilizing maximal matching algorithm. IEEE Trans. Infrmation and Systems E93-D(10), 2850–2853 (2010)
10. Manne, F., Mjelde, M., Pilard, L., Tixeuil, S.: A self-stabilizing 2/3-approximation algorithm for the maximum matching problem. Theoretical Computer Science 412(40), 5515–5526 (2011)
11. Tel, G.: Introduction to distributed algorithms. Cambridge university press (2000)

Dynamic Online Multiselection
in Internal and External Memory

Jérémy Barbay[1,*], Ankur Gupta[2,**], Srinivasa Rao Satti[3,***],
and Jonathan Sorenson[2]

[1] Departamento de Ciencias de la Computación (DCC)
Universidad de Chile
jeremy@jbarbay.cl
[2] Department of Computer Science and Software Engineering
Butler University
{agupta,jsorenso}@butler.edu
[3] School of Computer Science and Engineering
Seoul National University
ssrao@cse.snu.ac.kr

Abstract. We consider the dynamic version of the *online multiselection*
problem for internal and external memory, in which q selection queries
are requested on an unsorted array of N elements. Our internal memory
result is 1-competitive with the offline result of Kaligosi et al.[ICALP
2005]. In particular, we extend the results of Barbary et al.[ESA 2013]
by supporting arbitrary *insertions* and *deletions* while supporting online
select and *search* queries on the array. Assuming that the insertion of
an element is immediately preceded by a search for that element, we
show that our dynamic online algorithm performs an optimal number of
comparisons, up to lower order terms and an additive $O(N)$ term.

For the external memory model, we describe the first online multiselection algorithm that is $O(1)$-competitive. This result improves upon
the work of Sibeyn [Journal of Algorithms 2006] when $q > m$, where m is
the number of blocks that can be stored in main memory. We also extend
it to support searches, insertions, and deletions of elements efficiently.

1 Introduction

The *multiselection* problem asks for elements of rank r_i from the sequence $R =
(r_1, r_2, \ldots, r_q)$ on an unsorted array A of size N drawn from an ordered universe
of elements. We define $\mathcal{B}(S_q)$ as the information-theoretic lower bound on the
number of comparisons required in the comparison model to answer q queries,

* Supported by Project Regular Fondecyt number 1120054.
** Supported in part by the Butler Holcomb Awards grant and the Arete Initiative.
*** Supported by Basic Science Research Program through the National Research
Foundation of Korea (NRF) funded by the Ministry of Education, Science and
Technology (Grant number 2012-0008241).

M.S. Rahman and E. Tomita (Eds.): WALCOM 2015, LNCS 8973, pp. 199–209, 2015.
© Springer International Publishing Switzerland 2015

where $S_q = \{s_i\}$ denotes the queries ordered by rank. We define $\Delta_i = s_{i+1} - s_i$, where $s_0 = 0$ and $s_{q+1} = N$. Then,

$$\mathcal{B}(S_q) = \log N! - \sum_{i=0}^{q} \log(\Delta_i!) = \sum_{i=0}^{q} \Delta_i \log \frac{N}{\Delta_i} - O(N).^1$$

The *online multiselection* problem asks for elements of rank r_i, where the sequence R is given one element at a time. The lower bound $\mathcal{B}(S_q)$ also applies to *search* queries in the offline model, as well as both types of queries in the online model.

The *dynamic online multiselection* problem supports the *search*, *insert* and *delete* operations, described below:

- *select*(r), returns the position of the r-th element in A;
- *search*(a), returns the position of the largest element $y \leq a$ from A;
- *insert*(a), inserts a into A; and
- *delete*(i), deletes the ith sorted entry from A.

1.1   Previous Work

The offline multiselection problem has been well-studied for the internal memory model [DM81, Pro95, JM10]. Kaligosi et al. [KMMS05] give an optimal offline algorithm that uses $\mathcal{B}(S_q) + o(\mathcal{B}(S_q) + O(N)$ comparisons. In the external memory model [AV88] with parameters M and B, we define $n = N/B$ and $m = M/B$. Sibeyn [Sib06] solves external multiselection using $n + nq/m^{1-\epsilon}$ I/Os, where ϵ is any positive constant. The first term comes from creating a static index structure using n I/Os, and the reminder $nq/m^{1-\epsilon}$ comes from answering q searches using that index. In addition, this result requires the condition that $B \in \Omega(\log_m n)$. When $q = m$, Sibeyn's multiselection algorithm takes $O(nm^\epsilon)$ I/Os, whereas the optimum is $\Theta(n)$ I/Os. In fact, his bounds are within $\omega(\mathcal{B}_m(S_q))$, for any $q \geq m$, where $\mathcal{B}_m(S_q)$ is the lower bound on the number of I/Os required (see Section 4.1 for the definition of $\mathcal{B}_m(S_q)$).

Motwani and Raghavan [MR86] introduced the static online multiselection problem, where the selection and search queries arrive online, as a "Deferred Data Structure" for sorting (i.e., the input array is sorted over time, as queries are answered). They also described deferred data structures for some other problems in computational geometry. Barbay et al. [BGJ+13] described a simpler solution and refined their analysis so that it matches the offline results from Kaligosi et al. [KMMS05]. Ching et al. [YTC90] extended Motwani and Raghavan's solution [MR86] to add the support of the insertion and deletion operators, with optimal amortized complexity in the worst case over instances with a fixed number q of queries. Our solution is simpler, and our analysis finer, in the worst case over instances where the positions hit by the queries are fixed. To the best of our knowledge, there are no existing dynamic results for the multiselection problem in the external memory model.

---

[1] We use $\log_b a$ to refer to the base b logarithm of a. By default, we let $b = 2$.

1.2 Our Results

For the *dynamic online multiselection* problem in internal memory, we describe the *first* algorithm that supports a sequence R of q selection, search, insert, and delete operations, of which q' are search, insert, and delete, using $\mathcal{B}(S_q) + o(\mathcal{B}(S_q)) + O(N + q' \log N)$ comparisons.[2]

For the external memory model [AV88], we describe an online multiselection algorithm on an unsorted array stored on disk in n blocks, using $O(\mathcal{B}_m(S_q))$ I/Os, where $\mathcal{B}_m(S_q)$ is a lower bound on the number of I/Os required to support the given queries. This result improves upon the work of Sibeyn [Sib06] when $q > m$, where m is the number of blocks that can be stored in main memory. We also extend it to support search, insert, and delete operations using $O(\mathcal{B}_m(S_q) + q \log_B N)$ I/Os.

2 Background

Our dynamic online multiselection algorithm is an extension of the static algorithm of Barbay et al. [BGJ$^+$13]. To describe our algorithm, we first start by briefly describing their algorithm in this section. For complete details, we refer the reader to the description of the static solution [BGJ$^+$13].

2.1 Terminology

We call an element $A[i]$ at position i in array A a *pivot* if $A[1 \ldots i - 1] < A[i] \leq A[i + 1 \ldots n]$.

Query and Pivot Sets. Let R denote a sequence of q selection queries, ordered by time of arrival. Let $S_t = \{s_1, s_2, \ldots, s_t\}$ denote the first t queries from R, sorted by position. We also include $s_0 = 1$ and $s_{t+1} = n$ in S_t for convenience of notation, since the minimum and maximum are found during preprocessing. Let $P_t = \{p_i\}$ denote the set of k pivots found by the algorithm when processing S_t, sorted by position. Note that $p_1 = 1$, $p_k = n$, $V[p_i] = 1$ for all i, and $S_t \subseteq P_t$.

Pivot Tree and Recursion Depth. The pivots chosen by the algorithm form a binary tree structure, defined as the *pivot tree* T of the algorithm over time.[3] Pivot p_i is the parent of pivot p_j if, after p_i was used to partition an interval, p_j was the pivot used to partition either the right or left half of that interval. The root pivot is the pivot used to partition $A[2..n - 1]$ due to preprocessing. The *recursion depth*, $d(p_i)$, of a pivot p_i is the length of the path in the pivot tree from p_i to the root pivot. All leaves in the pivot tree are also selection queries, but it may be the case that a query is not a leaf.

---

[2] For the dynamic result, we make the (mild) assumption that the insertion of an element is immediately preceded by a search for that element. In that case, we show that our dynamic online algorithm performs an optimal number of comparisons, up to lower order terms and an additive $O(N)$ term.

[3] Intuitively, a pivot tree corresponds to a *recursion tree*, since each node represents one recursive call made during the quickselect algorithm [Hoa61].

Intervals. Each pivot was used to partition an interval in A. Let $I(p_i)$ denote the interval partitioned by pivot p_i, and let $|I(p_i)|$ denote its length. Such intervals form a binary tree induced by their pivots. If p_i is an ancestor of pivot p_j then $I(p_j) \subset I(p_i)$. The recursion depth of an array element is the recursion depth of the smallest interval containing that element, which in turn is the recursion depth of its pivot.

Gaps. Define the query gap $\Delta_i^{S_t} = s_{i+1} - s_i$ and similarly the pivot gap $\Delta_i^{P_t} = p_{i+1} - p_i$. By telescoping we have $\sum_{i=1}^q \Delta_i^{S_t} = \sum_{j=1}^k \Delta_j^{P_t} = n - 1$.

2.2 Description of the Static Algorithm

For the sake of completeness, we briefly outline the following lemma from Barbay et al. [BGJ$^+$13], which describes the optimal online multiselection algorithm for static data:

Lemma 1 (Static Online Multiselection [BGJ$^+$13]). *Given an unsorted array A of N elements, there exists an algorithm that supports a sequence R of q online selection queries using $\mathcal{B}(S_q)(1+o(1))+O(N)$ comparisons and $O(\mathcal{B}(S_q))$ time in the worst case.*

Barbay et al. [BGJ$^+$13] describe a variant of the static algorithm from Kaligosi et al. [KMMS05]. Both solutions consider *runs*, which are sorted sequences from A of length roughly $\ell = 1 + \lfloor \log(d(p)+1) \rfloor$ in the interval $I(p)$. They use a bitvector W to identify the endpoints of runs within each interval. Then, they compute the median μ of the medians of these sequences, and partition the runs based on μ. After partitioning, they recurse on the two sets of runs, sending *select* queries to the appropriate side of the recursion. To maintain the invariant on run length on the recursions, they merge short runs of the same size optimally until all but ℓ of the runs are again of length between ℓ and 2ℓ.

To perform the operation A.*select*(s), they first use bitvector V to identify the interval I containing s. If $|I| \le 4\ell^2$, they sort the interval I (making all elements of I pivots) and answer the query s. Otherwise, they compute the value of ℓ for the current interval, and proceed as in Kaligosi et al. [KMMS05] to answer the query s.

3 Optimal Online Dynamic Multiselection

In this section, we support insertions and deletions on the array, as well as selection and search queries. We are originally given the unsorted list A. To support *insert* and *delete* efficiently, we maintain newly-inserted elements in a separate data structure, and mark deleted elements in A. These *insert* and *delete* operations are occasionally merged to make the array A up-to-date. Let A$'$ denote the current array with length N'. We support two additional operations:

- *insert*(a), which inserts a into A$'$, and;
- *delete*(i), which deletes the ith sorted entry from A$'$.

3.1 Preliminaries

Our solution uses the *dynamic bitvector* of Hon et al. [HSS03]. This structure supports the following operations on a dynamic bitvector V. The $rank_b(i)$ operation tells the number of b bits up to the ith position in V. The $select_b(i)$ operation gives the position in V of the ith b bit. The $insert_b(i)$ operation inserts bit b in the ith position. The $delete(i)$ operation deletes the bit in the ith position. The $flip(i)$ operation flips the bit in the ith position.

Note that one can determine the ith bit of V by computing $rank_1(i) - rank_1(i-1)$. (For convenience, we assume that $rank_b(-1) = 0$.) The result of Hon et al. [HSS03, Theorem 1] can be re-stated as follows, for the case of maintaining a dynamic bit vector (the result of [HSS03] is stated for a more general case).

Lemma 2 ([HSS03]). *Given a bitvector V of length N, there exists a data structure that takes $N + o(N)$ bits and supports $rank_b$ and $select_b$ in $O(\log_t N)$ time, and insert, delete and flip in $O(t)$ time, for any t where $(\log N)^{O(1)} \leq t \leq N$. This structure assumes access to a precomputed table of size N^ϵ, for any fixed $\epsilon > 0$.*

All the pivots (and their positions) generated during *select, search, insert*, and *delete* operations on array A are maintained as in Barbay et al. [BGJ+13] using a bitvector V. In addition, we also maintain two bitvectors, each of length N': (i) an *insert bitvector* I such that $I[i] = 1$ if and only if $A'[i]$ is newly inserted, and (ii) a *delete bitvector* D such that if $D[i] = 1$, the ith element in A has been deleted. If a newly inserted item is deleted, it is removed from I directly. Both I and D are implemented as instances of the data structure described in Lemma 2.

We maintain the values of the newly inserted elements in a balanced binary search tree T. The inorder traversal of the nodes of T corresponds to the increasing order of their positions in A'. We support the following operations on this tree: (i) given an index i, return the element corresponding to the ith node in the inorder traversal of T, and (ii) insert/delete an element at a given inorder position. By maintaining the subtree sizes of the nodes in T, these operations can be performed in $O(\log N)$ time without having to perform any comparisons between the elements.

Our preprocessing steps are the same as in the static case. In addition, bitvectors I and D are each initialized to N **0**s. The tree T is initially empty.

After performing $|A|$ *insert* and *delete* operations, we merge all the elements in T with the array A, modify the bitvector B appropriately, and reset the bitvectors I and D (with all zeroes). This increases the amortized cost of the *insert* and *delete* operations by $O(1)$, without requiring additional comparisons.

3.2 Dynamic Online Multiselection

We now describe how to support the operators $A'.insert(a)$, $A'.delete(i)$, $A'.select(i)$, and $A'.search(a)$.

$A'.insert(a)$. First, we search for the appropriate unsorted interval $[\ell, r]$ containing a, using a binary search on the original (unsorted) array A. Now perform $A.search(a)$ on interval $[\ell, r]$ (choosing which subinterval to expand based on the insertion key a) until a's exact position j in A is determined. The original array A must have chosen as pivots the elements immediately to its left and right (positions $j-1$ and j in array A); hence, one never needs to consider newly-inserted pivots when choosing subintervals. Insert a in sorted order in T at position $\mathtt{I}.select_1(j)$ among all the newly-inserted elements. Calculate $j' = \mathtt{I}.select_0(j)$, and set a's position to $j'' = j' - \mathtt{D}.rank_1(j')$. Finally, we update our bitvectors by performing $\mathtt{I}.insert_1(j'')$ and $\mathtt{D}.insert_0(j'')$. Note that, apart from the $search$ operation, no other operation in the insertion procedure performs comparisons between the elements.

$A'.delete(i)$. Compute $i' = \mathtt{D}.select_0(i)$. If i' is newly-inserted (i.e., $\mathtt{I}[i'] = 1$), then remove the node (element) with inorder number $\mathtt{I}.rank_1(i')$ from T. Perform $\mathtt{I}.delete(i')$ and $\mathtt{D}.delete(i')$. If instead i' is an older entry, perform $\mathtt{D}.flip(i')$. In other words, we mark position i' in A as deleted even though the corresponding element may not be in its proper place.[4]

$A'.select(i)$. If $\mathtt{I}[i] = 1$, return the element corresponding to the node with inorder number $\mathtt{I}.rank_1(i)$ in T. Otherwise, compute $i' = \mathtt{I}.rank_0(i) - \mathtt{D}.rank_1(i)$, and return $A.select(i')$.

$A'.search(a)$. Search for the unsorted interval $[\ell, r]$ containing a using a binary search on the original (unsorted) array A. Then perform $A.search(a)$ on interval $[\ell, r]$ until a's exact position j is found. If a appears in A (which we discover through $search$), we need to check whether it has been deleted. We compute $j' = \mathtt{I}.select_0(j)$ and $j'' = j' - \mathtt{D}.rank_1(j')$. If $\mathtt{D}[j'] = 0$, return j''. Otherwise, it is possible that the item has been newly-inserted. Compute $p = \mathtt{I}.rank_1(j')$, which is the number of newly-inserted elements that are less than or equal to a. If $T[p] = a$, then return j''; otherwise, return failure.

We now analyze the above algorithm to show that the above algorithm achieves the comparison cost as in Theorem 1, and a running time as in Corollary 1.

Theorem 1 (Dynamic Online Multiselection). *Given an unsorted array A' of N elements, we provide a dynamic online algorithm that can support $q \in O(N)$ select, search, insert, and delete operations, of which r are search, insert, or delete, using at most $\mathcal{B}(S_q)(1 + o(1)) + O(N + r \log N)$ comparisons.*

Proof. Let N' denote the current length of the dynamic array A', after a sequence of queries and insertions. Let Q be the sequence of q selection operations performed (either directly or indirectly through other operations) on A', ordered by time of arrival. Let S_q be the queries of Q, ordered by position. We now analyze

---

[4] If a user wants to delete an item with value a, one could simply search for it first to discover its rank, and then delete it using this function.

the number of comparisons performed by a sequence of queries and *insert* and *delete* operations.

We consider the case when the number of *insert* and *delete* operations is less than N. In other words, we are between two re-buildings of our dynamic data structure. Recall that each of the r *search*, *insert*, and *delete* operations in the sequence will perform a constant number of search operations. To execute these searches, we require $O(r \log N')$ comparisons. Note that our algorithm does not perform any comparisons for *delete*(i) operations, until some other query is in the same interval as i. The deleted element will participate in the other costs (merging, pivot-finding, and partitioning) for these other queries, but its contribution can be bounded by $O(\log N)$, which we have as a credit.

Since a *delete* operation does not perform any additional comparisons beyond those needed to perform a *search*, we assume that all the updates are insertions in the rest of this section. Since each inserted element becomes a pivot immediately, it does not contribute to the comparison cost of any other *select* operation. Also, note that in the algorithm of Lemma 1, no pivot is part of a run and hence cannot affect the choice of any future pivot.

Since Q is essentially a set of q selection queries, we can bound its total comparison cost for selection queries by Lemma 1, which gives a bound of $\mathcal{B}(S_q)(1 + o(1)) + O(N)$. This proves the theorem.   □

By modifying Theorem 1 to account for the costs of the dynamic bit vector from Lemma 2, we obtain the following result.

Corollary 1. *Given a dynamic array $\mathbf{A}'$ of N original elements, there exists a dynamic online data structure that can support $q = O(N)$ select, search, insert, and delete operations, of which r are search, insert and delete and u of which are insert and delete, we provide a deterministic online algorithm that uses time within $O(\mathcal{B}(S_q) + q \log_t N + r \log N + ut)$, for any t where $(\log N)^{O(1)} \le t \le N$.*

4   External Online Multiselection

In the external memory model, we consider only two memory levels: the internal memory of size M, and the (unbounded) disk memory, which operates by reading and writing data in blocks of size B. We refer to the number of items of the input by N. For convenience, we define $n = N/B$ and $m = M/B$ as the number of blocks of input and memory, respectively. We make the reasonable assumption that $1 \le B \le M/2$. In this model, we assume that each I/O read or write is charged one unit of time, and that an internal memory operation is charged no units of time. To achieve the optimal sorting bound of $SortIO(N) \in \Theta(n \log_m n)$ in this setting, it is necessary to make the *tall cache* assumption [BF03]: $M \in \Omega(B^{1+\epsilon})$, for some constant $\epsilon > 0$, and we will make this assumption for the remainder of the paper.

Suppose we are given an unsorted array $\mathbf{A}$ of length N stored in $n = N/B$ blocks in the external memory. The techniques that we use in internal memory are not immediately applicable to the external memory model: in the extreme case where we have $q = N$ queries, the internal memory solution would require $O(n \log_2(n/m))$ I/Os. This compares poorly to the optimal $O(n \log_m n)$ I/Os performed by the optimal mergesort algorithm for external memory.

4.1 A Lower Bound for Multiselect in External Memory

As in the case of internal memory, the lower bound on the number of I/Os required to perform a given set of selection queries can be obtained by subtracting the number of I/Os required to sort the elements between the 'query gaps' from the sorting bound. More specifically, let $S_t = \{s_i\}$ be the first t queries from a query sequence R, sorted by position, and for $1 \leq i \leq t$, let $\Delta_i^{S_t} := s_{i+1} - s_i$ be the query gaps, as defined in Section 2.1. Then the lower bound on the number of I/Os required to support the queries in S_t is given by

$$\mathcal{B}_m(S_t) := n \log_m n - \sum_{i=0}^{t} \left(\Delta_i^{S_t} / B \right) \log_m \left(\Delta_i^{S_t} / B \right) - O(n),$$

where we assume that $\log_m \left(\Delta_i^{S_t} / B \right) = 0$ when $\Delta_i^{S_t} < mB = M$ in the above definition. Note that $\mathcal{B}_m(S_t) \in \Omega(n)$ for all $t \geq 1$.

4.2 Partitioning in External Memory

The main difference between our algorithms for internal and external memory is the partitioning procedure. In the internal memory algorithm, we partition the values according to a single pivot, recursing on the half that contains the answer. In the external memory algorithm, we modify this binary partition to a d-way partition, for some $d \in \Theta(m)$, by finding a sample of d "roughly equidistant elements." The next lemma describes how to find such a sample, and then partition the range of values into $d + 1$ subranges with respect to the sample.

As is usual in the external memory model [AV88], we assume that $B \in \Omega(\log_m n)$—which allows us to store a pointer to a memory block of the input using a constant number of blocks. This is similar to the word-size assumption for the transdichotomous word RAM model [FW93]. In addition, the algorithm of Sibeyn [Sib06] only works under this assumption, though this is not explicitly mentioned.

Lemma 3. *Given an unsorted array A containing N elements in external memory and an integer parameter $d < m/2$, one can perform a d-way partition in $O(n+d)$ I/Os, such that the size of each partition is in the range $[n/(2d), 3n/(2d)]$.*

Proof. Let $s = \lfloor \sqrt{m/4} \rfloor$. We perform the s-way partition described in [AV88] to obtain $s + 1$ super-partitions. We reapply the s-way partitioning method to each super-partition to obtain $d < m/2$ partitions in total.

Finally, our algorithm scans the data, keeping one input block and $d+1$ output blocks in main memory. An output block is written to external memory when it is full, or when the scan is complete. The algorithm performs n I/O to read the input, and at most $(n + d + 1)$ I/Os to write the output into $d + 1$ partitions, thus showing the result. $\qquad\square$

4.3 Algorithm Achieving $O(\mathcal{B}_m(S_q))$ I/Os

We now show that our lower bound is asymptotically tight, by describing an $O(1)$-competitive algorithm.

Theorem 2 (External Static Online Multiselection). *Given an unsorted array A occupying n blocks in external memory, we provide a deterministic algorithm that supports a sequence R of q online selection queries using $O(\mathcal{B}_m(S_q))$) I/Os under the condition that $B \in \Omega(\log_m n)$.*

Proof. Our algorithm uses the same approach as the simple internal memory algorithm described for the static version of the problem [BGJ+13], except that it chooses $d - 1$ pivots at once. In other words, each node v of the pivot tree T containing Δ_v elements has a branching factor of d. We subdivide its Δ_v elements into d partitions using Lemma 3. This requires $O(\delta_v + d)$ I/Os, where $\delta_v = \Delta_v/B$.

We also maintain the bitvector V of length N, as described before. For each $A.select(i)$ query, we access position $V[i]$. If $V[i] = 1$, return $A[i]$, else scan left and right from the ith position to find the endpoints of this interval I_i using $|I_i|/B$ I/Os. The analysis of the remaining terms follows directly from the internal memory algorithm, giving $O(\mathcal{B}_m(S_q)) + O(n) = O(\mathcal{B}_m(S_q))$ I/Os. $\qquad\square$

To add support for the search operator, instead of taking $O(\log N)$ time performing binary search on the blocks of V, we build a B-tree T maintaining all pivots from A. (During preprocessing, we insert $A[1]$ and $A[n]$ into T.) The B-tree T will be used to support *search* queries in $O(\log_B N)$ I/Os instead of $O(\log N)$ I/Os. We modify the proof of Theorem 2 to obtain the following result.

Corollary 2. *Given an unsorted array A occupying n blocks in external memory, we provide a deterministic algorithm that supports a sequence R of q online selection and search queries using $O(\mathcal{B}_m(S_q) + q \log_B N)$ I/Os under the condition that $B \in \Omega(\log_m n)$.*

Proof. The first term follows directly from the proof of Theorem 2. Now we explain the source of the second term, $q \log_B N$).

We build a B-tree T maintaining all pivots from A. (During preprocessing, we insert $A[1]$ and $A[n]$ into T.) Naively, for q queries, we must insert $qm \log_m N$ new pivots into T. The B-tree construction for these pivots would require $O(\min\{qm (\log_m N), N\}(\log_B N))$ I/Os, which is prohibitive.

Instead, we notice that the pivots for an individual query z are all inserted in some unsorted interval $I_z = [l, r]$, where l and r are consecutive leaves of the pivot tree T (in left-to-right level order). For z, we may spend $\log_B(\min\{qm(\log_m N), N\}) \in O(\log_B N)$ I/Os navigating to I_z using T. Our approach is to insert all $O(m \log_m N) = O((M/B) \log_m N) \subseteq O(M)$ pivots within I_z in a single batched manner. This process can easily be done in a bottom-up fashion by merging nodes in the tree T of an implicit B-tree T' for the $O(M)$ pivots using $O(m)$ I/Os.

Thus, we have $O(\min\{qm\log_m N, N\})$ pivots in T, and using the batched insertion process above, we only need $O(\min\{qm(\log_m N)/B, N/B\}) = O(\min\{qm, n\})$ I/Os. We must also add $O(q\log_B N)$ I/Os to navigate to the correct interval for each query.

For q queries, the algorithm takes $O(\mathcal{B}_m(S_q)) + O(n) + O(q\log_B N) = O(\mathcal{B}_m (S_q) + q\log_B N)$ I/Os, matching the result. $\qquad\square$

Combining the ideas from Corollary 2 and Theorem 1, we can dynamize the above algorithm. The proof follows from the fact that we can maintain the bit vectors I and D described in the multiselection algorithms of Section 3 using a B-tree in external memory.

Theorem 3 (External Dynamic Online Multiselection). *Given an unsorted array A occupying n blocks in external memory, we provide a deterministic algorithm that supports a sequence R of q online select, search, insert, and delete operations using $O(\mathcal{B}_m(S_q) + q\log_B N)$ I/Os under the condition that $B \in \Omega(\log_m n)$.*

Note that if $q \in O(\mathcal{B}_m(S_q)/\log_B N)$, then Corollary 2 and Theorem 3 require only $O(\mathcal{B}_m(S_q))$ I/Os, matching the bounds from Theorem 2. Hence, our results are asymptotically optimal when $\mathcal{B}_m(S_q)/q = O(\log_B N)$.

References

[AV88] Aggarwal, A., Vitter, J.S.: The input/output complexity of sorting and related problems. Commun. ACM 31(9), 1116–1127 (1988)

[BF03] Brodal, G., Fagerberg, R.: On the limits of cache-obliviousness. In: Proceedings of the ACM Symposium on Theory of Computing, pp. 307–315 (2003)

[BGJ+13] Barbay, J., Gupta, A., Jo, S., Rao, S.S., Sorenson, J.: Theory and implementation of online multiselection algorithms. In: Bodlaender, H.L., Italiano, G.F. (eds.) ESA 2013. LNCS, vol. 8125, pp. 109–120. Springer, Heidelberg (2013)

[DM81] Dobkin, D.P., Ian Munro, J.: Optimal time minimal space selection algorithms. J. ACM 28(3), 454–461 (1981)

[FW93] Fredman, M.L., Willard, D.E.: Surpassing the information theoretic bound with fusion trees. J. Comput. Syst. Sci. 47(3), 424–436 (1993)

[Hoa61] Hoare, C.A.R.: Algorithm 65: find. Commun. ACM 4(7), 321–322 (1961)

[HSS03] Hon, W.-K., Sadakane, K., Sung, W.-K.: Succinct data structures for searchable partial sums. In: Proceedings of the International Symposium on Algorithms and Computation, pp. 505–516 (2003)

[JM10] Jiménez, R.M., Martínez, C.: Interval Sorting. In: Abramsky, S., Gavoille, C., Kirchner, C., Meyer auf der Heide, F., Spirakis, P.G. (eds.) ICALP 2010. LNCS, vol. 6198, pp. 238–249. Springer, Heidelberg (2010)

[KMMS05] Kaligosi, K., Mehlhorn, K., Munro, J.I., Sanders, P.: Towards optimal multiple selection. In: Caires, L., Italiano, G.F., Monteiro, L., Palamidessi, C., Yung, M. (eds.) ICALP 2005. LNCS, vol. 3580, pp. 103–114. Springer, Heidelberg (2005)

[MR86] Motwani, R., Raghavan, P.: Deferred data structuring: Query-driven preprocessing for geometric search problems. In: Symposium on Computational Geometry, pp. 303–312 (1986)

[Pro95] Prodinger, H.: Multiple quickselect - Hoare's find algorithm for several elements. Inf. Process. Lett. 56(3), 123–129 (1995)

[Sib06] Sibeyn, J.F.: External selection. J. Algorithms 58(2), 104–117 (2006)

[YTC90] Ching, Y.-T., Mehlhorn, K., Smid, M.H.M.: Dynamic deferred data structuring. Information Processing Letters 35(1), 37–40 (1990)

Competitive Analysis
for Multi-objective Online Algorithms

Morten Tiedemann*, Jonas Ide, and Anita Schöbel

DFG RTG 1703, Institute for Numerical and Applied Mathematics,
University of Göttingen, Lotzestr. 16-18, D-37083 Göttingen, Germany
{m.tiedemann,j.ide,schoebel}@math.uni-goettingen.de

Abstract. So far, the concept of competitive analysis for online problems is in general applied to single-objective online problems. However, many online problems can be extended to multi-objective online problems in a natural way, but a uniform theory for the analysis of these problems is not provided in the literature. We expand the concept of competitive analysis to multi-objective online problems and achieve a consistent framework for the analysis of multi-objective online problems. Furthermore, we analyze the multi-objective time series search problem and present deterministic algorithms with best possible competitive ratios.

Keywords: competitive analysis, online optimization, multi-objective optimization, time series search.

1 Introduction

Online optimization is a helpful tool for various single-objective decision problems. However, online problems may also be of multi-objective nature. Imagine you want to sell your antique car and you are facing a sequence of offers by different people which you have to reject or accept immediately since the potential buyers are not willing to wait for a decision at a later time. On the one hand, you are eager to reach a high price, on the other hand, you want to know your antique car in safe keeping. Consequently, you evaluate each offer not only by the price, but also by your appreciation for the potential buyer. This leads to a bi-objective online problem.

In general, the concept of competitive analysis for online problems is only applied to single-objective online problems. However, the decision process of many online problems is subject to multiple objectives in real-world situations, but a uniform theory for the analysis of multi-objective online problems is not provided in the literature. In this work, we close this gap and expand the concept of competitive analysis to multi-objective online problems.

1.1 Previous Work

For an overview on the topic of competitive analysis for single-objective online problems, we refer to the textbook by Borodin and El-Yaniv [2], and, for an

---

* Corresponding author.

M.S. Rahman and E. Tomita (Eds.): WALCOM 2015, LNCS 8973, pp. 210–221, 2015.

overview on the topic of multi-objective optimization, we refer to the textbook by Ehrgott [3].

To the best of our knowledge, there exists no general definition of competitive analysis for multi-objective online problems. However, some approaches in the field of online optimization are related to multi-objective online optimization.

Resource augmentation refers to a relaxed notion of competitive analysis, in which the online player is allowed more resources than the adversary, see for example [7],[9]. Competitive ratios are then stated with respect to a fixed resource augmentation, which can be seen as a bi-objective online problem. Furthermore, for preemptive online scheduling problems, the trade-off between the competitive ratio and the cost of preemption is considered for example in [8], which is also a bi-objective setting. However, the competitive ratios considered for these problems are the classical single-objective competitive ratios, while we derive a notion for a multi-objective competitive ratio, i.e., for defining a competitive ratio for a multi-objective online problem.

Online algorithms are closely related to approximation algorithms. In the field of multi-objective optimization, a solution x of a multi-objective maximization problem is called a ρ-approximation of a solution x' if $f_i(x) \geq \rho \cdot f_i(x')$ for $i = 1, \ldots, n$, where f_i, $i = 1, \ldots, n$ are the components of the objective function and $0 < \rho \leq 1$. A set of feasible solutions X' is called a ρ-approximation of a set of efficient solutions if, for every feasible solution x, X' contains a feasible solution x' that is a ρ-approximation of x, see for example [1],[5]. In our approach, the output generated by an online algorithm is a single solution instead of a set of solutions, due to the online nature of the problem. Therefore, the competitiveness of a multi-objective online algorithm as introduced in this work is not directly deducible from the concept of ρ-approximation, but in some cases a close relation between our approach of competitive analysis for multi-objective online algorithms and multi-objective approximation algorithms is given.

1.2   Our Contribution

We introduce a general framework for the competitive analysis of multi-objective online problems which expands the known theory of competitive analysis for single-objective online problems to the multi-objective case. The fact that solutions to multi-objective optimization problems are sets rather than singletons as in the case of single objective optimization problems, requires a proper adaptation of the definition of competitiveness for multi-objective online problems.

For $c = (c_1, \ldots, c_n)^\mathsf{T}$, we define a multi-objective online algorithm ALG as c-competitive if, for each input sequence, there exists an efficient solution to the offline problem for which ALG is c_i-competitive (in the original sense) in the i-th component for $i = 1, \ldots, n$. Additionally, a multi-objective online algorithm ALG is labeled as *strongly* c-competitive if, for each input sequence, ALG is c_i-competitive (in the original sense) in the i-th component for $i = 1, \ldots, n$ for *all* efficient solutions to the offline problem. For $f : \mathbb{R}^n \mapsto \mathbb{R}$, the infimum over the set of all values $f(c)$ such that ALG is (strongly) c-competitive is then defined as the *(strong) competitive ratio with respect to f* of ALG.

Furthermore, we analyze the multi-objective time series search problem by means of the introduced notions of competitiveness with respect to functions $f_1(c) = \max_{i=1,\ldots,n} c_i$, $f_2(c) = \frac{1}{n}\sum_{i=1}^{n} c_i$, and $f_3(c) = \sqrt[n]{\prod_{i=1}^{n} c_i}$, and present deterministic algorithms featuring the best possible (strong) competitive ratios.

The rest of this paper is structured as follows. In Sect. 2, we formally introduce multi-objective online problems and the concept of competitive analysis for multi-objective online problems. In Sect. 3, the multi-objective time series search problem is defined and deterministic competitive online algorithms as well as matching lower bounds on the competitive ratios with respect to f_1, f_2, and f_3 achievable by any deterministic algorithm are presented. Finally, Sect. 4 contains a summary of our contribution and directions for future research.

2 Competitive Analysis for Multi-Objective Online Algorithms

In this section, we first introduce the notion of a multi-objective online problem and, secondly, define the concept of competitive analysis for multi-objective online problems.

2.1 Multi-Objective Online Problems

In the following, we define the concept of competitive analysis for multi-objective online problems with respect to minimization problems. If not mentioned otherwise, the definition for the corresponding maximization problem is analogous.

First of all, we define a *multi-objective optimization problem* $\mathcal{P}$ as a triple $(\mathcal{I}, \mathcal{X}, f)$, consisting of a set of inputs $\mathcal{I}$, a set of feasible outputs (or solutions) $\mathcal{X}(I) \in \mathbb{R}^n$ associated with every input $I \in \mathcal{I}$, and the objective function f given as $f : \mathcal{I} \times \mathcal{X} \mapsto \mathbb{R}^n_+$ where, for $\boldsymbol{x} \in \mathcal{X}(I)$, $f(I, \boldsymbol{x})$ represents the objective value of the solution x with respect to input $I \in \mathcal{I}$.

Given input $I \in \mathcal{I}$, an *algorithm* ALG for a multi-objective optimization problem $\mathcal{P}$ computes a feasible solution $\text{ALG}[I] \in \mathcal{X}(I)$. The objective associated with this feasible output is denoted by $\text{ALG}(I) = f(I, \text{ALG}[I])$. According to [3, p. 24], a feasible solution $\hat{\boldsymbol{x}} \in \mathcal{X}(I)$ is called efficient if there is no other $\boldsymbol{x} \in \mathcal{X}(I)$ such that $f(I, x) \preceq f(I, \hat{x})$, where $\preceq$ denotes a componentwise order, i.e., for $x, y \in \mathbb{R}^n$, $x \preceq y :\Leftrightarrow x_i \leq y_i$, for $i = 1, \ldots, n$, and $x \neq y$. An *optimal algorithm* OPT for $\mathcal{P}$ is such that, for all inputs $I \in \mathcal{I}$, $\text{OPT}[I]$ is the set of efficient solutions to $\mathcal{P}$, i.e., $\text{OPT}[I] = \{\boldsymbol{x} \in \mathcal{X}(I) \mid \boldsymbol{x} \text{ is an efficient solution to } \mathcal{P}\}$. The objective associated with a solution $\boldsymbol{x} \in \text{OPT}[I]$ is denoted by $\text{OPT}(\boldsymbol{x})$.

The definition of a multi-objective online problem is now given analogously to the definition of a single-objective online optimization problem in [2, p. 2]. Accordingly, *multi-objective online problems* are multi-objective optimization problems in which the input is revealed bit by bit and an output must be produced in an online manner, i.e., after each new bit of input a decision affecting the output must be made.

2.2 The Competitive Ratio and Competitiveness

The study of online problems is concerned with assessing the quality of corresponding online algorithms and, ultimately, the question of which is the best algorithm. We carry this leading question forward to multi-objective online problems. In the following, we list conditions that are supposed to be met by an appropriate measure for the quality of multi-objective online algorithms:

Condition 1: Worst Case Model. Just as in the case of competitive analysis for single-objective algorithms (cf. [6, p.4]), we aim for a worst case model for multi-objective competitive analysis that holds for any distribution in order to avoid the problems of probabilistic models.

Condition 2: Worst Case Ratio. Furthermore, a standard worst case analysis of multi-objective online algorithms leads to the same pitfall as in the single-objective case (cf. [6, p.3]): Due to the incomplete knowledge of the online algorithm, it is often possible to ensure that each decision made by an online algorithm is the worst possible decision with respect to all components. For example, consider the multi-objective time series search problem (see Sect. 3): If a sequence consisting only of the minimal price vector is revealed, the online player always ends up with the minimal price vector regardless of his strategy.

Therefore, following the underlying idea of competitive analysis, it is desirable to consider the ratio of the algorithm's performance and the optimal performance in every component on the same problem instance.

Condition 3: Independence from Efficient Solutions. Usually, the solution to a multi-objective optimization problem is given by a set of efficient solutions, compare, e.g., Ehrgott [3]. However, due to the online nature of our approach and the corresponding urge to obtain an autonomous algorithm, we assume a multi-objective online algorithm to compute a single solution instead of a set of solutions. The competitive ratio should nevertheless be independent of a particular solution chosen from the set of efficient solutions of the offline problem.

Condition 4: Total Order. So as to compare different multi-objective online algorithms, a total order on the competitive ratio of multi-objective online algorithms is necessary.

These requirements lead us to the following definition of c-competitiveness for multi-objective online algorithms:

Definition 1. *A multi-objective online algorithm* ALG *is c-competitive if for all finite input sequences I there exists an efficient solution $x \in \text{OPT}[I]$ such that*

$$\text{ALG}(I)_i \leq c_i \cdot \text{OPT}(x)_i + \alpha_i, \text{ for } i = 1, \ldots, n\,,$$

where $c = (c_1, \ldots, c_n)^\mathsf{T}$ and $\alpha \in \mathbb{R}^n$ is a constant vector independent of I.

Note that c is a vector instead of a scalar as in the classic definition of competitiveness for single-objective online algorithms. A multi-objective online algorithm which accomplishes this postulation even for all efficient solutions is called *strongly c-competitive*:

Definition 2. *A multi-objective online algorithm* ALG *is strongly c-competitive if for all finite input sequences* I *and all efficient solutions* $\boldsymbol{x} \in \mathrm{OPT}[I]$,

$$\mathrm{ALG}(I)_i \leq c_i \cdot \mathrm{OPT}(\boldsymbol{x})_i + \alpha_i, \ \text{for } i = 1, \dots, n,$$

where $c = (c_1, \dots, c_n)^\mathsf{T}$ *and* $\alpha \in \mathbb{R}^n$ *is a constant vector independent of* I.

Applying these definitions to single-objective problems results in the classical single-objective competitive ratio for both Definition 1 and Definition 2. Obviously, every strongly c-competitive multi-objective online algorithm is also c-competitive. For maximization problems, the inequalities in Definitions 1 and 2 are replaced by $\mathrm{ALG}(I)_i \geq 1/c_i \cdot \mathrm{OPT}(\boldsymbol{x})_i + \alpha_i$.

The definition of competitiveness for multi-objective online algorithms is a worst case ratio due to the consideration of all finite input sequences, as required by Conditions 1 and 2. Furthermore, the definition takes the set of all efficient offline solutions into account and hence does not rely on a particular efficient solution, as demanded by Condition 3.

In order to achieve a comparable competitive ratio of multi-objective online algorithms as demanded by Condition 4, a total order on the competitiveness of an online algorithm is necessary. This gives rise to the following definition of the competitive ratio for multi-objective online algorithms:

Definition 3. *Let* $f : \mathbb{R}^n \mapsto \mathbb{R}$. *The infimum over the set of all values* $f(c)$ *such that* ALG *is (strongly) c-competitive is called* the (strong) competitive ratio with respect to f *of* ALG *and is denoted by* $(\mathcal{R}_s^f(\mathrm{ALG}))$ $\mathcal{R}^f(\mathrm{ALG})$

The choice of the function f grants a certain degree of freedom that is left to the analyst of the online algorithm (in the style of the decision maker in the field of multi-objective optimization). However, f has to be chosen such that $f(c) \leq f(\hat{c})$ if $c_i \leq \hat{c}_i$ for $i = 1, \dots, n$ in order to guarantee a reasonable setting.

In this work, we consider three intuitive choices for the function f. First of all, consider f_1 given as $f_1(c) := \max_{i=1,\dots,n} c_i$. By this choice, the competitive ratio is guaranteed for each component of the objective function. We label this choice as *worst-component competitiveness*. Further possible choices for the function f are given by $f_2(c) := \frac{1}{n} \sum_{i=1}^n c_i$ and $f_3(c) := \sqrt[n]{\prod_{i=1}^n c_i}$. In these cases, the arithmetic and geometric mean value of the components' competitive ratios is taken, which is why these choices are labeled as *arithmetic-* and *geometric-mean-component competitiveness*.

The definitions of a (strongly) c-competitive algorithm and the (strong) competitive ratio with respect to f are given in the same way for *randomized* multi-objective online algorithms.

3   The Multi-objective Time Series Search Problem

In this section, the concept of competitive analysis for multi-objective online problems is applied to the classic time series search problem, where an online player is searching for the maximum (or minimum) price in a sequence of prices. At the beginning of each time period $t = 1, \ldots, T$, a price p_t is revealed to the online player and the player must decide whether to accept or reject the price p_t. When the player accepts a price p_t, the game ends and the return for the player is p_t.

Within the framework of competitive analysis, the time series search problem is initially investigated in [4] and competitive search algorithms are provided. Here, it is assumed that prices are chosen from the real interval $[m, M]$, where $0 < m \leq M$ and the online player can always end the game by accepting the minimum price m; otherwise the adversary is too powerful and there exist no competitive algorithms for the problem. The ratio $\varphi = M/m$ is defined as the fluctuation ratio of possible prices. The authors prove that, if only the fluctuation ratio φ is known to the online player, no better ratio than the trivial one of φ is achievable. Therefore, suppose that both m and M are known to the online player. Then, the best possible deterministic online algorithm is the reservation price policy RPP, where the algorithm accepts the first price greater than or equal to $p^\star = \sqrt{Mm}$, achieving a competitive ratio of $\sqrt{\varphi}$, as shown in [4].

The competitive ratio can be improved by means of randomized algorithms. Assume that $\varphi = 2^k$. The randomized algorithm EXPO presented in [4], which chooses the reservation price policy with reservation price $m2^l$, $l = 0, \ldots, k - 1$ with probability $1/k$ before the start of the game, is $\mathcal{O}\left(log(\varphi)\right)$-competitive, which is within a constant factor of the best possible competitive ratio. This result also holds when φ is not a power of 2. The best possible competitive ratio is achieved by algorithms that obey threat-based policies, cf. [4].

In order to investigate the time series search problem in a multi-objective setting, a price *vector* is introduced, i.e., at each time period $t = 1, \ldots, T$, a request $r_t \in \mathbb{R}^n_+$ is revealed to the online player, where $r_t = \left(p_t^1, \ldots, p_t^n\right)^{\mathsf{T}}$, and the player must decide whether to accept or reject r_t. When the player accepts a price vector r_t, the game ends and the return for the player is r_t. It is assumed that, for $i = 1, \ldots, n$, p_t^i is chosen from the real interval $[m_i, M_i]$, where $0 < m_i \leq M_i$. For $i = 1, \ldots, n$, the ratios $\varphi_i = M_i/m_i$ are defined as the fluctuation ratios of possible prices for the price component i. The online player can always end the game by accepting the minimum price vector $\left(m_1, \ldots, m_n\right)$. Without loss of generality, we assume for the fluctuation ratios that $M_1/m_1 \geq M_2/m_2 \geq \cdots \geq M_n/m_n$.

3.1   Worst-Component Competitive Analysis

In this section, a competitive analysis with respect to $f_1(c) = \max_{i=1,\ldots,n} c_i$ for the multi-objective time series search problem is presented, i.e., a worst-component competitive analysis. Consider the algorithm RPP-HIGH, which concentrates only on the component with the highest fluctuation ratio and achieves the best possible competitive ratio (see also Fig. 1):

---

for $t = 1, \ldots, T$ **do**
| Accept $r_t = \left(p_t^1, \ldots, p_t^n\right)^\mathsf{T}$ if $p_t^1 \geq \sqrt{m_1 M_1}$.
end

Algorithm 1. Multi-objective reservation price policy RPP-HIGH

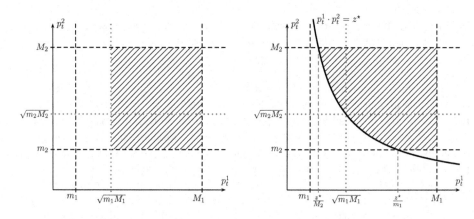

Fig. 1. Acceptance region of RPP-HIGH   **Fig. 2.** Acceptance region of RPP-MULT

Theorem 1. *The strong competitive ratio with respect to* $f_1(c) = \max_{i=1,\ldots,n} c_i$ *of* RPP-HIGH *is given by*

$$\mathcal{R}_s^{f_1}(\text{RPP-HIGH}) = \max \left\{ \sqrt{\frac{M_1}{m_1}}, \frac{M_2}{m_2} \right\}.$$

Proof. We distinguish two cases with respect to the sequence $\sigma = (r_1, \ldots, r_T)$ revealed by the adversary:

Case 1: there exists a request $r_{t'}$ **with** $p_{t'}^1 \geq \sqrt{m_1 M_1}$.
    In this case, the online player accepts the first request r_t with $p_t^1 \geq \sqrt{m_1 M_1}$. However, the adversary is able to reveal a further request r_j with $p_j^i = M_i$ for $i = 1, \ldots, n$, which is then the only efficient offline solution and, therefore, the optimal solution for the adversary, i.e., OPT $= \left(M_1, \ldots, M_n\right)^\mathsf{T}$.
    In the worst case with respect to all sequences, the request r_t accepted by the online player is such that $p_t^1 = \sqrt{m_1 M_1}$ and $p_t^i = m_i$ for $i = 2, \ldots, n$. According to Definition 3, the competitive ratio is in this case given by

$$\max \left\{ \frac{M_1}{\sqrt{m_1 M_1}}, \frac{M_2}{m_2}, \ldots, \frac{M_n}{m_n} \right\} = \max \left\{ \sqrt{\frac{M_1}{m_1}}, \frac{M_2}{m_2} \right\}, \qquad (1)$$

since $M_1/m_1 \geq M_2/m_2 \geq \cdots \geq M_n/m_n$.

Case 2: for all r_t, $t = 1, \ldots, T$, **we have** $p_t^1 < \sqrt{m_1 M_1}$.

In this instance, the online player does not accept any request and has to settle in the worst case for the lower bounds in each component, i.e. ALG = $(m_1, \ldots, m_n)^\mathsf{T}$. The adversary is able to offer (and accept) any request r_j for which the first price component is smaller than but arbitrarily close to $\sqrt{m_1 M_1}$, i.e., $p_j^1 = \sqrt{m_1 M_1} - \epsilon$, $\epsilon > 0$. For the other components the upper bound is chosen, i.e., $p_j^i = M_i$ for $i = 2, \ldots, n$. Ignoring the ϵ, the competitive ratio is in this case given by

$$\max \left\{ \frac{\sqrt{m_1 M_1}}{m_1}, \frac{M_2}{m_2}, \ldots, \frac{M_n}{m_n} \right\} = \max \left\{ \sqrt{\frac{M_1}{m_1}}, \frac{M_2}{m_2} \right\}, \qquad (2)$$

since $M_1/m_1 \geq M_2/m_2 \geq \cdots \geq M_n/m_n$.

The analysis above holds for all efficient solutions. Thus, the strong competitive ratio $\mathcal{R}_s^{f_1}$(RPP-HIGH) results in

$$\mathcal{R}_s^{f_1}(\text{RPP-HIGH}) = \max \left\{ \sqrt{\frac{M_1}{m_1}}, \frac{M_2}{m_2} \right\}. \qquad (3)$$

□

With respect to a worst-component competitive analysis, RPP-HIGH is the best possible deterministic algorithm for the multi-objective time series search problem which is proven by the following theorem:

Theorem 2. *No deterministic online algorithm for the multi-objective time series search problem can achieve a smaller worst-component competitive ratio than* $\max \left\{ \sqrt{M_1/m_1}, M_2/m_2 \right\}$.

Proof. If $\sqrt{M_1/m_1} \leq M_2/m_2$, the adversary offers a request

$$r_1 = (m_1, M_2, m_3, \ldots, m_n)^\mathsf{T}.$$

If the online player rejects r_1, no further requests are revealed, the online player has to settle for the lower bounds $(m_1, \ldots, m_n)^\mathsf{T}$, and the adversary accepts r_1. In this case, the competitive ratio is given by the trivial competitive ratio M_1/m_1. Otherwise, if the online player accepts r_1, the adversary reveals another request $r_2 = (M_1, \ldots, M_n)^\mathsf{T}$ and accepts this request. Thus, the competitive ratio is in this case given by

$$\max \left\{ \frac{m_1}{m_1}, \frac{M_2}{m_2}, \frac{m_3}{m_3}, \ldots, \frac{m_n}{m_n} \right\} = \frac{M_2}{m_2} =: \theta_1. \qquad (4)$$

If $\sqrt{M_1/m_1} > M_2/m_2$, the adversary offers a request

$$r_1 = (\sqrt{m_1 M_1}, m_2, \ldots, m_n)^\mathsf{T}.$$

If the online player rejects r_1, no further requests are revealed, the online player has to settle for the lower bounds $(m_1, \ldots, m_n)^\mathsf{T}$, and the adversary accepts r_1. In this case, the competitive ratio is given by

$$\max \left\{ \frac{\sqrt{m_1 M_1}}{m_1}, \frac{m_2}{m_2}, \ldots, \frac{m_n}{m_n} \right\} = \sqrt{\frac{M_1}{m_1}} =: \theta_2 . \tag{5}$$

Otherwise, if the online player accepts r_1, the adversary reveals another request $r_2 = (M_1, \ldots, M_n)^\mathsf{T}$ and accepts this request. Thus, the competitive ratio is in this case given by

$$\max \left\{ \frac{M_1}{\sqrt{m_1 M_1}}, \frac{M_2}{m_2}, \ldots, \frac{M_n}{m_n} \right\} = \sqrt{\frac{M_1}{m_1}} =: \theta_3 . \tag{6}$$

By means of θ_1, θ_2, and θ_3, no deterministic algorithm for the time series search problem can achieve a smaller worst-component competitive ratio ratio than

$$\max \{\theta_1, \theta_2, \theta_3\} = \max \left\{ \sqrt{\frac{M_1}{m_1}}, \frac{M_2}{m_2} \right\} = \mathcal{R}^{f_1}(\text{RPP-HIGH}) . \tag{7}$$

$\square$

With respect to a worst-component competitive analysis, only the component with the highest fluctuation ratio is decisive, and, therefore, the best possible deterministic single-objective policy applied to this component achieves the best competitive ratio.

3.2   Mean-Component Competitive Analysis

In this section, we first present a competitive analysis with respect to $f_2(c) = \frac{1}{n} \sum_{i=1}^{n} c_i$ for the multi-objective time series search problem. For the sake of simplicity, we consider the case $n = 2$, i.e., the bi-objective time series search problem. The algorithm RPP-MULT achieves the best possible arithmetic-mean-component competitive ratio (see also Fig. 2):

---

for $t = 1, \ldots, T$ **do**
 | Accept $r_t = (p_t^1, p_t^2)^\mathsf{T}$ if $p_t^1 \cdot p_t^2 \geq z^\star$, where $z^\star = \sqrt{m_1 M_1 m_2 M_2}$.
end

Algorithm 2. Multi-objective reservation price policy RPP-MULT

Theorem 3. *The strong competitive ratio with respect to $f_2(c) = \frac{1}{n} \sum_{i=1}^{n} c_i$ of* RPP-MULT *for the bi-objective time series search problem is given by*

$$\mathcal{R}_s^{f_2}(\text{RPP-MULT}) = \sqrt[4]{\frac{M_1 M_2}{m_1 m_2}} .$$

Proof. We distinguish two cases with respect to the sequence $\sigma = (r_1, \ldots, r_T)$ revealed by the adversary:

Case 1: there exists a request $r_{t'}$ with $p_t^1 \cdot p_t^2 \geq z^\star$.

In this case, the online player accepts the first request r_t with $p_t^1 \cdot p_t^2 \geq z^\star$, i.e., $\text{ALG} = \left(p_t^1, \, p_t^2\right)^\mathsf{T}$. However, the adversary is able to reveal a further request r_j with $p_j^i = M_i$ for $i = 1, 2$, which is then the only efficient offline solution and, therefore, the optimal solution for the adversary, i.e., $\text{OPT} = \left(M_1, \, M_2\right)^\mathsf{T}$.

In the worst case with respect to all sequences, the request r_t accepted by the online player is such that $p_t^1 \cdot p_t^2 = z^\star$. The set of all points in $[m_1, M_1] \times [m_2, M_2]$ satisfying $p_t^1 \cdot p_t^2 = z^\star$ is given by

$$\mathcal{I}_1 = \left\{ \left(x, \, \tfrac{z^\star}{x}\right)^\mathsf{T} \mid x \in [m_1, M_1] \text{ and } \frac{z^\star}{M_2} \leq x \leq \frac{z^\star}{m_2} \right\}. \tag{8}$$

According to Definition 3, the competitive ratio is defined as the infimum over the set of all values $f_2(c)$ such that

$$\text{ALG}_i \geq \frac{1}{c_i} \text{OPT}_i, \text{ for } i = 1, \dots, n,$$

where $\text{ALG} = \left(x, \, z^\star/x\right)^\mathsf{T}$, $x \in \mathcal{I}_1$, and $\text{OPT} = \left(M_1, \, M_2\right)^\mathsf{T}$. Since $f_2(c) = \frac{1}{n} \sum_{i=1,\dots,n} c_i$, the competitive ratio is in this case given by

$$\frac{1}{2} \max_{x \in \mathcal{I}_1} \left\{ \frac{M_1}{x} + \frac{M_2 x}{z^\star} \right\} = \frac{1}{2} \left(\frac{M_1}{\sqrt{(M_1 z^\star)/M_2}} + \frac{M_2 \sqrt{(M_1 z^\star)/M_2}}{z^\star} \right) = \sqrt[4]{\frac{M_1 M_2}{m_1 m_2}}. \tag{9}$$

Case 2: for all r_t, $t = 1, \dots, T$, we have $p_t^1 \cdot p_t^2 < z^\star$.

In this instance, the online player does not accept any request and has to settle in the worst case for the lower bounds in each component, i.e. $\text{ALG} = \left(m_1, \, m_2\right)^\mathsf{T}$.

The adversary is able to offer (and accept) any request r_j for which the product $p_t^1 \cdot p_t^2$ is smaller than but arbitrarily close to $z^\star$, i.e., $p_t^1 \cdot p_t^2 = z^\star - \epsilon$, $\epsilon > 0$. The set of efficient solutions for OPT is given by

$$\mathcal{I}_2 = \left\{ \left(x, \, \tfrac{z^\star - \epsilon}{x}\right)^\mathsf{T} \mid x \in [m_1, M_1] \text{ and } \frac{z^\star - \epsilon}{M_2} \leq x \leq \frac{z^\star - \epsilon}{m_2} \right\}. \tag{10}$$

Now, ignoring the ϵ, the competitive ratio is given by

$$\frac{1}{2} \max_{x \in \mathcal{I}_2} \left\{ \frac{x}{m_1} + \frac{z^\star}{m_2 x} \right\} = \frac{1}{2} \left(\frac{\sqrt{(m_1 z^\star)/m_2}}{m_1} + \frac{z^\star}{m_2 \sqrt{(m_1 z^\star)/m_2}} \right) = \sqrt[4]{\frac{M_1 M_2}{m_1 m_2}}. \tag{11}$$

By the analysis above, the arithemtic-mean-component competitive ratio results in

$$\mathcal{R}^{f_2}(\text{RPP-MULT}) = \sqrt[4]{\frac{M_1 M_2}{m_1 m_2}}. \tag{12}$$

This result holds for all efficient solutions since, in the first case, there is exactly one efficient solution for the adversary and, in the second case, we considered the maximum over all $x \in \mathcal{I}_2$. Consequently, we have $\mathcal{R}_s^{f_2}(\text{RPP-MULT}) = \mathcal{R}^{f_2}(\text{RPP-MULT})$.   □

Note that the arithmetic-mean-component competitive ratio of RPP-MULT is closely related to the competitive ratio of the corresponding single-objective algorithm RPP as given in [4]. In the following, we prove that RPP-MULT achieves the best possible arithmetic-mean-component competitive ratio.

Theorem 4. *No deterministic algorithm for the bi-objective time series search problem can achieve a smaller competitive ratio with respect to $f_2(c) = \frac{1}{n}\sum_{i=1}^{n} c_i$ than $\sqrt[4]{M_1 M_2 / m_1 m_2}$.*

Proof. The adversary offers a request r with

$$r = \left(\tilde{x}, \tfrac{z^\star}{\tilde{x}}\right)^\mathsf{T}, \text{ where } \tilde{x} = \sqrt{\frac{M_1 z^\star}{M_2}}.$$

If the online player accepts this request, another request $(M_1, M_2)^\mathsf{T}$ is revealed and the competitive ratio is given by

$$\mathcal{R}^{f_2}(\text{RPP-MULT}) = \frac{1}{2}\left(\frac{M_1}{\sqrt{(M_1 z^\star)/M_2}} + \frac{M_2\sqrt{(M_1 z^\star)/M_2}}{z^\star}\right) = \sqrt[4]{\frac{M_1 M_2}{m_1 m_2}}. \quad (13)$$

Otherwise, the adversary only reveals r until the online player has to settle for the lower bounds m_1 and m_2. The competitive ratio is then given by

$$\mathcal{R}^{f_2}(\text{RPP-MULT}) = \frac{1}{2}\left(\frac{\sqrt{(m_1 z^\star)/m_2}}{m_1} + \frac{z^\star}{\sqrt{(m_1 z^\star)/m_2}\,m_2}\right) = \sqrt[4]{\frac{M_1 M_2}{m_1 m_2}}. \quad (14)$$

Thus, no deterministic algorithm for the bi-objective time series search problem can achieve a smaller competitive ratio with respect to f than $\sqrt[4]{M_1 M_2 / m_1 m_2}$. Obviously, the same lower bound holds for the strong competitive ratio.   □

The competitive analysis with respect to $f_3(c) = \sqrt[n]{\prod_{i=1}^{n} c_i}$, i.e., a geometric-mean-component competitive analysis, yields the same results: For f_3, (9) and (11) are given by

$$\max_{x \in \mathcal{I}_1}\left\{\sqrt{\frac{M_1}{x}\frac{M_2 x}{z^\star}}\right\} = \sqrt[4]{\frac{M_1 M_2}{m_1 m_2}} \text{ and } \max_{x \in \mathcal{I}_2}\left\{\sqrt{\frac{x}{m_1}\frac{z^\star}{m_2 x}}\right\} = \sqrt[4]{\frac{M_1 M_2}{m_1 m_2}}. \quad (15)$$

For the lower bound, a request $r = \left(\tilde{x}, \tfrac{z^\star}{\tilde{x}}\right)^\mathsf{T}$, where $\tilde{x}$ is an arbitrary feasible value, leads to the same result as given in Theorem 4.

4   Conclusion and Future Research

In this paper, we introduced a general framework for the competitive analysis of multi-objective online problems which expands the known theory of competitive analysis for online problems in a straightforward manner. As an example, we demonstrated that the analysis of the multi-objective time series search problem by means of the introduced notions of competitiveness for multi-objective online problems yields results which are closely related to the single-objective algorithms and their competitive ratios as given in [4]. The analysis of mean-component competitive algorithms for the multi-objective time series search problem for dimensions $n \geq 3$ is subject to further research.

The concept of competitive analysis for multi-objective online problems seems highly promising and provides further insight into the nature of online problems. Questions for future research include the analysis of multi-objective counterparts of other well-known online problems such as the k-server problem or scheduling problems. Another direction for future research could be to study general relations between single-objective online problems and the corresponding multi-objective online problem or general propositions about the competitiveness of multi-objective online problems such as a multi-objective counterpart for Yao's principle.

Furthermore, the definition of the competitive ratio given in this work serves as a basis for further extensions such as a vector of competitive ratios with respect to different functions: For example, if the analyst of the online problem wants the worst component and the average of the components to be reasonably small at the same time, the vector of both competitive ratios could be analyzed in the sense of multi-objective optimization.

References

1. Bazgan, C., Gourvès, L., Monnot, J.: Approximation with a Fixed Number of Solutions of Some Biobjective Maximization Problems. In: Solis-Oba, R., Persiano, G. (eds.) WAOA 2011. LNCS, vol. 7164, pp. 233–246. Springer, Heidelberg (2012)
2. Borodin, A., El-Yaniv, R.: Online Computation and Competitive Analysis. Cambridge University Press (1998)
3. Ehrgott, M.: Multicriteria Optimization. Springer (2005)
4. El-Yaniv, R., Fiat, A., Karp, R.M., Turpin, G.: Optimal search and one-way trading online algorithms. Algorithmica 30(1), 101–139 (2001)
5. Erlebach, T., Kellerer, H., Pferschy, U.: Approximating Multiobjective Knapsack Problems. Management Sciences 48(12), 1603–1612 (2002)
6. Fiat, A. (ed.): Online Algorithms 1996. LNCS, vol. 1442. Springer, Heidelberg (1998)
7. Kalyanasundaram, B., Pruhs, K.: Speed is as powerful as clairvoyance. Journal of the ACM 47(4), 617–643 (2000)
8. Motwani, R., Phillips, S., Torng, E.: Non-clairvoyant scheduling. Theoretical Computer Science 130, 17–47 (1994)
9. Phillips, C.A., Stein, C., Torng, E., Wein, J.: Optimal time-critical scheduling via resource augmentation. In: Proceedings of the 29th ACM Symposium on the Theory of Computing (STOC), pp. 140–149 (1997)

Simultaneous Drawing of Planar Graphs with Right-Angle Crossings and Few Bends[*]

Michael A. Bekos[1], Thomas C. van Dijk[2], Philipp Kindermann[2], and Alexander Wolff[2]

[1] Wilhelm-Schickard-Institut für Informatik, Universität Tübingen, Germany
bekos@informatik.uni-tuebingen.de
[2] Lehrstuhl für Informatik I, Universität Würzburg, Germany
http://www1.informatik.uni-wuerzburg.de/en/staff

Abstract. Given two planar graphs that are defined on the same set of vertices, a *RAC simultaneous drawing* is a drawing of the two graphs where each graph is drawn planar, no two edges overlap, and edges of one graph can cross edges of the other graph only at right angles. In the geometric version of the problem, vertices are drawn as points and edges as straight-line segments. It is known, however, that even pairs of very simple classes of planar graphs (such as wheels and matchings) do not always admit a geometric RAC simultaneous drawing.

In order to enlarge the class of graphs that admit RAC simultaneous drawings, we allow edges to have bends. We prove that any pair of planar graphs admits a RAC simultaneous drawing with at most six bends per edge. For more restricted classes of planar graphs (e.g., matchings, paths, cycles, outerplanar graphs, and subhamiltonian graphs), we significantly reduce the required number of bends per edge. All our drawings use quadratic area.

1 Introduction

A simultaneous embedding of two planar graphs embeds each graph in a planar way—using the same vertex positions for both embeddings. Edges of one graph are allowed to intersect edges of the other graph. There are two versions of the problem: In the first version, called *Simultaneous Embedding with Fixed Edges* (SEFE), edges that occur in both graphs must be embedded in the same way in both graphs (and hence, cannot be crossed by any other edge). In the second version, called *Simultaneous Embedding*, these edges can be drawn differently for each of the graphs. Both versions of the problem have a geometric variant where edges must be drawn using straight-line segments.

Simultaneous embedding problems have been extensively investigated over the last few years, starting with the work of Brass et al. [6] on simultaneous straight-line drawing problems. Bläsius et al. [5] recently surveyed the area. For example, it is possible to decide in linear time whether a pair of graphs admits a SEFE or not, if the common

---

[*] This research was supported by the ESF EuroGIGA project GraDR (DFG grant Wo 758/5-1). The work of M.A. Bekos is implemented within the framework of the Action "Supporting Postdoctoral Researchers" of the Operational Program "Education and Lifelong Learning" (Action's Beneficiary: General Secretariat for Research and Technology), and is co-financed by the European Social Fund (ESF) and the Greek State.

M.S. Rahman and E. Tomita (Eds.): WALCOM 2015, LNCS 8973, pp. 222–233, 2015.

graph is biconnected [1]. When actually drawing these simultaneous embeddings, a natural choice is to use straight-line segments. Only very few graphs can be drawn in this way, however, and some existing results require exponential area. For instance, there exist a tree and a path which cannot be drawn simultaneously with straight-line segments [2], and the algorithm for simultaneously drawing a tree and a matching [8] does not provide a polynomial area bound. For the case of edges with bends, that is, polygonal edges, Erten and Kobourov [10] showed that three bends per edge and quadratic area suffice for any pair of planar graphs (without fixed edges), and that one bend per edge suffices for pairs of trees. Kammer [11] reduced the number of bends to two for the general planar case. In these results, however, the *crossing angles* can be very small.

We suggest a new approach that overcomes the aforementioned problems. We insist that crossings occur at right angles, thereby "taming" them. We do this while drawing all vertices and all bends on a grid of size $O(n) \times O(n)$ for any n-vertex graph, and we can still draw any pair of planar graphs simultaneously. We do not consider the problem of fixed edges. In a way, our results give a measure for the geometric complexity of simultaneous embeddability for various pairs of graph classes, some of which can be combined more easily (that is, with fewer bends) than others.

Brightwell and Scheinermann [7] proved that the problem of simultaneously drawing a (primal) embedded graph and its dual always admits a solution if the input graph is a triconnected planar graph. Erten and Kobourov [9] presented an $O(n)$-time algorithm that computes simultaneous drawings of a triconnected planar graph and its dual on an integer grid of size $O(n^2)$, where n is the total number of vertices in the graph and its dual. However, these drawings can have non-right angle crossings.

In this paper, we study the *RAC simultaneous (*RacSim*) drawing problem*. Let $G_1 = (V, E_1)$ and $G_2 = (V, E_2)$ be two planar graphs on the same vertex set. We say that G_1 and G_2 admit a RacSim drawing if we can place the vertices on the plane such that (i) each edge is drawn as a polyline, (ii) each graph is drawn planar, (iii) there are no edge overlaps, and (iv) crossings between edges in E_1 and E_2 occur at right angles.

Argyriou et al. [3] introduced and studied the geometric version of RacSim drawing. In particular, they proved that any pair of a cycle and a matching admits a geometric RacSim drawing on a grid of quadratic size, while there exists a pair of a wheel and a cycle that does not admit a geometric RacSim drawing. The problem that we study was left as an open problem.

Our Contribution. Our main result is that any pair of planar graphs admits a RAC simultaneous drawing with at most six bends per edge. We can compute such drawings in linear time. For pairs of subhamiltonian graphs and pairs of outerplanar graphs, we need four bends and three bends per edge, respectively; see Section 2. Then, we turn our attention to pairs of more restricted graph classes where we can guarantee one bend per edge or two bends per edge; see Sections 3 and 4, respectively. Table 1 lists our results. The main approach of all our algorithms is to find linear orders on the vertices of the two graphs and then to compute coordinates for the vertices based on these orders. All crossings in our drawings appear between horizontal and vertical edge segments. We call the non-rectilinear edge segments *slanted* segments. All our drawings fit on a grid of quadratic size. Due to lack of space, some proofs are only sketched. The corresponding detailed proofs are given in the full version [4].

Table 1. A short summary of our results

Graph classes			Number of bends	Ref.
planar	+	planar	$6 + 6$	Thm. 1
subhamiltonian	+	subhamiltonian	$4 + 4$	Cor. 1
outerplanar	+	outerplanar	$3 + 3$	Thm. 2
cycle	+	cycle	$1 + 1$	Thm. 3
caterpillar	+	cycle	$1 + 1$	Thm. 4
four matchings			$1 + 1 + 1 + 1$	Thm. 5
tree	+	matching	$1 + 0$	Thm. 6
wheel	+	matching	$2 + 0$	Thm. 7
outerpath	+	matching	$2 + 1$	Thm. 8

2   RacSim Drawings of General Graphs

In this section, we study general planar graphs and show how to efficiently construct RacSim drawings with few bends per edge in $O(n) \times O(n)$ area. We prove that two planar graphs on a common set of n vertices admit a RacSim drawing with six bends per edge (Theorem 1). For pairs of subhamiltonian graphs, we lower the number of bends per edge to 4 (Corollary 1) and for pairs of outerplanar graphs to 3 (Theorem 2). Recall that the class of subhamiltonian graphs is equivalent to the class of 2-page book-embeddable graphs, and the class of outerplanar graphs is equivalent to the class of 1-page book-embeddable graphs.

Central to our approach is an algorithm by Kaufmann and Wiese [12] that embeds any planar graph such that vertices are mapped to points on a horizontal line (called *spine*) and each edge crosses the spine at most once; see Fig. 1(a). We introduce a *dummy vertex* at each spine crossing. This yields a linear order of the (original and dummy) vertices with the property that every edge is either above or below the spine. For our problem, in order to determine the locations of the (original and dummy) vertices of the two given graphs, we basically use the linear order induced by one graph for the x-coordinates and the order induced by the other graph for the y-coordinates. (We reserve additional rows and columns for routing the edges of the first and second graph, respectively.) Let R be the bounding box of the vertex positions. Then, for the first graph, we route, in a planar fashion, the above/below-edges using short slanted segments and long vertical segments inside R as well as horizontal segments above/below R; see Fig. 1(c). We treat the second graph analogously, but turn the drawing by $90°$. As the following theorem assures, the resulting simultaneous drawing has only right-angle crossings; see Fig. 1(d). Note that the algorithm of Kaufmann and Wiese has been used for simultaneous drawing problems before [10].

Theorem 1. *Two planar graphs on a common set of n vertices admit a RacSim drawing on an integer grid of size $(14n - 26) \times (14n - 26)$ with six bends per edge. The drawing can be computed in $O(n)$ time.*

Proof. Let $\mathcal{G}_1 = (V, E_1)$ and $\mathcal{G}_2 = (V, E_2)$ be the given planar graphs. For $m = 1, 2$, let ξ_m be an embedding of $\mathcal{G}_m$ according to the algorithm of Kaufmann and Wiese,

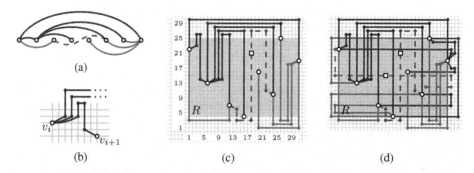

Fig. 1. (a) A drawing of a planar graph by Kaufmann and Wiese [12], (b) reserving additional columns between two vertices, (c) the graph in (a) drawn by our algorithm with at most six bends per edge, and (d) the RACSIM drawing of two planar graphs by our algorithm. The edges that cross the spine are drawn dashed; the dummy vertices on these edges are drawn as squares.

and denote by A_m and B_m the edges that are drawn completely above and below the spine in ξ_m, respectively. Further, let $\mathcal{G}'_m = (V'_m, E'_m) = (V \cup V_m, A_m \cup B_m)$ be the resulting graph, where V_m is the set of dummy vertices of $\mathcal{G}_m$.

We now show how to determine the x-coordinates of the vertices in V'_1; the y-coordinates of the vertrices in V'_2 are determined analogously. The missing y-coordinates of the dummy vertices in V_1 and the missing x-coordinates of the dummy vertices in V_2 are arbitrary (as long as they are inside R).

Let n'_1 be the number of vertices in V'_1 and let $v_1, \ldots, v_{n'_1}$ be the linear order of the vertices in V'_1 along the spine in ξ_1. We place v_1 in the first column. Between any two consecutive vertices v_i and v_{i+1}, we reserve several columns for the bends of the edges incident to v_i and v_{i+1}, in the following order; see Fig. 1(b):

(i) a column for the first bend on all edges leaving v_i in A_2;
(ii) a column for each edge $(v_i, v_j) \in E'_1$ with $j > i$;
(iii) a column for each edge $(v_k, v_{i+1}) \in E'_1$ with $k \leq i$;
(iv) a column for the last bend on all edges entering v_{i+1} in B_2.

Note that, for (ii) and (iii), we can save some columns because an edge in A_1 and an edge in B_1 can use the same column for their bend. Further, we may save the additional column of (i) and (iv) if no such edges exist. Now, we draw $\mathcal{G}'_1$ and $\mathcal{G}'_2$ with at most four bends per edge such that all edge segments of $\mathcal{G}'_1$ in R are either vertical or of y-length exactly 1, and all edge segments of $\mathcal{G}'_2$ in R are either horizontal or of x-length exactly 1; see Fig. 1(d).

First, we draw the edges $(v_i, v_j) \in A_1$ with $i < j$ in a nested order: When we place the edge (v_i, v_j), then there is no edge $(v_k, v_l) \in A_1$ with $k \leq i$ and $l \geq j$ that has not already been drawn. Recall that the first column to the right and the first column to the left of every vertex is reserved for the edges in E_1; hence, we assume that they are already used. We draw (v_i, v_j) with at most four bends as follows. We start with a slanted segment that has its endpoint in the row above v_i and in the first unused column that does not lie to the left of v_i. We follow with a vertical segment to the top that leaves R. We add a horizontal segment above R. In the last unused column that does not lie to the right of v_j, we add a vertical segment that ends one row above v_j. We close the edge with a slanted segment that has its endpoint in v_j. We draw the edges in B_1 symmetrically with the horizontal segment below R.

Note that this algorithm always uses the top and the bottom port of a vertex v, if there is at least one edge incident to v in A_1 and B_1, respectively. There is exactly one edge incident to each dummy vertex t in A_1 and B_1, respectively. Thus, the edges incident to t only use the top and the bottom port. We create a drawing of $\mathcal{G}_1$ and $\mathcal{G}_2$ with at most 6 bends per edge by removing the dummy vertices from the drawing.

We now show that combining the drawings of $\mathcal{G}_1$ and $\mathcal{G}_2$ yields a RACSIM drawing. By construction, all segments of E_1 inside R are either vertical segments or slanted segments of y-length 1, and all segments of E_2 inside R are either horizontal segments or slanted segments of x-length 1. Thus, the slanted segments cannot overlap. Furthermore, all crossings inside R occur between a horizontal and a vertical segment, and thus form right angles. Also, there are no segments in E_1 that lie to the left or to the right of R, and there are no segments in E_2 that lie above or below R. Hence, there are no crossings outside R, and the drawing is a RACSIM drawing.

We now count the columns used by the drawing. For the leftmost and the rightmost vertex, we reserve one additional column for its incident edges in E_2; for the remaining vertices, we reserve two such columns. For each edge in E_1, we need up to three columns: one for each endpoint of the slanted segment at each vertex and one for the vertical segment that crosses the spine, if it exists. Note that at least one edge per vertex does not need a slanted segment. For each edge in E_2, we need at most one column for the vertical segment to the side of R. Since there are at most $3n - 6$ edges, we need at most $3n - 2 + 3 \cdot (3n - 6) - n + 3n - 6 = 14n - 26$ columns. By symmetry, we need the same number of rows.

Since the algorithm of Kaufmann and Wiese runs in $O(n)$ time, our algorithm also runs in $O(n)$ total time.   □

We can improve the results of Theorem 1 for subhamiltonian graphs. Recall that a subhamiltonian graph has a 2-page book embedding, in which no edges cross the spine. Since such edges are the only ones that need six bends, we can reduce the number of bends per edge to four. Further, the number of columns and rows are reduced by one per edge. This yields the following corollary.

Corollary 1. *Two subhamiltonian graphs on a common set of n vertices admit a* RAC-SIM *drawing on an integer grid of size* $(11n - 32) \times (11n - 32)$ *with four bends per edge.*

Theorem 2. *Two outerplanar graphs on a common set of n vertices admit a* RACSIM *drawing on an integer grid of size* $(7n - 10) \times (7n - 10)$ *with three bends per edge.*

Proof. Let $\mathcal{O}_1 = (V, E_1)$ and $\mathcal{O}_2 = (V, E_2)$ be the given outerplanar graphs. First, we create a 1-page book embedding for $\mathcal{O}_1$ and $\mathcal{O}_2$. This gives us the order of the x-coordinates and y-coordinates, respectively. It follows by Corollary 1 that, by using the algorithm described in the proof of Theorem 1, we obtain a RACSIM drawing with at most four bends per edge. We will now show how to adjust the algorithm to reduce the number of bends by one.

It follows by Nash-Williams' formula [13] that every outerplanar graph has arboricity 2, that is, it can be decomposed into two forests. We embed both graphs on two pages with one forest per page. Let A_1 and B_1 be the two forests $\mathcal{O}_1$ is decomposed into. We will draw the edges of A_1 above the spine and the edges B_1 below the spine. By rooting the trees in A_1 in arbitrary vertices, we can direct each edge such that every vertex has exactly one incoming edge. Recall that, in the drawing produced in Theorem 1, one edge per vertex can use the top port. We adjust the algorithm such that every directed edge (v, w) enters the vertex w from the top port. To do so, we draw the edge as follows. We start with a slanted segment of y-length exactly 1. We follow with a vertical segment to the top. We proceed with a horizontal segment that ends directly above w and finish the edge with a vertical segment that enters w from the top port. We use the same approach for the edges in B_1, using the bottom port. We treat the second outerplanar graph $\mathcal{O}_2$ analogously, but turn the drawing by $90°$.

Since every port of a vertex is only used once, the drawing has no overlaps. We now analyze the number of columns used. For every vertex except for the leftmost and rightmost, we again reserve two additional columns for the edges in E_2; for the remaining two vertices, we reserve one additional column. However, the edges in E_1 now only need one column for the bend of the single slanted segment. For every edge in E_2, we need up to one column for the vertical segment to the side of R. Since there are at most $2n - 4$ edges, our drawing needs $3n - 2 + 2n - 4 + 2n - 4 = 7n - 10$ columns. Analogously, we can show that the algorithm needs $7n - 10$ rows.   □

3 RACSIM Drawings with One Bend Per Edge

In this section, we study simple classes of planar graphs and show how to efficiently construct RACSIM drawings with one bend per edge in quadratic area. In particular, we prove that two cycles or four matchings on a common set of n vertices admit a RACSIM drawing on an integer grid of size $2n \times 2n$; see Theorems 3 and 5, respectively. If the input to our problem is a caterpillar and a cycle, then we can compute a RACSIM drawing with one bend per edge on an integer grid of size $(2n-1) \times 2n$; see Theorem 4. For a tree and a cycle, we can construct a RACSIM drawing with one bend per tree edge and no bends in the matching edges on an integer grid of size $n \times (n-1)$; see Theorem 6.

Lemma 1. *Two paths on a common set of n vertices admit a RACSIM drawing on an integer grid of size $2n \times 2n$ with one bend per edge. The drawing can be computed in $O(n)$ time.*

Proof. Let $\mathcal{P}_1 = (V, E_1)$ and $\mathcal{P}_2 = (V, E_2)$ be the two input paths. Following standard practices from the literature (see, e.g., Brass et al. [6]), we draw $\mathcal{P}_1$ x-monotone and $\mathcal{P}_2$ y-monotone. This ensures that the drawing of both paths will be planar. We will now describe how to compute the exact coordinates of the vertices and how to draw the edges of $\mathcal{P}_1$ and $\mathcal{P}_2$, such that all crossings are at right angles and, more importantly, no edge segments overlap.

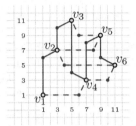
(a) two paths: $\mathcal{P}_1$ (solid) and $\mathcal{P}_2$ (dashed)

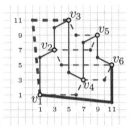
(b) two cycles: $\mathcal{C}_1$ (solid) and $\mathcal{C}_2$(dashed)

Fig. 2. RACSIM drawings with one bend per edge

For $m = 1, 2$ and any vertex $v \in V$, let $\pi_m(v)$ be the position of v in $\mathcal{P}_m$. Then, v is drawn at the point $(2\pi_1(v) - 1, 2\pi_2(v) - 1)$; see Fig. 2a. It remains to determine, for each edge $e = (v, v')$, where it bends. First, assume that $e \in E_1$ and e is directed from its left to its right endpoint. Then, we place the bend at $v' - (2, \operatorname{sgn}(y(v') - y(v)))$. Second, assume that $e \in E_2$ and e is directed from its bottom to its top endpoint. Then, we place the bend at $v' - (\operatorname{sgn}(x(v') - x(v)), 2)$.

Clearly, the area required by the drawing is $(2n - 1) \times (2n - 1)$. The edges of $\mathcal{P}_1$ leave the left endpoint vertically and enter the right endpoint with a slanted segment of x-length 1 and y-length 2. Similarly, the edges of $\mathcal{P}_2$ leave the the bottom endpoint horizontally and enter the top endpoint with a slanted segment of x-length 2 and y-length 1. Hence, the slanted segments cannot be involved in crossings or overlaps. Since $\mathcal{P}_1$ and $\mathcal{P}_2$ are x- and y-monotone, respectively, it follows that all crossings must involve a vertical edge segment of $\mathcal{P}_1$ and a horizontal edge segment of $\mathcal{P}_2$, which clearly yields right angles at the crossing points. □

We say that an edge uses the bottom/left/right/top *port* of a vertex if it enters the vertex from the bottom/left/right/top.

Theorem 3. *Two cycles on a common set of n vertices admit a RACSIM drawing on an integer grid of size $2n \times 2n$ with at most one bend per edge. The drawing can be computed in $O(n)$ time.*

Proof. Let $\mathcal{C}_1 = (V, E_1)$ and $\mathcal{C}_2 = (V, E_2)$ be the two input cycles, and let $v \in V$ be an arbitrary vertex. We temporarily delete one edge $(v, w_1) \in E_1$ from $\mathcal{C}_1$ and $(v, w_2) \in E_2$ from $\mathcal{C}_2$ (refer to the bold-drawn edges of Figure 2b). This way, we obtain two paths $\mathcal{P}_1 = \langle v, \ldots, w_1 \rangle$ and $\mathcal{P}_2 = \langle v, \ldots, w_2 \rangle$. We employ the algorithm desribed in Lemma 1 to construct a RACSIM drawing of $\mathcal{P}_1$ and $\mathcal{P}_2$ on an integer grid of size $(2n - 1) \times (2n - 1)$. Since v is the first vertex in both paths, it is placed at the bottom-left corner of the bounding box containing the drawing. Since w_1 and w_2 are the last vertices in $\mathcal{P}_1$ and $\mathcal{P}_2$, respectively, w_1 is placed on the right side, and w_2 on the top side of the bounding box. By construction, the bottom port of w_1 and the left port of w_2 are both unoccupied. Hence, the edges (v, w_1) and (v, w_2) that form $\mathcal{C}_1$ and $\mathcal{C}_2$ can be drawn with a single bend at points $(2n - 1, 0)$ and $(0, 2n - 1)$, respectively; see Figure 2b. Clearly, none of them is involved in crossings, while the total area of the drawing gets larger by a single unit in each dimension. □

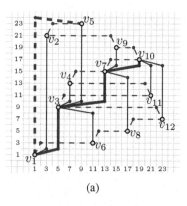
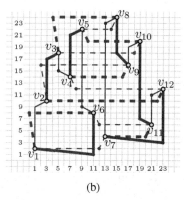

(a)   (b)

Fig. 3. RACSIM drawings with one bend per edge of: (a) a caterpillar $\mathcal{A}$ (solid; its spine is drawn bold) and a cycle $\mathcal{C}$ (dashed); (b) four matchings $\mathcal{M}_1$ (solid-plain), $\mathcal{M}_2$ (solid-bold), $\mathcal{M}_3$ (dashed-plain) and $\mathcal{M}_4$ (dashed-bold)

Theorem 4. *A caterpillar and a cycle on a common set of n vertices admit a RACSIM drawing on an integer grid of size $(2n - 1) \times 2n$ with one bend per edge. The drawing can be computed in $O(n)$ time.*

Proof. We denote by $\mathcal{A} = (V, E_\mathcal{A})$ and $\mathcal{C} = (V, E_\mathcal{C})$ the caterpillar and the cycle, respectively. Let $v_1, v_2, \ldots, v_n$ be the vertex set of $\mathcal{A}$ ordered as follows (see Fig. 3a). Starting from an endpoint of the spine of $\mathcal{A}$, we traverse the spine such that we first visit all legs incident to a spine vertex before moving on to the next spine vertex. This order defines the x-order of the vertices in the output drawing.

As in the proof of Theorem 3, we temporarily delete an edge of $\mathcal{C}$ incident to v_1 (see the bold dashed edge in Fig. 3a) and obtain a path $\mathcal{P} = (V, E_\mathcal{P})$. For any vertex $v \in V$, let $\pi(v)$ be the position of v in $\mathcal{P}$. The map π determines the y-order of the vertices in our drawing. For $i = 1, 2, \ldots, n$, we draw v_i at point $(2i - 1, 2\pi(v_i) - 1)$. It remains to determine, for each edge $e = (v, v')$, where it bends. First, assume that $e \in E_\mathcal{P}$ and e is directed from its bottom to its top endpoint. Then, we place the bend at $v + (\mathrm{sgn}(x(v') - x(v)), 2)$. Second, assume that $e \in E_\mathcal{A}$ and e is directed from its left to its right endpoint. Then, we place the bend at $(x(v'), y(v) + \mathrm{sgn}(y(v') - y(v)))$.

The approach described above ensures that $\mathcal{P}$ is drawn y-monotone, hence planar. The spine of $\mathcal{A}$ is drawn x-monotone. The legs of a spine vertex of $\mathcal{A}$ are drawn to the right of their parent spine vertex and to the left of the next vertex along the spine. Hence, $\mathcal{A}$ is drawn planar as well. The slanted segments of $\mathcal{A}$ are of y-length 1, while the slanted segments of $\mathcal{P}$ are of x-length 1. Thus, they cannot be involved in crossings, which implies that all crossings form right angles.

It remains to draw the edge e in $E_\mathcal{C} \setminus E_\mathcal{P}$. Recall that e is incident to v_1, which lies at the bottom-left corner of the bounding box containing our drawing. Let v_j be the other endpoint of e. Since $\pi(v_j) = n$, vertex v_j lies at the top side of the bounding box. The top port of v_1 is not used, so we draw the first segment of e vertically, bending at $(1, 2n)$; see the bold dashed edge in Fig. 3a.

Clearly, the total area required by the drawing is $(2n - 1) \times 2n$.   □

Theorem 5. *Four matchings on a common set of n vertices admit a RACSIM drawing on an integer grid of size $2n \times 2n$ with at most one bend per edge. The drawing can be computed in $O(n)$ time.*

Proof. Let $\mathcal{M}_1 = (V, E_1)$, $\mathcal{M}_2 = (V, E_2)$, $\mathcal{M}_3 = (V, E_3)$ and $\mathcal{M}_4 = (V, E_4)$ be the input matchings. W.l.o.g, we assume that all matchings are perfect; otherwise, we augment them to perfect matchings. Let $\mathcal{M}_{1,2} = (V, E_1 \cup E_2)$ and $\mathcal{M}_{3,4} = (V, E_3 \cup E_4)$. Since $\mathcal{M}_1$ and $\mathcal{M}_2$ are defined on the same vertex set, $\mathcal{M}_{1,2}$ is a 2-regular graph. Thus, each connected component of $\mathcal{M}_{1,2}$ corresponds to a cycle of even length which alternates between edges of $\mathcal{M}_1$ and $\mathcal{M}_2$; see Fig. 3b. The same holds for $\mathcal{M}_{3,4}$. We will determine the x-coordinates of the vertices from $\mathcal{M}_{1,2}$, and the y-coordinates from $\mathcal{M}_{3,4}$.

We start with choosing an arbitrary vertex $v \in V$. Let $\mathcal{C}$ be the cycle of $\mathcal{M}_{1,2}$ containing vertex v. We determine the x-coordinates of the vertices of $\mathcal{C}$ by traversing $\mathcal{C}$ in some direction, starting from vertex v. For each vertex u in $\mathcal{C}$, let $\pi_1(u)$ be the discovery time of u according to this traversal, with $\pi_1(v) = 0$. Then, we set $x(u) = 2\pi_1(u) + 1$. Next, we determine the y-coordinates of the vertices of all cycles $\mathcal{C}_1, \ldots, \mathcal{C}_k$ of $\mathcal{M}_{3,4}$ that have at least one vertex with a determined x-coordinate, ordered as follows. For $i = 1, \ldots, k$, let a_i be the *anchor* of $\mathcal{C}_i$, that is, the vertex with the smallest determined x-coordinate of all vertices in $\mathcal{C}_i$. Then, $x(a_1) < \ldots < x(a_k)$. In what follows, we start with the first cycle $\mathcal{C}_1$ of the computed order and determine the y-coordinates of its vertices. To do so, we traverse $\mathcal{C}_1$ in some direction, starting from its anchor vertex a_1. For each vertex u in $\mathcal{C}_1$, let $\pi_2(u)$ be the discovery time of u according to this traversal, with $\pi_2(a_1) = 0$. Then, we set $y(u) = 2\pi_2(u) + 1$. We proceed analogously with the remaining cycles $\mathcal{C}_i$, $i = 2, \ldots, k$, setting $\pi_2(a_i) = \max_{u \in \mathcal{C}_{i-1}} \pi_2(u) + 1$.

Now, there are no vertices with only one determined x-coordinate. However, there might exist vertices with only one determined y-coordinate. If this is the case, we repeat the aforementioned procedure to determine the x-coordinates of the vertices of all cycles of $\mathcal{M}_{1,2} \setminus \mathcal{C}$ that have at least one vertex with a determined y-coordinate, but without determined x-coordinates. If there are no vertices with only one determined coordinate left, either all coordinates are determined, or we restart this procedure with another arbitrary vertex that has no determined coordinates. Thus, our algorithm guarantees that the x- and y-coordinate of all vertices are eventually determined.

Note that, for each cycle in $\mathcal{M}_{1,2}$, there is exactly one edge $e = (v, v')$, called *closing edge*, with $\pi_1(v') > \pi_1(v) + 1$. Analogously, for each cycle in $\mathcal{M}_{3,4}$, there is exactly one closing edge $e = (u, u')$ with $\pi_2(u') > \pi_2(u) + 1$.

It remains to determine, for each edge $e = (v, v')$, where it bends. First, assume that $e \in E_1 \cup E_2$ and e is directed from its left to its right endpoint. If e is not a closing edge, we place the bend at $v' - (2, \text{sgn}(y(v') - y(v)))$. Otherwise, we place the bend at $(x(v'), y(v) - 1)$. Second, assume that $e \in E_3 \cup E_4$ and e is directed from its bottom to its top endpoint. If e is not a closing edge, we place the bend at $v' - (\text{sgn}(x(v') - x(v)), 2)$. Otherwise, we place the bend at $(x(v) - 1, y(v'))$; see Fig. 3b.

Our choice of coordinates guarantees that the x-coordinates of the cycles of $\mathcal{M}_{1,2}$ and the y-coordinates of the cycles of $\mathcal{M}_{3,4}$ form disjoint intervals. Thus, the area below a cycle of $\mathcal{M}_{1,2}$ and the area to the left of a cycle of $\mathcal{M}_{3,4}$ are free from vertices. Hence, the slanted segments of the closing edges cannot have a crossing that violates the RAC restriction. Clearly, the total area required by the drawings is $2n \times 2n$. ☐

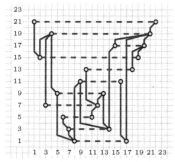
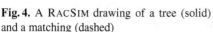

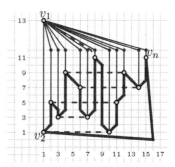

Fig. 4. A RACSIM drawing of a tree (solid) and a matching (dashed)

Fig. 5. A RACSIM drawing of a wheel (solid; its rim is drawn bold) and a matching (dashed)

Theorem 6. *A tree and a matching on a common set of n vertices admit a RACSIM drawing on an integer grid of size $n \times (n-1)$ with one bend per tree-edge, and no bends in the edges of the matching. The drawing can be computed in $O(n)$ time.*

Sketch of Proof. We inductively place each matching edge in one row. In every step, we decide whether to add the next matching edge to the stack at the top or at the bottom. We determine the x-coordinates of the matching with the help of a specific post-order visit; see Fig. 4. A detailed proof is given in the full version [4]. □

4   RACSIM Drawings with Two Bends Per Edge

In this section, we study more complex classes of planar graphs, and show how to efficiently construct RACSIM drawings with two bends per edge in quadratic area. In particular, we prove that a wheel and a matching on a common set of n vertices admit a RACSIM drawing on an integer grid of size $(1.5n - 1) \times (n + 2)$ with two bends per edge and no bends, respectively; see Theorem 7. If the input to our problem is an outerpath—that is, an outerplanar graph whose weak dual is a path—and a matching, then a RACSIM drawing with two bends per edge and no bends, respectively, is also possible on an integer grid of size $(3n - 2) \times (3n - 2)$; see Theorem 8.

Theorem 7. *A wheel and a matching on a common set of n vertices admit a RACSIM drawing on an integer grid of size $(1.5n - 1) \times (n + 2)$ with two bends per edge and no bends, respectively. The drawing can be computed in $O(n)$ time.*

Proof. We denote the wheel by $W = (V, E_W)$ and the matching by $M = (V, E_M)$. A wheel can be decomposed into a cycle, called *rim*, a *center* vertex, and a set of edges that connect the center to the rim, called *spikes*. Let $V = \{v_1, v_2, \ldots, v_n\}$, such that v_1 is the center of W and $C = \langle v_2, v_3, \ldots, v_n, v_2 \rangle$ is the rim of W. Thus, $E_W = \{(v_i, v_{i+1}) \mid i = 1, \ldots, n-1\} \cup \{(v_n, v_2)\} \cup \{(v_1, v_i) \mid i = 2, \ldots, n\}$. Let $M' = (V, E_{M'})$ be the matching M without the edge incident to v_1.

We first compute the x-coordinates of the vertices, such that $C - \{(v_n, v_2)\}$ is x-monotone (if drawn with straight-line edges). More precisely, for $i = 2, \ldots, n$ we set

$x(v_i) = 2i - 3$. The y-coordinates of the vertices are computed based on the matching $\mathcal{M}'$, as follows. Let $E_{\mathcal{M}'} = \{e_1, \ldots, e_k\}$ be the matching edges with v_2 incident to e_1. For $i = 1, \ldots, k$, we assign the y-coordinate $2i - 1$ to the endpoints of e_i. Next, we assign the y-coordinate $2k + 1$ to the vertices incident to the rim without a matching edge in $\mathcal{M}'$. Finally, the center v_1 of $\mathcal{W}$ is located at point $(1, 2k + 3)$.

It remains to determine, for each edge $e \in E_{\mathcal{W}}$, where it bends, as $\mathcal{M}'$ is drawn bendless. First, let $e = (v_1, v_i)$, $i = 3, \ldots, n$ be a spike. Then, we place the bend at $(x(v_j), 2k + 2)$. Since both v_1 and v_2 are located in column 1, we can save the bend of the spike (v_1, v_2). Second, let $e = (v_i, v_{i+1})$, $i = 2, \ldots, n-1$ be an edge of the rim $\mathcal{C}$. If $y(v_{i+1}) > y(v_i)$, we place the bend at $(x(v_{i+1}), y(v_i) + 1)$. If $y(v_{i-1}) > y(v_i) > y(v_{i+1})$, we place the bend at $(x(v_{i+1}), y(v_i) - 1)$. If $y(v_i) > y(v_{i-1}), y(v_{i+1})$, the bottom port at v_i is already used. Thus, we draw the edge with two bends at $(x(v_{i+1}), y(v_i) - 1)$ and $(x(v_{i+1}), y(v_{i+1}) + 1)$. Finally, let $e = (v_n, v_2)$ be the remaining edge of the rim. Then, we place the bend at $(2n - 2, 0)$. See Fig. 5 for an illustration.

Our approach ensures that $\mathcal{C} - \{(v_n, v_2)\}$ is drawn x-monotone, hence planar. The last edge (v_n, v_2) of $\mathcal{C}$ outside of the bounding box containing the vertices; thus, it is crossing-free. Further, the spikes are not involved in crossings with the rim, as they are outside of the bounding box containing the rim edges. Hence, $\mathcal{W}$ is drawn planar. On the other hand, all edges of $\mathcal{M}'$ are drawn as horizontal, non-overlapping line segments. Thus, $\mathcal{M}'$ is drawn planar as well. The slanted segments of $\mathcal{W} - (v_n, v_2)$ are of y-length 1. So, they cannot be crossed by the edges of $\mathcal{M}'$. As the edge (v_n, v_2) is not involved in crossings, it follows that all crossings between $\mathcal{W}$ and $\mathcal{M}'$ form right angles.

Finally, we have to insert the matching edge (v_1, v_i) in $E_{\mathcal{M}} \setminus E'_{\mathcal{M}}$. Since v_i is not incident to a matching edge in $\mathcal{M}'$, it is placed above all matching edges. Then, $(v_1, v_i) \in \mathcal{W}$ does not cross a matching edge, so we can use this edge as a double edge.

We will now prove the area bound of the drawing algorithm. To that end, we remove all columns that contain neither a vertex, nor a bend. First, we count the rows used. Since we remove the matching edge incident to v_1, the matching $\mathcal{M}'$ has $k \leq n/2 - 1$ matching edges. We place the bottommost vertex in row 1 and the topmost vertex, that is, vertex v_1, in row $2k + 3$. We add one extra bend in row 0 for the edge (v_n, v_2). Thus, our drawing uses $2k + 3 + 1 \leq n + 2$ rows. Next, we count the columns used. The vertices $v_2, \ldots, v_n$ are each placed in their own column. Every spike has exactly one bend in the column of a vertex. An edge (v_i, v_{i+1}) of rim $\mathcal{W}$ has exactly one bend in a vertex column, except for the case that $y(v_i) > y(v_{i-1}), y(v_{i+1})$, in which it needs an extra bend between v_i and v_{i+1}, $i = 1, \ldots, n-1$. Clearly, there can be at most $n/2 - 1$ vertices satisfying this condition. Since the edge (v_n, v_2) uses an extra column to the right of v_n, our drawing uses $(n - 1) + (n/2 - 1) + 1 = 1.5n - 1$ columns. $\square$

Theorem 8. *An outerpath and a matching on a common set of n vertices admit a* RAC-SIM *drawing on an integer grid of size $(3n - 2) \times (3n - 2)$ with two bends per edge and one bend, respectively. The drawing can be computed in $O(n)$ time.*

Sketch of Proof. We augment the outerpath to a maximal outerpath. Removing its outer cycle, the result is a caterpillar, which determines the x-coordinates of the vertices as outlined in Thm. 4. Then, the y-coordinates are computed similar to Thm. 7 such that the matching is planar. A detailed proof is given in the full version [4]. $\square$

5 Conclusions and Open Problems

The results presented in this paper raise several questions that remain open.
1. Is it possible to reduce the number of bends per edge for the classes of graphs that we presented in this paper? What additional non-trivial classes of graphs admit a RACSIM drawing with better-than-general number of bends?
2. As a variant of the problem, it may be possible to reduce the required number of bends per edge by relaxing the strict constraint that intersections must be right-angled and instead ask for drawings that have close-to-optimal crossing resolution.
3. What is the computational complexity of the general problem: Given two or more planar graphs on the same set of vertices and an integer k, is there a RACSIM drawing in which each graph is drawn with at most k bends per edge, and the crossings are at right angles?

References

1. Angelini, P., Battista, G.D., Frati, F., Patrignani, M., Rutter, I.: Testing the simultaneous embeddability of two graphs whose intersection is a biconnected or a connected graph. J. Discrete Algorithms 14, 150–172 (2012)
2. Angelini, P., Geyer, M., Kaufmann, M., Neuwirth, D.: On a tree and a path with no geometric simultaneous embedding. J. Graph Algorithms Appl. 16(1), 37–83 (2012)
3. Argyriou, E.N., Bekos, M.A., Kaufmann, M., Symvonis, A.: Geometric RAC simultaneous drawings of graphs. J. Graph Algorithms Appl. 17(1), 11–34 (2013)
4. Bekos, M.A., van Dijk, T.C., Kindermann, P., Wolff, A.: Simultaneous drawing of planar graphs with right-angle crossings and few bends. Arxiv report (2014), http://arxiv.org/abs/1408.3325
5. Bläsius, T., Kobourov, S.G., Rutter, I.: Simultaneous embedding of planar graphs. In: Tamassia, R. (ed.) Handbook of Graph Drawing and Visualization, ch. 11, pp. 349–381. CRC Press (2013)
6. Brass, P., Cenek, E., Duncan, C.A., Efrat, A., Erten, C., Ismailescu, D.P., Kobourov, S.G., Lubiw, A., Mitchell, J.S.: On simultaneous planar graph embeddings. Comput. Geom. Theory Appl. 36(2), 117–130 (2007)
7. Brightwell, G., Scheinerman, E.R.: Representations of planar graphs. SIAM J. Discrete Math. 6(2), 214–229 (1993)
8. Cabello, S., van Kreveld, M., Liotta, G., Meijer, H., Speckmann, B., Verbeek, K.: Geometric simultaneous embeddings of a graph and a matching. J. Graph Algorithms Appl. 15(1), 79–96 (2011)
9. Erten, C., Kobourov, S.G.: Simultaneous embedding of a planar graph and its dual on the grid. Theory Comput. Syst. 38(3), 313–327 (2005)
10. Erten, C., Kobourov, S.G.: Simultaneous embedding of planar graphs with few bends. J. Graph Algorithms Appl. 9(3), 347–364 (2005)
11. Kammer, F.: Simultaneous embedding with two bends per edge in polynomial area. In: Arge, L., Freivalds, R. (eds.) SWAT 2006. LNCS, vol. 4059, pp. 255–267. Springer, Heidelberg (2006)
12. Kaufmann, M., Wiese, R.: Embedding vertices at points: Few bends suffice for planar graphs. J. Graph Algorithms Appl. 6(1), 115–129 (2002)
13. Nash-Williams, C.: Decomposition of finite graphs into forests. J. London Math. Soc. 39, 12 (1964)

An Improved Algorithm for Parameterized Edge Dominating Set Problem

Ken Iwaide and Hiroshi Nagamochi

Department of Applied Mathematics and Physics,
Graduate School of Informatics,
Kyoto University, Japan
{iwaide,nag}@amp.i.kyoto-u.ac.jp

Abstract. An edge dominating set of a graph $G = (V, E)$ is a subset $M \subseteq E$ of edges such that each edge in $E \setminus M$ is incident to at least one edge in M. In this paper, we consider the parameterized edge dominating set problem which asks us to test whether a given graph has an edge dominating set with size bounded from above by an integer k or not, and we design an $O^*(2.2351^k)$-time and polynomial-space algorithm. This is an improvement over the previous best time bound of $O^*(2.3147^k)$. We also show that a related problem: the parameterized weighted edge dominating set problem can be solved in $O^*(2.2351^k)$ time and polynomial space.

1   Introduction

An *edge dominating set* of a graph $G = (V, E)$ is a subset $M \subseteq E$ of edges in the graph such that each edge in $E \setminus M$ is incident with at least one edge in M. The *edge dominating set problem* (EDS) is to find a minimum edge dominating set of a given graph. The problem is one of the basic problems highlighted by Garey and Johnson [4] in their work on NP-completeness. Yanakakis and Gavril [14] showed that EDS is NP-hard even in planar or bipartite graphs of maximum degree 3. Randerath and Schiermeyer [7] designed an $O^*(1.4423^m)$-time and polynomial-space algorithm for EDS, where $m = |E|$ and O^* notation suppresses all polynomially bounded factors. The result was improved to $O^*(1.4423^n)$ by Raman *et al.* [6], where $n = |V|$. Considering the treewidth of the graph, Fomin *et al.* [3] obtained an $O^*(1.4082^n)$-time and exponential-space algorithm. With the measure and conquer method, van Rooij and Bodlaender [8] designed an $O^*(1.3226^n)$-time and polynomial-space algorithm and an improved $O^*(1.3160^n)$-time and polynomial-space algorithm was presented by Xiao and Nagamochi [12]. For EDS in graphs of maximum degree 3, the best algorithm is an $O^*(1.2721^n)$-time and polynomial-space algorithm due to Xiao and Nagamochi [13].

The *parameterized edge dominating set problem* (PEDS) is, given a graph $G = (V, E)$ with an integer k, to decide whether there is an edge dominating set of size up to k. It is known that there is an FPT algorithm for PEDS; we can design an algorithm with the running time $f(k)poly(n)$ to solve the problem, where $f(k)$ is a function of k and $poly(n)$ is a polynomial of the number of vertices in G. For PEDS, an $O^*(2.6181^k)$-time and polynomial-space algorithm was given by Fernau [2]. Fomin *et al.* [3] obtained

M.S. Rahman and E. Tomita (Eds.): WALCOM 2015, LNCS 8973, pp. 234–245, 2015.
© Springer International Publishing Switzerland 2015

an $O^*(2.4181^k)$-time and exponential-space algorithm based on dynamic programming on treewidth-bounded graphs. With the measure and conquer method, Binkele-Raible and Fernau [1] designed an $O^*(2.3819^k)$-time and polynomial-space algorithm. Xiao et al. [10] give an $O^*(2.3147^k)$-time and polynomial-space branching algorithm. For PEDS in graphs of maximum degree 3, the best parameterized algorithm is an $O^*(2.1479^k)$-time and polynomial-space algorithm due to Xiao and Nagamochi [11].

EDS and PEDS are related to the *vertex cover problem*. A *vertex cover* of a graph is a set of vertices such that each edge of the graph is incident to at least one vertex in the set. The set of endpoints of all edges in any edge dominating set is a vertex cover. To find an edge dominating set of a graph, we may enumerate vertex covers of the graph and construct edge dominating sets from the vertex covers. Many previous algorithms are based on enumeration of vertex covers. We enumerate candidates of such edge dominating sets by branching on a vertex: fixing it as a vertex incident on at least one edge in an edge dominating set with a bounded size or not. In the $O^*(2.3147^k)$-time algorithm to PEDS, Xiao et al. [10] observed that branching on vertices in a local structure called "2-path component" is the most inefficient among branchings on other local structures, and that reducing the number of branchings on 2-path components leads to an improvement over the time complexity. For this, they retained branching on 2-path components until no other structure remains, and effectively skipped subinstances that will not deliver edge dominating sets with a bounded size by systematically treating the set of 2-path components. In this paper, identifying new local structures, called "bi-claw," "leg-triangle" and "tri-claw components" and establishing a refined lower bound on the size of edge dominating sets, we design an $O^*(2.2351^k)$-time and polynomial-space algorithm.

Section 2 gives some terminologies and notations and introduces our branching operations of our algorithm. After Section 3 describes our algorithm that consists of three major stages, Section 4 analyzes the time complexity by deriving an upper bound on the number of all subinstances. Section 5 discusses a weighted variant of PEDS. Section 6 makes some concluding remarks. For space limitation, the proofs of lemmata are available in the full version of the paper [5].

2 Preliminaries

2.1 Terminology and Notation

For non-negative integers $k_1, k_2, \ldots, k_m$, a multinomial coefficient $\frac{(\sum_{i=1}^m k_i)!}{k_1! \cdots k_m!}$ is denoted by $\binom{\sum_{i=1}^m k_i}{k_1, \ldots, k_m}$.

Lemma 1. *Let $k_1, k_2, \ldots, k_m$ be non-negative integers, where $m \geq 1$. Then for any positive reals $\gamma_1, \gamma_2, \ldots, \gamma_m$ such that $\sum_{i=1}^m 1/\gamma_i \leq 1$, it holds that*

$$\binom{\sum_{i=1}^m k_i}{k_1, k_2, \ldots, k_m} \leq \prod_{i=1}^m \gamma_i^{k_i}.$$

The set of vertices and edges in a graph H is denoted by $V(H)$ and $E(H)$, respectively. For a vertex v in a graph, let $N(v)$ denote a set of neighbors of v and let $N[v]$

denote a set of v and its neighbors (i.e., $N[v] = \{v\} \cup N(v)$). A vertex of degree d is called a *degree-d* vertex. The degree of a vertex v in a graph H is denoted by $d(v; H)$. For a set F of edges, we use $V(F)$ to denote a set of vertices incident on at least one edge in F, and we say that F *covers* a vertex set $S \subseteq V$ if $V(F) \supseteq S$. For a subset $S \subseteq V$ of vertices, $G[S]$ denote the subgraph of G induced by S. A cycle of length ℓ is called an *ℓ-cycle*, and is denoted by the sequence $v_1 v_2 \ldots v_\ell$ of vertices in it, where the cycle contains edges $v_1 v_2, \ldots, v_{\ell-2} v_{\ell-1}$ and $v_\ell v_1$. A connected component containing only one vertex is called *trivial*. We define five types of connected components as follows:

a *clique component*, a connected component that is a complete subgraph;

- a *2-path component*, a connected component consisting of a degree-2 vertex u_1 and its two degree-1 neighbors $u_0, u_2 \in N(u_1)$, denoted by $u_0 u_1 u_2$, as illustrated in Fig. 1(a);

- a *bi-claw component*, a connected component consisting of two adjacent degree-3 vertices u_1 and v_1 and their four degree-1 neighbors $u_0, u_2 \in N(u_1)$ and $v_0, v_2 \in N(v_1)$, denoted by $(u_0 u_1 u_2)(v_0 v_1 v_2)$, as illustrated in Fig. 1(b);

- a *legged triangle component* (or *leg-triangle component*), a connected component consisting of two adjacent degree-3 vertices u_1 and v_1, their two degree-1 neighbors $u_0 \in N(u_1)$ and $v_0 \in N(v_1)$ and one common degree-2 neighbor $w \in N(u_1) \cap N(v_1)$, denoted by $u_0(u_1 w v_1)v_0$, as illustrated in Fig. 1(c); and

- a *tri-claw component*, a connected component consisting of three degree-3 vertices u_1, v_1 and w_1, their six degree-1 neighbors $u_0, u_2 \in N(u_1)$, $v_0, v_2 \in N(v_1)$ and $w_0, w_2 \in N(w_1)$ and their common degree-3 neighbor $t \in N(u_1) \cap N(v_1) \cap N(w_1)$, denoted by $t(u_0 u_1 u_2)(v_0 v_1 v_2)(w_0 w_1 w_2)$, as illustrated in Fig. 1(d).

The last four types of components, 2-path, bi-claw, leg-triangle and tri-claw components are called *bad components* collectively.

2.2 Instances with Covered and Discarded Vertices

Throughout our algorithm, we do not modify a given graph $G = (V, E)$ or a parameter k, but fix vertices to *covered* vertices or *discarded* vertices so that a pair of the sets C and D of covered and discarded vertices gives an instance (C, D) that asks to find an edge dominating set M of G such that $C \subseteq V(M) \subseteq V \setminus D$. We call such an edge dominating set a (C, D)-*eds* for short. An instance (C, D) is called *feasible* if it admits a (C, D)-eds, and is called *k-feasible* if it admits a (C, D)-eds M of size $|M| \leq k$. We call vertices in $V \setminus (C \cup D)$ *undecided* and denote by U the set of undecided vertices.

We use two kinds of fundamental branching operations. One is to branch on an undecided vertex $v \in U$ in (C, D): fix v as a new covered vertex in the first branch or as a new discarded vertex in the second branch. This is based on the fact that there is a (C, D)-eds M with $v \in V(M)$ or there is no such (C, D)-eds. Then we also fix all the vertices in $N(v)$ as covered vertices in the second branch, since any edge $e = vw$ incident to v needs to be incident to an edge dominating set at the vertex w. The other is to branch on a 4-cycle $v_0 v_1 v_2 v_3$ over undecided vertices: fix vertices v_0 and v_2 as new covered vertices or fix vertices v_1 and v_3 as new covered vertices. This is based on the fact that for any edge dominating set M, the set $V(M)$ is a vertex cover and one of $\{v_0, v_2\}$ and $\{v_1, v_3\}$ is contained in any vertex cover [9]. Van Rooij and Bodlaender [8] found the following solvable case.

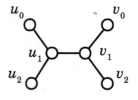

(a) A 2-path component $u_0u_1u_2$

(b) A bi-claw component $(u_0u_1u_2)(v_0v_1v_2)$

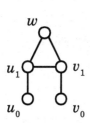

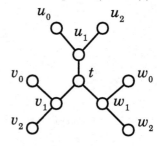

(c) A leg-triangle component $u_0(u_1wv_1)v_0$

(d) A tri-claw component
$t(u_0u_1u_2)(v_0v_1v_2)(w_0w_1w_2)$

Fig. 1. The four types of bad components

Lemma 2. [8] *A minimum (C, D)-eds of an instance (C, D) such that $G[U]$ contains only clique components can be found in polynomial time.*

We denote by U_1 the set of vertices of all clique components in $G[U]$, and let $U_2 = U \setminus U_1$. An instance (C, D) is called a *leaf instance* if $U_2 = \emptyset$. By Lemma 2, we only need to select vertices from U_2 to apply branching operations until all instances become leaf instances.

The next lower bound on the size of (C, D)-edses is immediate since for each clique component Q in $G[U]$, it holds that $|V(Q) \cap V(M)| \geq |V(Q)| - 1$.

Lemma 3. *For any (C, D)-eds M in a graph G, it holds that*

$$|V(M)| \geq |C| + \sum\{|V(Q)| - 1 \mid \text{clique components } Q \text{ in } G[U]\}.$$

Based on this, we define the *measure* μ of an instance (C, D) to be

$$\mu(C, D) = 2k - |C| - \sum\{|V(Q)| - 1 \mid \text{clique components } Q \text{ in } G[U]\}.$$

We do not need to generate any instances (C, D) with $\mu(C, D) < 0$ since they are not k-feasible. In this paper, we introduce the following new lower bound.

Lemma 4. *Let M be a (C, D)-eds in a graph G. Then for any subset $S \subseteq C$ it holds that*

$$|M| \geq \sum\{\lceil |V(H)|/2 \rceil \mid \text{components } H \text{ in } G[S]\} \geq \lceil |S|/2 \rceil.$$

Branching on a bad component H in $G[U_2]$ means to keep branching on vertices in $U_2 \cap V(H)$ until all vertices in $V(H)$ are contained in $C \cup D \cup U_1$. We treat a series of such branchings as an operation of branching on H that generates r new instances defined as follows. For each type of a bad component H, we define the number r and $C^{(j)}(H)$ (resp., $D^{(j)}(H)$), $j = 1, 2, \ldots, r$ to be a set of vertices of H fixed as covered (resp., discarded) vertices in the j-th branch:

For a 2-path component $H_1 = u_0 u_1 u_2$, by branching on u_1, we can branch on H_1 into $r = 2$ branches:

1. $C^{(1)}(H_1) = \{u_1\}$ and $D^{(1)}(H_1) = \emptyset$; and
2. $C^{(2)}(H_1) = \{u_0, u_2\}$ and $D^{(2)}(H_1) = \{u_1\}$.

For a bi-claw component $H_2 = (u_0 u_1 u_2)(v_0 v_1 v_2)$, where at least one of adjacent vertices u_1 and v_1 must be in $V(M)$ of any (C, D)-eds M, we can branch on this component into $r = 3$ branches:

1. $C^{(1)}(H_2) = \{u_1, v_1\}$ and $D^{(1)}(H_2) = \emptyset$;
2. $C^{(2)}(H_2) = \{u_0, u_2, v_1\}$ and $D^{(2)}(H_2) = \{u_1\}$; and
3. $C^{(3)}(H_2) = \{u_1, v_0, v_2\}$ and $D^{(3)}(H_2) = \{v_1\}$.

For a leg-triangle component $H_3 = u_0(u_1 w v_1)v_0$, where at least one of adjacent vertices u_1 and v_1 must be in $V(M)$ of any (C, D)-eds M, we can branch on this component into $r = 3$ branches:

1. $C^{(1)}(H_3) = \{u_1, v_1\}$ and $D^{(1)}(H_3) = \emptyset$;
2. $C^{(2)}(H_3) = \{u_0, v_1, w\}$ and $D^{(2)}(H_3) = \{u_1\}$; and
3. $C^{(3)}(H_3) = \{u_1, v_0, w\}$ and $D^{(3)}(H_3) = \{v_1\}$.

For a tri-claw component $H_4 = t(u_0 u_1 u_2)(v_0 v_1 v_2)(w_0 w_1 w_2)$, we can branch on u_1, v_1 and w_1 sequentially to generate the following $r = 8$ branches:

1. $C^{(1)}(H_4) = \{u_1, v_1, w_1\}$ and $D^{(1)}(H_4) = \emptyset$;
2. $C^{(2)}(H_4) = \{t, u_0, u_2, v_1, w_1\}$ and $D^{(2)}(H_4) = \{u_1\}$;
3. $C^{(3)}(H_4) = \{t, u_1, v_0, v_2, w_1\}$ and $D^{(3)}(H_4) = \{v_1\}$;
4. $C^{(4)}(H_4) = \{t, u_1, v_1, w_0, w_2\}$ and $D^{(4)}(H_4) = \{w_1\}$;
5. $C^{(5)}(H_4) = \{t, u_0, u_2, v_0, v_2, w_1\}$ and $D^{(5)}(H_4) = \{u_1, v_1\}$;
6. $C^{(6)}(H_4) = \{t, u_1, v_0, v_2, w_0, w_2\}$ and $D^{(6)}(H_4) = \{v_1, w_1\}$;
7. $C^{(7)}(H_4) = \{t, u_0, u_2, v_1, w_0, w_2\}$ and $D^{(7)}(H_4) = \{u_1, w_1\}$; and
8. $C^{(8)}(H_4) = \{t, u_0, u_2, v_0, v_2, w_0, w_2\}$ and $D^{(8)}(H_4) = \{u_1, v_1, w_1\}$.

For each of the above branch, we define two kinds of values α and β which will be summed up to give lower bounds on the size of a (C', D')-eds of a leaf instance (C', D'). For each (i, j), let

$$\alpha_{i,j} = |C^{(j)}(H_i)| \text{ and } \beta_{i,j} = \sum \{\lceil |V(T)|/2 \rceil \mid \text{components } T \text{ in } G[C^{(j)}(H_i)]\}.$$

Observe that $\beta_{i,j}$ is a lower bound on the size of a $(C^{(j)}(H_i), \emptyset)$-eds by Lemma 4. For $(i, j) \in \{(1, 1), (1, 2), (2, 2), (2, 3), (3, 2), (4, 8)\}$, the graph $G[C^{(j)}(H_i)]$ contains only isolated vertices, and $\beta_{i,j} = |C^{(j)}(H_i)| = \alpha_{i,j}$. For other (i, j), the graph $G[C^{(j)}(H_i)]$ consists of exactly one nontrivial component of size $p \in \{2, 3\}$ and $|C^{(j)}(H_i)| - p$ isolated vertices, and $\beta_{i,j} = \lceil p/2 \rceil + (|C^{(j)}(H_i)| - p) = |C^{(j)}(H_i)| - 1 = \alpha_{i,j} - 1$.

In this paragraph, we introduce criteria in choosing 4-cycle/vertices to branch on used in our algorithm. For a subset $S \subseteq U_2$ of vertices, we let q_S and b_S denote the sum of $|V(Q)| - 1$ over all clique components Q and the number of bad components newly generated by removing S from $G[U_2]$, respectively. A 4-cycle $v_0 v_1 v_2 v_3$ in $G[U_2]$ is called *admissible* if $b_{\{v_0,v_2\}} + b_{\{v_1,v_3\}} \leq 1$. A vertex v in $G[U_2]$ such that $b_v = x$ and $b_{N[v]} = y$ is called an (x, y)-*vertex*. A vertex v in $G[U_2]$ is called *optimal* if it satisfies a condition (c-i) below with the minimum i over all vertices in $G[U_2]$:

(c-1) v is a degree-3 $(0, 0)$-vertex;

(c-2) v is a degree-2 (x, y)-vertex with $x + y \leq 1$ and $q_v \geq 1$;

(c-3) (i) v is in an admissible 4-cycle;

(ii) v is a degree-d (x, y)-vertex such that $2 \leq d \leq 3$, $x + y \leq 1$ and $q_v + q_{N[v]} \geq 4 - d$;

(iii) v is a degree-d (x, y)-vertex such that $2 \leq d \leq 3$, $x + y \leq 1$, $q_{N[v]} = 3 - d$ and removing each of v and $N[v]$ produces no new 2-path component; or

(iv) v is a degree-3 $(0, 1)$-vertex such that $G[U_2 \setminus \{v\}]$ contains at least one degree-3 $(0, 0)$-vertex and removing $N[v]$ produces exactly one new 2-path component;

(c-4) v is a degree-2 vertex with $q_v = 1$;

(c-5) v is a degree-3 vertex; and

(c-6) v is a degree-2 vertex.

3   The Algorithm

Given a graph G and an integer k, our algorithm returns **TRUE** if it admits an edge dominationg set of size $\leq k$ or **FALSE** otherwise. The algorithm is designed to be a procedure that returns **TRUE** if a given instance (C, D) is k-feasible or **FALSE** otherwise, by branching on a vertex/4-cycle/bad component in (C, D) to generate new smaller instances $(C^{(1)}, D^{(1)}), \ldots, (C^{(r)}, D^{(r)})$, to each of which the procedure is recursively applied. The procedure is initially given an instance $(\emptyset, \emptyset)$, and always returns **FALSE** whenever $\mu(C, D) < 0$ holds.

Our algorithm takes three stages. The first stage keeps branching on vertices of degree ≥ 4, and retains the set $\mathcal{B}$ of all the produced bad components without branching on them. The second stage keeps branching on optimal vertices of degree ≤ 3, immediately branching on any newly produced bad component before it chooses the next optimal vertex to branch on. The third stage generates leaf instances by fixing all undecided vertices in the bad components in $\mathcal{B}$, where we try to decrease the number of leaf instances to be generated based on some lower bound on the size of solutions of leaf instances. To derive the lower bounds in the third stage, we let C_i store all vertices fixed to covered vertices during branching operations in the i-th stage. Formally EDSSTAGE1 is described as follows.

Algorithm. EDSSTAGE1(C, D)

Input: A graph $G = (V, E)$ with an integer k, and subsets C and D of V (initially, $C = D = \emptyset$).

Output: **TRUE** if (C, D) is k-feasible or **FALSE** otherwise.

1: **if** $\mu(C, D) < 0$ **then**

2:      **return FALSE**

3: **else if** there is a vertex v of degree ≥ 4 in $G[U_2]$ **then**

```
4:     return EDSSTAGE1(C ∪ {v}, D) ∨ EDSSTAGE1(C ∪ N(v), D ∪ {v})
5: else
6:     C₁ := C; C₂ := ∅;
7:     Let B store all bad components in G[U₂];
8:     return EDSSTAGE2(C₁, C₂, B, D)
9: end if
```

For a given instance (G, k) of PEDS, let $\mathcal{I}_1$ denote the set of all instances constructed immediately after the first stage. Let $V(\mathcal{B})$ denote the set of vertices in the bad components in $\mathcal{B}$. Given an instance $(C_1, C_2, \mathcal{B}, D) \in \mathcal{I}_1$, the second stage EDSSTAGE2 fixes all vertices in $U_2 \setminus V(\mathcal{B})$ to covered/discarded vertices by repeatedly branching on optimal vertices or any newly produced bad component in $G[U_2 \setminus V(\mathcal{B})]$ if it exists. During the second stage, the sets C_1 and $\mathcal{B}$ obtained in the first stage never change. When no vertex is left in $U_2 \setminus V(\mathcal{B})$, we switch to the third stage. Formally EDSSTAGE2 is described as follows.

Algorithm. EDSSTAGE2$(C_1, C_2, \mathcal{B}, D)$

```
Input:  A graph G = (V, E) with an integer k, disjoint subsets C₁, C₂, D ⊆ V and a set of
        bad components B in G[U₂].
Output: TRUE if (C₁ ∪ C₂, D) is k-feasible or FALSE otherwise.
 1: if μ(C₁ ∪ C₂, D) < 0 then
 2:     return FALSE
 3: else if there is a 2-path component H₁ in G[U₂ \ V(B)] then
 4:     return ⋁ⱼ₌₁,₂ EDSSTAGE2(C₁, C₂ ∪ C⁽ʲ⁾(H₁), B, D ∪ D⁽ʲ⁾(H₁))
 5: else if there is a bi-claw component H₂ in G[U₂ \ V(B)] then
 6:     return ⋁₁≤ⱼ≤₃ EDSSTAGE2(C₁, C₂ ∪ C⁽ʲ⁾(H₂), B, D ∪ D⁽ʲ⁾(H₂))
 7: else if there is a leg-triangle component H₃ in G[U₂ \ V(B)] then
 8:     return ⋁₁≤ⱼ≤₃ EDSSTAGE2(C₁, C₂ ∪ C⁽ʲ⁾(H₃), B, D ∪ D⁽ʲ⁾(H₃))
 9: else if there is a tri-claw component H₄ in G[U₂ \ V(B)] then
10:     return ⋁₁≤ⱼ≤₈ EDSSTAGE2(C₁, C₂ ∪ C⁽ʲ⁾(H₄), B, D ∪ D⁽ʲ⁾(H₄))
11: else if U₂ \ V(B) ≠ ∅ then
12:     Choose an optimal vertex v in G[U₂ \ V(B)];
13:     if v is in an admissible 4-cycle v₀v₁v₂v₃ of condition (c-4) then
14:         return  EDSSTAGE2(C₂ ∪ {v₀, v₂}, D, B, C₁) ∨ EDSSTAGE2(C₁, C₂ ∪
            {v₁, v₃}, B, D)
15:     else
16:         return EDSSTAGE2(C₁, C₂∪{v}, B, D) ∨ EDSSTAGE2(C₁, C₂∪N(v), B, D∪{v})
17:     end if
18: else /* Now U₂ = V(B) */
19:     return EDSSTAGE3(C₁, C₂, B, D)
20: end if
```

Let $\mathcal{I}_2$ denote the set of all instances constructed immediately after the second stage. Consider an instance $I = (C_1, C_2, \mathcal{B}, D) \in \mathcal{I}_2$, where the graph $G[U_2]$ consists of the bad components in $\mathcal{B}$ retained at the first stage. Let $\mathcal{B}_1$ (resp., $\mathcal{B}_2, \mathcal{B}_3$ and $\mathcal{B}_4$) be the sets of 2-path (resp., bi-claw, leg-triangle and tri-claw) components in $\mathcal{B}$, and $y_i = |\mathcal{B}_i|$, $i = 1, 2, 3, 4$ in $I \in \mathcal{I}_2$. To obtain a leaf instance from the instance I, we need to fix all vertices in $V(\mathcal{B})$. The number of all leaf instances that can be constructed from the instance $I \in \mathcal{I}_2$ is $\prod_{i=1}^{4} r_i^{y_i} = 2^{y_1} \cdot 3^{y_2} \cdot 3^{y_3} \cdot 8^{y_4}$, where r_i is the number of subinstances generated by branching on a bad component $H \in \mathcal{B}_i$.

In the third stage, we avoid constructing of some "k-infeasible" leaf instances among all leaf instances. For a leaf instance $I' = (C' = C_1 \cup C_2 \cup C_3, D')$ obtained from the instance $I \in \mathcal{I}_2$, where C_3 denotes the set of undecided vertices in $V(\mathcal{B})$ that are fixed to covered vertices in I', we let $w_{i,j}$ be the number of bad components in $\mathcal{B}_i$ to which the j-th branch is applied to generate I', and call the vector w with these 16 entries $w_{i,j}$ the *occurrence vector* of I'. Note that $\sum_{i,j} \alpha_{i,j} w_{i,j} = |C_3|$ holds, and that $\sum_{i,j} \beta_{i,j} w_{i,j}$ is a lower bound on the size of (C_3, D')-eds by Lemma 4, since no edge in G joins two components in $\mathcal{B}$. We derive two necessary conditions for a vector w to be the occurrence vector of a k-feasible leaf instance $I' = (C', D')$. One is that $2k \geq 2|M| \geq |V(M)| \geq |C_1| + |C_2| + |C_3|$, i.e.,

$$2k \geq |C_1| + |C_2| + \sum_{i,j} \alpha_{i,j} w_{i,j}. \tag{1}$$

Observe that there is no edge between C_3 and C_2 in I', since any vertex in C_2 is contained in some component in $G[U_2 \setminus V(\mathcal{B})]$ during an execution of EDSSTAGE2. Hence $\sum_{i,j} \beta_{i,j} w_{i,j} + \lceil |C_2|/2 \rceil$ is a lower bound on the size of a $(C_3 \cup C_2, D')$-eds by Lemma 4, and another necessary condition is given by

$$k \geq |C_2|/2 + \sum_{i,j} \beta_{i,j} w_{i,j}. \tag{2}$$

Note that the number $\ell(w)$ of leaf instances I' whose occurrence vectors are given by w is

$$\ell(w) = \binom{y_1}{w_{1,1}, w_{1,2}} \binom{y_2}{w_{2,1}, w_{2,2}, w_{2,3}} \binom{y_3}{w_{3,1}, w_{3,2}, w_{3,3}} \binom{y_4}{w_{4,1}, w_{4,2}, \ldots, w_{4,8}}. \tag{3}$$

For each instance $I = (C_1, C_2, \mathcal{B}, D) \in \mathcal{I}_2$, the third stage EDSSTAGE3 generates an occurrence vector w satisfying the conditions (1) and (2) and $\sum_j w_{i,j} = y_i$, $1 \leq i \leq 4$, and constructs all leaf instances $I' = (C_1 \cup C_2 \cup C_3, D')$ from $I \in \mathcal{I}_2$ with the vector w, before it returns TRUE if one of the leaf instances is k-feasible or FALSE otherwise. Formally EDSSTAGE3 is described as follows.

Algorithm. EDSSTAGE3$(C_1, C_2, \mathcal{B}, D)$

Input: A graph $G = (V, E)$ with an integer k, disjoint subsets $C_1, C_2, D \subseteq V$ and a set of bad components $\mathcal{B}$ in $G[U_2]$.
Output: TRUE if $(C_1 \cup C_2, D)$ is k-feasible or FALSE otherwise.
1: Let $\mathcal{B}_1$ (resp. $\mathcal{B}_2, \mathcal{B}_3$ and $\mathcal{B}_4$) be a set of 2-path (resp. bi-claw, leg-triangle and tri-claw) components in $\mathcal{B}$, and $y_i := |\mathcal{B}_i|$, $i = 1, 2, 3, 4$;
2: **for** each occurrence vector w that satisfies the conditions (1) and (2) and $\sum_j w_{i,j} = y_i$, $1 \leq i \leq 4$ **do**
3:     **for** each combination of partitions of $\mathcal{B}_1, \mathcal{B}_2, \mathcal{B}_3$ and $\mathcal{B}_4$ into
$$\mathcal{B}_1^{(1)} \cup \mathcal{B}_1^{(2)} = \mathcal{B}_1, \mathcal{B}_2^{(1)} \cup \mathcal{B}_2^{(2)} \cup \mathcal{B}_2^{(3)} = \mathcal{B}_2, \mathcal{B}_3^{(1)} \cup \mathcal{B}_3^{(2)} \cup \mathcal{B}_3^{(3)} = \mathcal{B}_3, \text{ and}$$

$\mathcal{B}_4^{(1)} \cup \mathcal{B}_4^{(2)} \cup \cdots \cup \mathcal{B}_4^{(8)} = \mathcal{B}_4$ such that $|\mathcal{B}_j^{(j)}| = w_{i,j}$ for all i and j; **do**

4:     **for** each $j = 1, 2$ and each 2-path component $H_1 \in \mathcal{B}_1^{(j)}$ **do**
5:         $C_3 := C^{(j)}(H_1); D := D \cup D^{(j)}(H_1)$
6:     **end for**;
7:     **for** each $j = 1, 2, 3$ and each bi-claw component $H_2 \in \mathcal{B}_2^{(j)}$ **do**
8:         $C_3 := C^{(j)}(H_2); D := D \cup D^{(j)}(H_2)$
9:     **end for**;
10:    **for** each $j = 1, 2, 3$ and each leg-triangle component $H_3 \in \mathcal{B}_3^{(j)}$ **do**
11:        $C_3 := C^{(j)}(H_3); D := D \cup D^{(j)}(H_3)$
12:    **end for**;
13:    **for** each $j = 1, 2, \ldots, 8$ and each tri-claw component $H_4 \in \mathcal{B}_4^{(j)}$ **do**
14:        $C_3 := C^{(j)}(H_4); D := D \cup D^{(j)}(H_4)$
15:    **end for**; /* Now $U_2 = \emptyset$ and $(C_1 \cup C_2 \cup C_3, D)$ is a leaf instance */
16:    Test whether $(C = C_1 \cup C_2 \cup C_3, D)$ is k-feasible or not by computing a minimum (C, D)-eds by Lemma 2
17:   **end for**
18: **end for**;
19: **if** there is a k-feasible leaf instance $(C_1 \cup C_2 \cup C_3, D)$ in the for loop **then**
20:    **return** TRUE
21: **else**
22:    **return** FALSE
23: **end if**

4 The Analysis

For a given instance (G, k) of PEDS, let $\mathcal{I}_i$, $i = 1, 2, 3$ be the set of all instances constructed immediately after the i-th stage during the execution of EDSSTAGE1$(\emptyset, \emptyset)$, where $\mathcal{I}_3$ is the set of all leaf instances, which correspond to the leaf nodes of the search tree of the execution. To analyze the time complexity of our algorithm, it suffices to derive an upper bound on $|\mathcal{I}_3|$.

Lemma 5. *For any non-negative integer x_1, the number of instances $I = (C_1, \emptyset, \mathcal{B}, D) \in \mathcal{I}_1$ with $|C_1| = x_1$ is $O(1.380278^{x_1})$.*

Lemma 6. *For any non-negative integer x_2 and an instance $I = (C_1, \emptyset, \mathcal{B}, D) \in \mathcal{I}_1$, the number of instances $I' = (C_1, C_2, \mathcal{B}, D') \in \mathcal{I}_2$ with $|C_2| = x_2$ that can be generated from I is $O(1.494541^{x_2})$.*

From these, we obtain the next.

Lemma 7. *For any non-negative integers x_1 and x_2, the number of instances $(C_1, C_2, \mathcal{B}, D) \in \mathcal{I}_2$ such that $|C_1| = x_1$ and $|C_2| = x_2$ is $O(1.380278^{x_1} \cdot 1.494541^{x_2})$.*

Note that the number of combinations (x_1, x_2) for $(|C_1|, |C_2|)$ is $O(n^2)$. For a given instance $(C_1, C_2, \mathcal{B}, D) \in \mathcal{I}_2$, the number of possible occurrence vectors w satisfying the conditions (1) and (2) and $\sum_j w_{i,j} = y_i$, $1 \leq i \leq 4$ is also bounded by a polynomial of n. To show that our algorithm runs in $O^*(2.2351^k)$ time, it suffices to prove that the number of leaf instances generated from an instance $I = (C_1, C_2, \mathcal{B}, D) \in \mathcal{I}_2$

with specified size $|C_1| = x_1$ and $|C_2| = x_2$ and a specified occurrence vector $\boldsymbol{w}$ is $O^*(2.2351^k)$. Let $\mathcal{I}_3(x_1, x_2, \boldsymbol{w})$ denote the set of all such leaf instances. By Lemma 7 and (3), we see that $|\mathcal{I}_3(x_1, x_2, \boldsymbol{w})| = O(1.380278^{x_1} \cdot 1.494541^{x_2} \cdot \ell(\boldsymbol{w}))$.

In what follows, we derive an upper bound on $O(1.380278^{x_1} \cdot 1.494541^{x_2} \cdot \ell(\boldsymbol{w}))$ under the constraints (1) and (2). For this, we merge some entries in $\boldsymbol{w}$ into ten numbers by $z_{1,1} = w_{1,1}$, $z_{1,2} = w_{1,2}$, $z_{2,1} = w_{2,1}$, $z_{2,2} = w_{2,2} + w_{2,3}$, $z_{3,1} = w_{3,1}$, $z_{3,2} = w_{3,2} + w_{3,3}$, $z_{4,1} = w_{4,1}$, $z_{4,2} = w_{4,2} + w_{4,3} + w_{4,4}$, $z_{4,3} = w_{4,5} + w_{4,6} + w_{4,7}$ and $z_{4,4} = w_{4,8}$. Then $\ell(\boldsymbol{w})$ is restated as

$$
\binom{z_{1,1}+z_{1,2}}{z_{1,1}, z_{1,2}} \cdot \binom{z_{2,1}+z_{2,2}}{z_{2,1}, z_{2,2}} \cdot 2^{z_{2,2}} \cdot \binom{z_{3,1}+z_{3,2}}{z_{3,1}, z_{3,2}} \cdot 2^{z_{3,2}} \cdot \binom{z_{4,1}+z_{4,2}+z_{4,3}+z_{4,4}}{z_{4,1}, z_{4,2}, z_{4,3}, z_{4,4}} \cdot 3^{z_{4,2}+z_{4,3}},
$$

which is bounded from above by an exponential function in Lemma 1

$$
\gamma_{1,1}^{z_{1,1}} \gamma_{1,2}^{z_{1,2}} \cdot \gamma_{2,1}^{z_{2,1}} \gamma_{2,2}^{z_{2,2}} \cdot \gamma_{3,1}^{z_{3,1}} \gamma_{3,2}^{z_{3,2}} \cdot \gamma_{4,1}^{z_{4,1}} \gamma_{4,2}^{z_{4,2}} \gamma_{4,3}^{z_{4,3}} \gamma_{4,4}^{z_{4,4}}
$$

for any positive reals $\gamma_{1,1}, \gamma_{1,2}, \gamma_{2,1}, \gamma_{2,2}, \gamma_{3,1}, \gamma_{3,2}, \gamma_{4,1}, \gamma_{4,2}, \gamma_{4,3}$ and $\gamma_{4,4}$ such that $1/\gamma_{1,1} + 1/\gamma_{1,2} \leq 1$, $1/\gamma_{2,1} + 2/\gamma_{2,2} \leq 1$, $1/\gamma_{3,1} + 2/\gamma_{3,2} \leq 1$ and $1/\gamma_{4,1} + 3/\gamma_{4,2} + 3/\gamma_{4,3} + 1/\gamma_{4,4} \leq 1$. Then we have

$$
|\mathcal{I}_3(x_1, x_2, \boldsymbol{w})| = O(1.380278^{x_1} \cdot 1.494541^{x_2} \gamma_{2,1}^{z_{2,1}} \gamma_{2,2}^{z_{2,2}} \gamma_{3,1}^{z_{3,1}} \gamma_{3,2}^{z_{3,2}} \gamma_{4,1}^{z_{4,1}} \gamma_{4,2}^{z_{4,2}} \gamma_{4,3}^{z_{4,3}} \gamma_{4,4}^{z_{4,4}}),
$$

which is bounded by

$$
O(\max\{1.380278^{1/c_1}, 1.494541^{1/c_2}, \gamma_{11}^{1/c_{1,1}}, \gamma_{12}^{1/c_{1,2}}, \gamma_{21}^{1/c_{2,1}}, \gamma_{22}^{1/c_{2,2}},
$$
$$
\gamma_{31}^{1/c_{3,1}}, \gamma_{32}^{1/c_{3,2}}, \gamma_{41}^{1/c_{4,1}}, \gamma_{42}^{1/c_{4,2}}, \gamma_{43}^{1/c_{4,3}}, \gamma_{44}^{1/c_{4,4}}\}^k) \qquad (4)
$$

for any constants c_1, c_2 and $\{c_{i,j}\}$ such that

$$
k \geq c_1 x_1 + c_2 x_2 + c_{1,1} z_{1,1} + c_{1,2} z_{1,2} + c_{2,1} z_{2,1} + c_{2,2} z_{2,2}
$$
$$
+ c_{3,1} z_{3,1} + c_{3,2} z_{3,2} + c_{4,1} z_{4,1} + c_{4,2} z_{4,2} + c_{4,3} z_{4,3} + c_{4,4} z_{4,4}. \qquad (5)
$$

Conditions (1) and (2) are restated as

$$
k \geq x_1/2 + x_2/2 + (z_{1,1} + 2z_{1,2})/2 + (2z_{2,1} + 3z_{2,2})/2
$$
$$
+ (2z_{3,1} + 3z_{3,2})/2 + (3z_{4,1} + 5z_{4,2} + 6z_{4,3} + 7z_{4,4})/2; \qquad (6)
$$

$$
k \geq x_2/2 + (z_{1,1} + 2z_{1,2}) + (z_{2,1} + 3z_{2,2})
$$
$$
+ (z_{3,1} + 2z_{3,2}) + (3z_{4,1} + 4z_{4,2} + 5z_{4,3} + 7z_{4,4}). \qquad (7)
$$

As a linear combination of (6) and (7) with λ and $(1 - \lambda)$, we get (5) for constants $c_1 = \lambda/2, c_2 = 1/2, c_{1,1} = 1 - \lambda/2, c_{1,2} = 2 - \lambda, c_{2,1} = 1, c_{2,2} = 3 - 3\lambda/2, c_{3,1} = 1, c_{3,2} = 2 - \lambda/2, c_{4,1} = 3 - 3\lambda/2, c_{4,2} = 4 - 3\lambda/2, c_{4,3} = 3 - 2\lambda$ and $c_{4,4} = 7 - 7\lambda/2$.

From (4), we obtain $|\mathcal{I}_3(x_1, x_2, \boldsymbol{w})| = O(2.2351^k)$ by setting $\lambda = 0.80142$, $\gamma_{1,1} = 1.61804$, $\gamma_{1,2} = 2.61804$, $\gamma_{2,1} = 2.10457$, $\gamma_{2,2} = 3.81068$, $\gamma_{3,1} = 2.23510$, $\gamma_{3,2} = 3.61931$, $\gamma_{4,1} = 3.60818$, $\gamma_{4,2} = 7.36647$, $\gamma_{4,3} = 11.29854$ and $\gamma_{4,4} = 19.96819$. This establishes the next theorem.

Theorem 1. *Algorithm* EDSSTAGE1, *accompanied by Algorithm* EDSSTAGE2 *and* EDSSTAGE3, *can solve the parameterized edge dominating set problem in* $O^*(2.2351^k)$ *time and polynomial space.*

5   A Related Problem: The Parameterized Weighted Edge Dominating Set Problem

We also consider a weighted variant of PEDS. The *weighted edge dominating set problem* (WEDS) is, given a graph $G = (V, E)$ with an edge weight function $\omega : E \to \mathbb{R}_{\geq 0}$, to find an edge dominating set M of minimum total weight $\omega(M) = \sum_{e \in M} \omega(e)$. The *parameterized weighted edge dominating set problem* (PWEDS) is, given a graph $G = (V, E)$ with an edge weight function $\omega : E \to \mathbb{R}_{\geq 1}$ and a positive real k, to test whether there is an edge dominating set M such that $\omega(M) \leq k$. We show that a modification of our algorithm for PEDS can solve PWEDS in the same time and space complexities as our algorithm does PEDS.

For PWEDS we use the same terminologies and notations as for PEDS; for example, an instance of PWEDS is also denoted by (C, D). Van Rooij and Bodlaender [8] found the following solvable case for a weighted variant of EDS.

Lemma 8. *[8] A minimum (C, D)-eds of an instance (C, D) of WEDS such that $G[U]$ contains only clique components of size ≤ 3 can be found in polynomial time.*

Based on this lemma, for PWEDS we modify U_1 to be the set of vertices of clique components of size ≤ 3 in $G[U]$. We call our algorithm to which this modification is applied a modified algorithm. This modification brings the following corollary.

Corollary 1. *The modified algorithm can solve the parameterized weighted edge dominating set problem in $O^*(2.2351^k)$ time and polynomial space.*

Proof. We first show the correctness. If an edge dominating set M of G is k-feasible, i.e., $\omega(M) \leq k$, then it holds that $|V(M)| \leq 2k$ and $|M| \leq k$ since $\omega(e) \geq 1$ for any edge $e \in E$. This ensures the correctness of the measure $\mu(C, D)$ and the conditions (1) and (2) for an instance (C, D) of the weighted variant. Therefore we can solve PWEDS by the same branching method as PEDS.

Second we show the time complexity is the same as PEDS. Only difference between our algorithm for PEDS and one for PWEDS is treatment of clique components of size ≥ 4. In what follows, we describe the treatment by the modified algorithm and it guarantees that the time complexity is $O^*(2.2351^k)$. For a clique component H of size ≥ 5 of an instance (C, D), the degree of a vertex of H in $G[U_2]$ is $|V(H)| - 1 \geq 4$, on which therefore the modified algorithm branches in the first stage. For a clique component H of size 4 of an instance (C, D), a vertex of H satisfies condition (c-2), on which therefore the algorithm branches in the second stage.   □

6   Conclusion

In this paper, we have presented an $O^*(2.2351^k)$-time and polynomial-space algorithm to PEDS. The algorithm retains bad components produced at the first stage for branching on vertices of degree ≥ 4, and branching on the remaining undecided vertices not in clique components by choosing 4-cycles/vertices to branch on carefully. Based on our new lower bound on the size of (C, D)-edses, we derived an upper bound on the number of leaf instances generated in the third stage. We have also shown that a modi-fication of our algorithm can solve PWEDS in the same time and space complexities as PEDS.

For a possible achievement of further improved algorithms, it is still left to modify the first stage of our algorithm to branch on vertices of degree ≤ 4 in the second stage and to identify several new components as bad components.

References

1. Binkele-Raible, D., Fernau, H.: Enumerate and Measure: Improving Parameter Budget Man-agement. In: Raman, V., Saurabh, S. (eds.) IPEC 2010. LNCS, vol. 6478, pp. 38–49. Springer, Heidelberg (2010)
2. Fernau, H.: EDGE DOMINATING SET: Efficient Enumeration-Based Exact Algorithms. In: Bodlaender, H.L., Langston, M.A. (eds.) IWPEC 2006. LNCS, vol. 4169, pp. 142–153. Springer, Heidelberg (2006)
3. Fomin, F., Gaspers, S., Saurabh, S., Stepanov, A.: On Two Techniques of Combining Branch-ing and Treewidth. Algorithmica 54(2), 181–207 (2009)
4. Garey, M.R., Johnson, D.S.: Computers and Intractability: A Guide to The Theory of NP-Completeness. Freeman, San Francisco (1979)
5. Iwaide, K., Nagamochi, H.: An Improved Algorithm for Parameterized Edge Dominating Set Problem. Technical Report 2014-004 Kyoto University (2014), http://www.amp.i.kyoto-u.ac.jp/tecrep/
6. Raman, V., Saurabh, S., Sikdar, S.: Efficient Exact Algorithms through Enumerating Maxi-mal Independent Sets and Other Techniques. Theory of Computing Systems 42(3), 563–587 (2007)
7. Randerath, B., Schiermeyer, I.: Exact Algorithms for Minimum Dominating Set. Technical Report zaik 2005-501, Universität zu Köln, Cologne, Germany (2005)
8. van Rooij, J.M.M., Bodlaender, H.L.: Exact Algorithms for Edge Domination. Algorith-mica 64(4), 535–563 (2012)
9. Xiao, M.: A Simple and Fast Algorithm for Maximum Independent Set in 3-Degree Graphs. In: Rahman, M. S., Fujita, S. (eds.) WALCOM 2010. LNCS, vol. 5942, pp. 281–292. Springer, Heidelberg (2010)
10. Xiao, M., Kloks, T., Poon, S.-H.: New Parameterized Algorithms for the Edge Dominating Set Problem. TCS 511, 147–158 (2013)
11. Xiao, M., Nagamochi, H.: Parameterized Edge Dominating Set in Cubic Graphs. In: Atallah, M., Li, X.-Y., Zhu, B. (eds.) FAW-AAIM 2011. LNCS, vol. 6681, pp. 100–112. Springer, Heidelberg (2011)
12. Xiao, M., Nagamochi, H.: A Refined Exact Algorithm for Edge Dominating Set. In: Agrawal, M., Cooper, S.B., Li, A. (eds.) TAMC 2012. LNCS, vol. 7287, pp. 360–372. Springer, Heidelberg (2012)
13. Xiao, M., Nagamochi, H.: Exact Algorithms for Annotated Edge Dominating Set in Graphs with Degree Bounded by 3. IEICE Transactions on Information and Systems E96-D(3), 408–418 (2013)
14. Yanakakis, M., Gavril, F.: Edge Dominating Set in Graphs. SIAM J. Appl. Math. 38(3), 364–372 (1980)

On Bar $(1, j)$-Visibility Graphs

(Extended Abstract)

Franz J. Brandenburg[1], Niklas Heinsohn[2],
Michael Kaufmann[2], and Daniel Neuwirth[1]

[1] Fakultät für Informatik und Mathematik, Universität Passau, Germany
{brandenb,neuwirth}@informatik.uni-passau.de
[2] Wilhelm-Schickard-Institut für Informatik, Universität Tübingen, Germany
{heinsohn,mk}@informatik.uni-tuebingen.de

Abstract. A graph is called a bar $(1, j)$-visibility graph if its vertices can be represented as horizontal vertex-segments (bars) and each edge as a vertical edge-segment connecting the bars of the end vertices such that each edge-segment intersects at most one other bar and each bar is intersected by at most j edge-segments. Bar $(1, j)$-visibility refines bar 1-visibility in which there is no bound on the number of intersections of bars.

We construct gadgets which show structural properties of bar $(1, j)$-visibility graphs, study bounds on the maximal number of edges and show that there is an infinite hierarchy of bar $(1, j)$-visibility graphs. Finally, we prove that it is NP-complete to test whether a graph is bar $(1, \infty)$-visible.

1 Introduction

Planar graphs are an important topic in graph theory, combinatorics, and in particular in graph drawing. Planar graphs with n vertices have at most $3n - 6$ edges, and triangulated planar graphs meet this upper bound. Planar graphs admit straight-line drawings in the plane. A (weak) visibility representation is another way to draw a planar graph. Here the vertices are displayed as horizontal vertex-segments, called bars, and there is a vertical line of sight between the bars of the end vertices for each edge [15,18].

Developing a theory of graph drawing beyond planarity has received increasing interest in recent years. This is partly motivated by applications of network visualization, where it is important to compute readable drawings of non-planar graphs, and from cognitive experiments of Huang et al. [11], which indicate that large angle crossings have no negative impact on the human understanding of graph drawings. Recent approaches are k-planarity [14], bar k-visibility [6], rectangle visibility [12], k-quasi planarity [2] and right angle crossings (RAC) [7]. Studies of such classes of graphs address the maximal number of edges of graphs of size n, complete and complete bipartite graphs, the complexity of the recognition problem, and containment relations.

Recently, bar k-visibility, and especially bar 1-visibility representation gained new progress by the works of Brandenburg [4], Evans et al. [8] and Sultana et al. [16].

M.S. Rahman and E. Tomita (Eds.): WALCOM 2015, LNCS 8973, pp. 246–257, 2015.

We refine the model and introduce bar (k, j)-visibility representations of graphs, where k is the maximal number of intersections per edge-segment and j is the maximal number of intersections per vertex-segment. Using the new definition, (weak) bar visibility is bar $(0, 0)$-visibility, bar k-visibility used in Dean et al. [6] is bar (k, ∞)-visibility and 1-visibility used in Brandenburg [4] is bar $(1, 1)$-visibility. The respective graphs are called bar (k, j)-visibility graphs, where we will focus only on $k = 1$.

First, we investigate the density of bar $(1, j)$-visibility graphs for $j \leq 4$, where the density is the number of edges as a function of the number of vertices of edge-maximal graphs. A graph G is *edge-maximal* for a class of graphs $\mathcal{G}$ if the addition of any edge e violates the defining properties of $\mathcal{G}$ such that the graph $G + e \notin \mathcal{G}$. The density of planar graphs is $3n - 6$, while $4n - 8$ is the upper bound of the density of 1-planar graphs [3,14,9]. However, there are sparse edge-maximal 1-planar graphs with only $2.64n$ edges [5]. The upper bound on the number of edges is $6.5n$ for quasi-planar graphs [1], $4n - 10$ for RAC graphs [7], and $6n - 20$ for bar 1-visibility graphs [6] and these bounds can be reached for 1-planar, RAC and bar 1-visibility graphs.

We study the density of bar $(1, j)$-visibility graphs and show that the known upper bounds for bar $(1, 2)$- and bar $(1, 3)$-visibility representations can almost be reached, while the general upper bound of $6n - 20$ can be realized already by bar $(1, 4)$-visibility graphs. Our results are surveyed in Table 1. We also show that there are edge-maximal bar $(1, \infty)$-visibility graphs with only $5n + O(1)$ edges, such there is a range between $5n + O(1)$ to $6n - 20$ for the density of edge-maximal bar $(1, \infty)$-visibility graphs, which parallels the situation of 1-planar graphs.

The paper is organized as follows: First, we present existential upper bounds for the number of edges of bar $(1, 2)$- and $(1, 3)$-visibility graphs. In Section 3, structural properties of bar $(1, j)$-visibility graphs are revealed by our core gadget called *ship*. We show that it has a very limited bar-visibility representation. For all our following results, the basic gadget is heavily being used. In Section 4, we consider maximal bar $(1, j)$-visible graphs and derive bounds on the number of edges. We give an infinite hierarchy of bar $(1, j)$-visibility graphs, i.e. for $j = 1, 2, \ldots$, graphs that are bar $(1, j)$-visible, but not bar $(1, j - 1)$-visible, as well as a class of edge-maximal bar $(1, \infty)$-visible graphs with only $5n - 12$ edges. Finally, we sketch the NP-completeness of the recognition problem of bar $(1, \infty)$-visibility. We conclude with a summary and open problems.

Table 1. Overview of density bounds of bar $(1, j)$-visibility graphs

	upper bounds	example
$(0, 0)$	$3n - 6$	$3n - 6$
$(1, 1)$	$4n - 8$	$4n - 8$
$(1, 2)$	$5n - 10$	$5n - 12$ (this paper)
$(1, 3)$	$6n - 20$	$6n - 21$ (this paper)
$(1, j), j \geq 4$	$6n - 20$	$6n - 20$ (this paper)

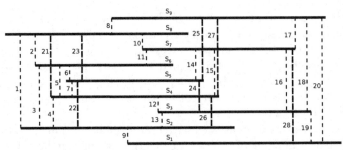

Fig. 1. The basic graph structure G'_9 with 9 vertex-segments, 20 planar edges (numbered 1 to 20 from left to right) and 8 crossing edge-segments (21 to 28). The bar $(1,2)$-visibility representation of G'_9 has $28 = 5n - 17$ edges.

2   Upper Edge Bounds for Bar $(1, j)$-Visibility Graphs

Upper bounds on the number of edges of bar $(1, j)$-visibility graphs can easily be obtained from the observation that the two outermost bars cannot be crossed at all, and that there are at most $3n - 6$ edges that are not involved in a crossing. Hence, bar $(1,1)$-visible graphs have at most $4n - 8$ edges, while bar $(1,2)$- and bar $(1,3)$-visible graphs have at most $5n - 10$ and $6n - 12$ edges, respectively. The bound for $(1,1)$-visible graphs is tight, since the complete graph on seven vertices without one edge i.e. $K_7 - e$ is $(1,1)$-visible and can be augmented by $n - 7$ vertices of degree four [4], and the $6n - 12$ bound for bar $(1,3)$-visibility is close to the $6n - 20$ bound for bar $(1, \infty)$-visible graphs from [6], but clearly an overestimation. We improve these bounds and make them (nearly) tight.

Lemma 1. *For every odd $n \geq 9$ there is a graph G_n with $5n - 12$ edges, which has a bar $(1, 2)$-visibility representation.*

Proof. We construct our graph G_n with $n = 2k+1$ vertices and $n \geq 9$, by placing first the vertex-segments and then adding edge-segments. For each vertex v_i in G_n let $s_i = (l_i, r_i, y_i)$ be the corresponding vertex-segment where l_i and r_i denote the x-coordinates of the left and right endpoints, and y_i is its y-coordinate. We use indexing in increasing order s.t. for all v_i, we have $y_i = i$.

At first we present the basic structure G'_9 with $n = 9$ vertex-segments and $28 = 5n - 17$ edge segments. Then we introduce the inductive way to add two vertex-segments that induce 10 new edge-segments and as a last step we extend the two outermost vertex-segments to reach the upper bound of $5n - 12$.

The basic structure of 9 vertex-segments has 8 crossing edge-segments and 20 planar edges are placed close to the right or left end points of the vertex-segments going up and downwards. Thus, G'_9 in Figure 1 has $28 = 5 \cdot 9 - 17$ edge-segments.

In each inductive step, we will add two vertex segments to the basic structure G'_n to end up with $n + 2$ vertex-segments. Let $s_1, \ldots, s_n$ be segments of G'_n increasingly ordered by their y-coordinates with $5n - 17$ edge-segments.

If s_2 and s_{n-1} are located on the left side, we will move all right endpoints $> r_2$ by two units to the right, while the left endpoints stay fixed such that all

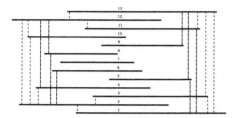
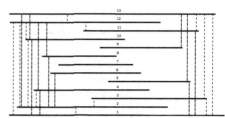

Fig. 2. G'_{13} is shown after the first two inductive steps. Note that 13 vertex-segments drawn, but for visual reasons only the 20 new induced edge-segments.

Fig. 3. The final bar $(1, 2)$-visibility representation with n=13 vertex-segments and $5n - 12$ edge-segments. The segments s_1 and s_{13} are enlarged.

edge segments are stretched by two units. Thus, $r_2 + 1$ and $r_2 + 2$ are coordinates without existing endpoints and the stretched vertex-segments are extended by two units. The new vertex-segment $s_0 = (l_{n-1} - 1, r_2 + 2, 0)$ will be placed below s_2 and $s_{n+1} = (l_{n-1} - 2, r_2 + 1, n + 1)$ above s_{n-1}. Note that $l_{n-1} - 1$ and $l_{n-1} - 2$ were unused coordinates too. Now we add the following planar edge-segments $(s_0, s_1), (s_0, s_2), (s_0, s_{n-1}), (s_0, s_{n+1})$ and $(s_{n+1}, s_n), (s_{n+1}, s_{n-1})$. Additionally we add two edge-segments crossing s_2, namely $(s_0, s_4), (s_0, s_{n-3})$, as well as two edge-segments crossing s_{n-1}, namely $(s_{n+1}, s_{n-3}), (s_{n+1}, s_2)$.

Vice versa, if s_2 and s_{n-1} are located on the right side, we will stretch all left endpoints $< l_{n-1}$ by 2 units to the left. The new vertex-segment $s_0 = (l_{n-1} - 1, r_2 + 2, 0)$ will be placed again below s_2 and the segment $s_{n+1} = (l_{n-1} - 2, r_2 + 1, n + 1)$ above s_{n-1}. The planar edge-segments are $(s_0, s_1), (s_0, s_2), (s_0, s_{n+1})$ and $(s_{n+1}, s_n), (s_{n+1}, s_{n-1}), (s_{n+1}, s_1)$. The two edge-segments crossing s_2 are $(s_0, s_4), (s_0, s_{n-1})$, while the two edge-segments crossing s_{n-1} are (s_{n+1}, s_{n-3}), (s_{n+1}, s_4). In both cases, we add ten new edge-segments when inserting two vertex-segments and the resulting representation has $5(n+2) - 17$ edge-segments, cf. Figure 2. At the very end, we extend the basic structure G'_n to obtain G_n.

Let again be $s_1, ..., s_n$ the vertex-segments of the drawing numbered from bottom to top. Extending s_1 and s_n to the left and right to the minimal and maximal x-coordinates of the drawing, we have to consider two cases again: If s_1 and s_n are located on the left, we can add the edges $(s_1, s_4), (s_1, s_{n-1})$ crossing s_2 as well as $(s_n, s_{n-3}), (s_n, s_4)$ crossing s_{n-1} and the planar edge (s_n, s_2). Otherwise we can add the following edges: $(s_1, s_4), (s_1, s_{n-3})$ crossing s_2 as well as $(s_n, s_{n-3}), (s_n, s_2)$ crossing s_{n-1} and the planar edge (s_1, s_{n-1}).

Thus, we can add 5 more edge-segments and end up with a bar $(1, 2)$-visibility representation with n vertex-segments and $5n - 12$ edge-segments. The extension of the outermost vertex-segments is illustrated in Figure 3 for the case, where s_1 and s_n are located on the right. □

Next, we extend this to a bar $(1, 3)$-visibility representation.

Lemma 2. *For every odd $n \geq 9$ there is a graph G_n with n vertices and $6n - 21$ edges, which has a bar $(1, 3)$-visibility representation.*

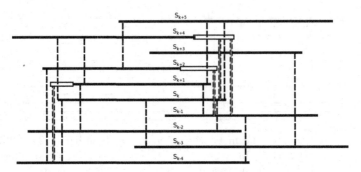

Fig. 4. We show the crossings in the center of the bar $(1,3)$-visibility representation. We have drawn only edge-segments with visible start- and end-segments. The changes in the length is shown in rectangles, while dashed rectangles indicate the new edges. The number of planar edges stays the same.

Proof. We will start with the same basic structure of $n = 9$ vertex-segments (Figure 1) with two additional edge-segments, namely $(1,4)$ and $(6,9)$. Thus the visibility representation of the basic structure with $n = 9$ vertex-segments has $30 = 6n - 24$ edge-segments. The inductive step remains exactly the same when locating the two new segments s_0 and s_{n+1}. The only difference are two additional edge-segments, namely (s_0, s_3) and (s_{n+1}, s_{n-2}), which can be drawn by crossing s_1 and s_n. Note, that these edges would have otherwise occurred in the last expanding-step $((s_1, s_4)$ and $(s_n, s_{n-3}))$. Hence the expanding step, which also stays exactly the same does add three edges, instead of five. Thus we end up with an bar $(1,3)$-visibility representation with $6n - 21$ edge-segments. $\square$

Lemma 3. *For every odd $n \geq 9$ there is a graph with $6n - 20$ edges, which has a bar $(1,4)$-visibility representation.*

Proof. Using the bar $(1,3)$-visibility graph with $6n - 21$ edges from the previous lemma, we reconfigure the vertex-segments to achieve a bar $(1,4)$-visibility representation with $6n - 20$ edges. This requires three steps given below. Figure 4 presents resulting bar $(1,4)$-visibility representation.

1. extend l_{k+1} to $l_{k+2} + 0.5$. This interrupts the edge-segment (s_k, s_{k+4}) and enables the edge-segment (s_{k+1}, s_{k-4}) which has not been drawn already. Note that s_{k-2} is now crossed by four edge-segments.
2. extend r_{k+2} to $r_{k+1} + 0.5$. This interrupts the edge-segment (s_{k+1}, s_{k+5}) and enables the edge-segment (s_{k+2}, s_{k-1}) crossing s_k.
3. extend r_{k+4} to $r_k + 0.5$. This interrupts the edge-segment (s_k, s_{k+5}) and enables the edge-segment (s_{k+4}, s_{k-1}) crossing s_{k+3}. Additionally we can add the edge-segment (s_k, s_{k+4}) crossing s_{k+3} again.

Thus, we have added one edge ending up with a bar $(1,4)$-visibility representation with n vertex-segments and $6n - 20$ edge-segments. Note that in the case $n = 9$, the segment s_{k-4} is replaced and s_{k-3} is enlarged, the additional edge is also possible. $\square$

3   Construction of the Main Tool: The Ship-Gadget

It is well-known that a 3-connected planar graph has an unique embedding on the sphere, which is unique in the plane if the outer face is determined. Beyond planarity such a stabilizing property may not hold. There are maximal 1-planar graphs with $4n - 8$ edges with two embeddings where some crossing and non-crossing edges are exchanged [17]. To circumvent such problems Korzhik and Mohar [13] constructed so-called U-graphs which are 1-planar graphs with an almost unique drawing. In the same spirit we construct the ship gadget $S(r)$ which is parameterized by the length of the chain and has a distinct bar $(1, \infty)$-visibility representation. A ship $S(6)$ is shown in Fig. 5 in a straight line drawing and in Fig. 6 in a bar visibility representation.

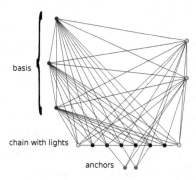

Fig. 5. A straight-line drawing of a ship $S(6)$ with the basis b_1, b_2, b_3 (in blue) and h_1, h_2 (in grey), the lights (l_1, l_2 in yellow), the chain ($c_1, ..., c_6$ in black) and the two anchors (a_1, a_2 in red)

A *ship* $S(r)$ consists of $r + 9$ vertices and $5r + 24$ edges. It has three basis vertices $\{b_1, b_2, b_3\}$ (blue), two hull vertices $\{h_1, h_2\}$ (grey), two light vertices $\{l_1, l_2\}$ (yellow), two anchor vertices $\{a_1, a_2\}$ (red), and r chain vertices $\{c_1, ..., c_r\}$ (black). The chain vertices connect l_1 and l_2 by a chain, and we call them odd and even chain vertices according to the index. The basis, hull vertices and light vertices form a $K_7 - e$, where edge $e = (l_1, l_2)$ is missing. Each chain vertex is connected to each of $\{b_1, b_3, b_3\}$ and alternating to one of h_1 and h_2, which each time form a K_5. Finally, the anchor vertices are connected to chain vertices in the middle. For $m = \lfloor \frac{r}{2} \rfloor$ there are edges $(a_1, h_1), (a_2, h_2)$ and $(a_1, c_{m-1}), (a_2, c_m), (a_1, c_{m+1}), (a_2, c_{m+2})$. In addition, the edges $(a_1, b_1), (a_1, b_3), (a_2, b_3), (a_2, b_2)$ may be added.

It shall be shown that in any bar $(1, \infty)$-visibility representation of a ship $S(r)$ the following property holds, if r is chosen to be large enough:

Theorem 1. *In any bar $(1, \infty)$-visibility representation of a ship $S(r)$ with $r \geq 640$, the bars are vertically ordered by $(h_1, a_1, b_1, b_2, b_3, a_2, h_2)$. There is a subsequence including the four chain vertices adjacent to the two anchors whose* *bars are alternating just above and below the bar of b_2. The bars of h_1, b_1, b_2, b_3, h_2 completely cover the bars of the chain vertices from the above subsequence and bars of a_1 and a_2, and they do not cover the bars of the lights l_1 and l_2.*

We can only provide the main steps of the proof.

The projection of the vertex-segment s representing v forms an interval, which will be denoted by $I(s)$ or less accurately by $I(v)$. A necessary condition for an edge between vertices v_i and v_j is that the intervals $I(v_i)$ and $I(v_j)$ have a non-empty intersection. By Helly's Theorem each set of convex structures in an Euclidean space with pairwise intersections has a common intersection. Hence, the vertex-segments of K_k have a non-empty common interval.

The stated properties of any bar $(1, \infty)$-visibility representation of a ship $S(r)$ are based on the following observations.

Lemma 4. *The intervals of the basis and hull vertices share an interval I_c of common x-coordinates.*

Lemma 5. *The intervals of the lights $I(l_1)$ and $I(l_2)$ intersect the common interval I_c, but are not contained in I_c.*

Proof. Each light together with the basis and the hull forms K_6, which by Helly's Theorem implies a common interval. If $I(l_1) \subseteq I_c$ then the intervals of at least three basis and hull vertices are all above (or below) $I(l_1)$ and the edge-segment to the interval with the largest distance from $I(l_1)$ must intersect at least two other intervals, which contradicts bar $(1, \infty)$-visibility. The case of l_2 is similar. □

For the existence of such odd and even chain segments we employ topological arguments and consider a visibility representation as a polyline drawing, where edges might intersect vertices.

We assume that we have r chain vertices, where r is a large enough constant.

Lemma 6. *There are at least $r - 26$ chain vertices c_i , $1 \leq i \leq r$ with an edge (c_i, b) to some basis vertex b, whose edge-segment must intersect the interval of another basis vertex b'.*

Proof. Consider the three basis vertices $\{b_1, b_2, b_3\}$ and maximize the number s of chain vertices that can be added without intersecting the interval of a basis vertex. By case checking, we can bound $s \leq 26$. □

Lemma 7. *There exists a subset C_S of odd (even) chain vertices of size at least $\frac{r-26}{24}$ and there are two basis vertices b and b' such that for each $c_i \in C_S$ the edge (c_i, h) to the corresponding hull vertex h has to intersect the interval of b and the edge (b'', c_i) to the third vertex b'' has to intersect the interval of b'.*

Proof. From Lemma 6, we have at least $r - 26$ chain vertices with edges that intersect the interval of at least one basis vertex. We can subdivide them into 12 classes, intersecting only one of the vertices b_1, b_2, b_3 and ending at another one, and 6 intersecting two basis vertices b_1 and b_2, b_2 and b_3, and b_1 and b_3 and ending at some other vertices. The largest of these classes has at least $\frac{r-26}{12}$ vertices. For this class, we prove the claimed property by case checking. □

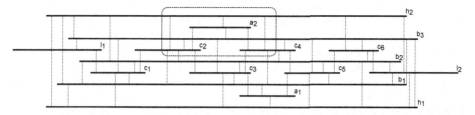

Fig. 6. A bar $(1, \infty)$-visibility representation of the ship-gadget with 6 chain-nodes $(c_1, ..., c_6)$

By geometric arguments, we can prove the following two lemmas

Lemma 8. *The subsequence C_S of the chain implies that in any bar $(1, \infty)$-visibility representation of shipe $S(r)$, h_1, b_1, b_2, b_3 and h_2 are placed in this order such that the even chain segments are between b_1 and b_2 and the odd chain segments are between b_2 and b_3.*

Lemma 9. *The subset C_S is in the common interval I_c.*

For the 11 other classes of chain segments, we can argue that they contain at most 13 vertices each. Furthermore, we observe that those vertices have to be consecutive and include the light vertices. Hence we conclude that almost all but a constant number c of chain elements are within the subset C_S.

We estimate this constant c from above roughly by 312, observing that we have to distinguish odd and even chain segments. So we have at most $2 \cdot 11 \cdot 13$ segments plus the additional 26 segments from Lemma 6 = 312 elements, which might be not in the subset C_S. Since we do not know exactly where those 312 elements are located (either at the beginning or at the end of the bar $(1, \infty)$-visibility representation), we ensure that the middle part of the sequence of chain elements must be in the subset C_S by choosing $r = 2 \cdot 312 + 16$. We conclude

Lemma 10. *Choosing $r = 640$ for ship $S(r)$ ensures that at least 16 subsequent chain segments in the middle of the chain are in interval I_C.*

Proof. (of Theorem 1) Since there is a subsequence of at least 16 chain vertices $c_p, \ldots, c_q$, whose intervals are completely contained in the common interval I_c, the intervals of the basis vertices must be ordered $(h_1, b_1, b_2, b_3, h_2)$, and the intervals of $c_p, \ldots, c_q$ alternate directly above and below the interval of b_2.

Now, the intervals of the anchors must be placed between the intervals of the hull and the next basis vertex. Otherwise, one of the edges (a_1, c_m) or (c_{m+1}, b_2) would intersect two intervals, and similarly for a_2.

The intervals of the anchors are contained in I_c, since $I(a_1) \subseteq I(c_m) \cup I(c_{m+2}) \subseteq I_c$, and similarly for $I(a_2)$. $\qquad \square$

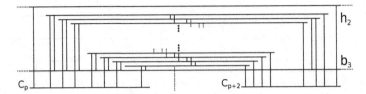

Fig. 7. The region of the anchor-segment a_2 of a ship $S(r)$. Each anchor α_i has two edge-segments intersecting b_3 $((\alpha_i, c_p), (\alpha_i, c_{p+2}))$ and induces 5 edge -segments. The two topmost anchor-segments can be connected to h_2.

4   Hierarchy and Sparsity of Bar $(1,j)$-Visibility Graphs

Using the ship we establish an infinite hierarchy of bar $(1,j)$-visibility graphs and show that such graphs can be sparser than the previously given upper bounds.

Theorem 2. *For every $j \geq 2$ there is a bar $(1,j)$-visibility graph which has no bar $(1, j-1)$-visibility representation.*

Proof. We apply the argument from Lemma 6 to the graph $K_{3,n}$. After placing at most 26 vertices, we have to intersect one basis vertex. Hence after placing at most 30 vertices, we have to intersect one basis vertex at least twice, and for $n = 26 + 3 \cdot j + 1$, we have to intersect at least one basis vertex j times. Hence, for every j there is an n such that $K_{3,n}$ has a bar $(1,j)$-visibility representation, and there is no bar $(1, j-1)$-visibility representation for $K_{3,n}$. ☐

Corollary 1. *There is an infinite hierarchy of classes of bar $(1,j)$-visibility graphs.*

Theorem 3. *There are edge-maximal bar $(1, \infty)$-visibility graphs with n vertices and $5n + O(1)$ edges.*

Proof. We extend the ship-gadget $S(r)$ and add edges to make it edge-maximal. Then we replace anchor a_2 by a set of anchors $\alpha_1, \ldots \alpha_t$ for $t > 0$. In any visibility representation, $\alpha_1, \ldots \alpha_t$ are in the same place of a_2 between h_2 and b_3 and in the common interval I_c and maintain the same connections.

Let a_2 be connected to the chain vertices c_p and c_{p+2}. We add the edges $(\alpha_i, b_3), (\alpha_i, c_p), (\alpha_i, c_{p+2})$ for $1 \leq i \leq t$. For $1 < i < t$ add the edges (α_{i-1}, α_i) and $(\alpha_{i-1}, \alpha_{i+1})$, and finally add $(\alpha_t, h_2), (\alpha_{t-1}, h_2)$. The edge (α_1, b_2) may be added as well. See Fig. 7 for an illustration.

Due to the edges (α_i, c_p) and (α_i, c_{p+2}), which intersect the bar of b_3, in any bar $(1, \infty)$-visibility representation the interval $I(\alpha_i)$ is properly included in the interval of $I(\alpha_{i+1})$ and the anchors properly nest. Hence, the edges $(\alpha_{i-1}, \alpha_{i+1})$ intersect the bar of α_i for $1 < i < t$. Thereby the position of the anchors is fixed and they are in place of the anchor a_2 in any bar $(1, \infty)$-visibility representation.

The obtained graph is edge-maximal, since no further edges can be added the bars of vertices from $S(r)$ and from the new anchors and between the new anchors. The ship $S(r)$ without the anchors has a fixed number of edges, and each new anchor α_i increases the number of edges by 5. ☐

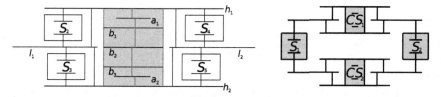

Fig. 8. A chain-ship (left), where the lights l_1 and l_2 are secured by four help-ships $S_2...S_5$. A ring (right) consisting of two chain-ships CS_1 and CS_2, whose lights are connected via ships S_1, S_2 with each other.

5   NP-Completeness of Bar $(1, \infty)$-Visibility

We only sketch the ideas here. The recognition problem asks if a graph G has a bar $(1, \infty)$-visibility representation. The problem is in NP since we can guess a bar $(1, \infty)$-visibility representation consisting of polynomial many segments on a grid of polynomial size and check its correctness in polynomial time. To prove the NP-completeness we use a reduction from the NAE-3-SAT problem [10] and heavily use the ship gadget and the fact that in any bar $(1, \infty)$-visibility representation the anchors are placed between the intervals of the hull and outer basis vertices and are completely covered by the intervals of the hull vertices. Hence, an edge from an anchor to the outside must intersect the interval from the hull.

We combine five ships to a so-called chain ship, which prevents a nesting of ships. A ring is built from two chain chips and two ships such that two rings must be disjoint in any bar $(1, \infty)$-visibility representation. Both constructs are presented in Figure 8. Further two consecutive rings are connected by edges, which forces the rings to nest in each other. For an instance α of NAE-3-SAT of n variables and m clauses we construct graph G_α consisting of a sequence of $n + 4m + 3$ nested rings, with a ring for each variable and four rings for each clause and three protection rings. As gadgets of the variable and the literals in each clause we use chain ships in the corresponding rings. For each variable x and a literal x or $\bar{x}$ in a clause C, where x occurs there is a path of a fixed length between the light vertices l_1 of their gadgets, which consists of edge segments which are all intersected in any bar $(1, \infty)$-visibility representation.

A clause gadget consists of four rings (R_1, R_2, R_3, R_4) and three additional segments s_1, s_2, s_3. For $i = 1, 2, 3$ s_i is connected to the l_1 segments of R_i by a path of length $4 - i$ and the anchors of the chain ships of R_3 and R_4. The constructions of the paths forces that at most two l_1s of different rings in one clause gadget are on the same side. This corresponds that two literals of a clause have the same value and ensures the condition for the clauses of NAE-3-SAT. Additionally the connection between the variables and the literals gadgets guarantees the consistent value. A comprehensive example is given in Fig. 9.

Theorem 4. *A graph G_α has a bar 1-visibility representation if and only if the CNF 3-SAT formula α has a valid NAE-3-SAT assignment.*

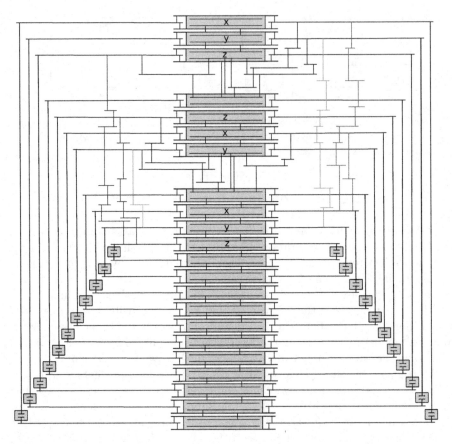

Fig. 9. An reduction for a CNF of NAE-3-SAT with $c_1 = (x \vee y \vee x)$ and $c_2 = (\bar{x} \vee \bar{y} \vee z)$ with $x, z = 1$ and $y = 0$. The different colored paths connect the variable chain ships from the variable group in the middle with the chain ships of their corresponding clauses.

6   Conclusion and Open Problems

We refined bar 1-visibility graphs by a new parameter on the maximal number of intersections of vertex-segments and started a detailed study of such graphs. We established upper bounds on the number of edges of such graphs and showed an infinite hierarchy, and proved NP-completeness for bar $(1, \infty)$-visibility representations using our powerful ship-gadget. More open problems remain:

1. Are there sparser edge-maximal bar $(1, j)$-visibility graphs?
2. Are there any bar $(1, 3)$-visibility graphs with $6n - 20$ edges?
3. Is bar $(1, j)$-visibility NP-complete if j is fixed?
4. Does a k-planar graph G have a bar (k, k)-visibility representation?
5. Are there 2-planar graphs without a bar $(1, \infty)$-visibility representation?

Acknowledgement. We wish to thank an anonymous referee for her/his useful suggestions. This work was supported in part by DFG grant Br 835/18.

References

1. Ackerman, E., Tardos, G.: On the maximum number of edges in quasi-planar graphs. Journal of Combinatorial Theory, Series A 114(3), 563–571 (2007)
2. Agarwal, P.K., Aronov, B., Pach, J., Pollack, R., Sharir, M.: Quasi-planar graphs have a linear number of edges. Combinatorica 17(1), 1–9 (1997)
3. Bodendiek, R., Schumacher, H., Wagner, K.: Über 1-optimale graphen. Mathematische Nachrichten 117(1), 323–339 (1984)
4. Brandenburg, F.-J.: 1-visibility representations of 1-planar graphs. J. Graph Algorithms Appl. 18(3), 421–438 (2014)
5. Brandenburg, F.-J., Eppstein, D., Gleißner, A., Goodrich, M.T., Hanauer, K., Reislhuber, J.: On the density of maximal 1-planar graphs. In: Didimo, W., Patrignani, M. (eds.) GD 2012. LNCS, vol. 7704, pp. 327–338. Springer, Heidelberg (2013)
6. Dean, A.M., Evans, W., Gethner, E., Laison, J.D., Safari, M.A., Trotter, W.T.: Bar k-visibility graphs. J. Graph Algorithms Appl. 11(1), 45–59 (2007)
7. Didimo, W., Eades, P., Liotta, G.: Drawing graphs with right angle crossings. Theor. Comput. Sci. 412(39), 5156–5166 (2011)
8. Evans, W.S., Kaufmann, M., Lenhart, W., Liotta, G., Mchedlidze, T., Wismath, S.K.: Bar 1-visibility graphs and their relation to other nearly planar graphs. CoRR, abs/1312.5520 (2013)
9. Fabrici, I., Madaras, T.: The structure of 1-planar graphs. Discrete Mathematics 307(7-8), 854–865 (2007)
10. Garey, M.R., Johnson, D.S.: Computers and Intractability; A Guide to the Theory of NP-Completeness. W. H. Freeman & Co., New York (1990)
11. Huang, W., Hong, S.-H., Eades, P.: Effects of crossing angles. In: PacificVis, pp. 41–46 (2008)
12. Hutchinson, J.P., Shermer, T.C., Vince, A.: On representations of some thickness-two graphs. Comput. Geom. 13(3), 161–171 (1999)
13. Korzhik, V.P., Mohar, B.: Minimal obstructions for 1-immersions and hardness of 1-planarity testing. In: Tollis, I.G., Patrignani, M. (eds.) GD 2008. LNCS, vol. 5417, pp. 302–312. Springer, Heidelberg (2009)
14. Pach, J., Tóth, G.: Graphs drawn with few crossings per edge. Combinatorica 17(3), 427–439 (1997)
15. Rosenstiehl, P., Tarjan, R.E.: Rectilinear planar layouts and bipolar orientations of planar graphs. Discrete & Computational Geometry 1, 343–353 (1986)
16. Sultana, S., Rahman, M. S., Roy, A., Tairin, S.: Bar 1-visibility drawings of 1-planar graphs. In: Gupta, P., Zaroliagis, C. (eds.) ICAA 2014. LNCS, vol. 8321, pp. 62–76. Springer, Heidelberg (2014)
17. Suzuki, Y.: Re-embeddings of maximum 1-planar graphs. SIAM Journal on Discrete Mathematics 24(4), 1527–1540 (2010)
18. Tamassia, R., Tollis, I.G.: A unified approach a visibility representation of planar graphs. Discrete & Computational Geometry 1, 321–341 (1986)

Simultaneous Time-Space Upper Bounds for Red-Blue Path Problem in Planar DAGs*

Diptarka Chakraborty and Raghunath Tewari

Department of Computer Science & Engineering,
Indian Institute of Technology, Kanpur, India
{diptarka,rtewari}@cse.iitk.ac.in

Abstract. In this paper, we consider the RedBluePath problem, which states that given a graph G whose edges are colored either red or blue and two fixed vertices s and t in G, is there a path from s to t in G that alternates between red and blue edges. The RedBluePath problem in planar DAGs is NL-complete. We exhibit a polynomial time and $O(n^{\frac{1}{2}+\epsilon})$ space algorithm (for any $\epsilon > 0$) for the RedBluePath problem in planar DAG. We also consider a natural relaxation of RedBluePath problem, denoted as EvenPath problem. The EvenPath problem in DAGs is known to be NL-complete. We provide a polynomial time and $O(n^{\frac{1}{2}+\epsilon})$ space (for any $\epsilon > 0$) bound for EvenPath problem in planar DAGs.

1 Introduction

A fundamental problem in computer science is the problem of deciding reachability between two vertices in a directed graph. This problem characterizes the complexity class non-deterministic logspace (NL) and hence is an important problem in computational complexity theory. Polynomial time algorithms such as Breadth First Search (BFS) algorithm and Depth First Search (DFS) give a solution to this problem, however they require linear space as well. On the other hand, Savitch showed that reachability can be solved by an $O(\log^2 n)$ space algorithm, however that takes $\Theta(n^{\log n})$ time [1]. The readers may refer to a survey by Wigderson [2] to know more about the reachability problem.

It is an important open question whether these two bounds can be achieved by a single algorithm. In other words can we exhibit a polynomial time and $O(\log^k n)$ space algorithm for the reachability problem in directed graphs. Barnes, Buss, Ruzzo and Schieber made some progress in this direction by giving the first polynomial time and sub-linear space algorithm. They showed that directed reachability can be solved by an $O(n/2^{k\sqrt{\log n}})$ space and polynomial time algorithm [3], by cleverly combining BFS and Savitch's algorithm. Till now this is the best known simultaneous time-space bound known for the directed reachability problem in this direction. Recently, Imai et. al. [4] have improved the BBRS bound for the class of directed planar graph. They gave a polynomial

---

* Research supported by Research-I Foundation.

M.S. Rahman and E. Tomita (Eds.): WALCOM 2015, LNCS 8973, pp. 258–269, 2015.
© Springer International Publishing Switzerland 2015

time and $O(n^{\frac{1}{2}+\epsilon})$ space algorithm by efficiently constructing a *planar separator* and applying a divide and conquer strategy. In a recent work, their result has been extended to the class of *high-genus* and *H-minor-free* graphs [5].

An interesting generalization of the reachability problem, is the RedBluePath problem. Given a directed graph where each edge is colored either Red or Blue, the problem is to decide if there is a (simple) directed path between two specified vertices that alternates between red and blue edges. Kulkarni showed that the RedBluePath problem is NL-complete even when restricted to planar DAGs [6]. Unfortunately, no sublinear $(O(n^{1-\epsilon})$, for any $\epsilon > 0)$ space and polynomial time algorithm is known for RedBluePath problem in planar DAGs.

A natural relaxation is the EvenPath problem, which asks if there is a (simple) directed path of even length between two specified vertices in a given directed graph. In general, EvenPath problem is NP-complete [7], but for planar graphs, it is known to be in P [8]. It is also known that for DAGs, this problem is NL-complete. Datta *et. al.* showed that for planar DAGs, EvenPath problem is in UL. However, no sublinear $(O(n^{1-\epsilon})$, for any $\epsilon > 0)$ space and polynomial time algorithm is known for this problem also.

In this paper, we provide a sublinear space and polynomial time bound for both the RedBluePath and EvenPath problem. Our main idea is the use of a space efficient construction of separators for planar graphs [4]. We then devise a modified DFS approach on a smaller graph to solve the RedBluePath problem. As a consequence, we show that a similar bound exists for the directed reachability problem for a class of graphs that is a superset of planar graphs. Using similar approach, we design an algorithm to detect the presence of an odd length cycle in a directed planar graph, which serves as a building block to solve the EvenPath problem.

Our Contributions

The first contribution of this paper is to give an improved simultaneous time-space bound for the RedBluePath problem in planar DAGs.

Theorem 1. *For any constant* $0 < \epsilon < \frac{1}{2}$, *there is an algorithm that solves* RedBluePath *problem in planar DAGs in polynomial time and* $O(n^{\frac{1}{2}+\epsilon})$ *space.*

We first construct a separator for the underlying undirected graph and perform a DFS-like search on the separator vertices.

Using the reduction given in [6] and the algorithm stated in the above theorem, we get an algorithm to solve the directed reachability problem for a fairly large class of graphs as described in Section 3, that takes polynomial time and $O(n^{\frac{1}{2}+\epsilon})$ space. Thus we are able to beat the BBRS bound for such a class of graphs. One such class is all k-planar graphs, where $k = O(\log^c n)$, for some constant c and this is a strict superset of the set of planar graphs.

In this paper, we also establish a relation between EvenPath problem in a planar DAG and the problem of finding odd length cycle in a directed planar graph and thus we argue that both of these problems have the same simultaneous time-space complexity. We use two colors Red and Blue to color the vertices of the given graph

and then use the color assigned to the vertices of the separator to detect the odd
length cycle. The conflicting assignment of color to the same vertex in the separator will lead to the presence of an odd length cycle. Here also we use the recursive
approach to color the vertices and as a base case we use BFS to solve the problem
of detecting odd length cycle in each small component. Thus we have the following
result regarding the EvenPath problem.

Theorem 2. *For any constant* $0 < \epsilon < \frac{1}{2}$, *there is an algorithm that solves*
EvenPath *problem in planar DAGs in polynomial time and* $O(n^{\frac{1}{2}+\epsilon})$ *space.*

The rest of the paper is organized as follows. In Section 2, we give some basic
definitions and notations that we use in the rest of the paper. In Section 3, we
prove our main results regarding the RedBluePath and EvenPath problem.

2   Preliminaries

A graph $G = (V, E)$ consists of a set of vertices V and a set of edges E where
each edge can be represented as an ordered pair (u, v) in case of directed graph
and as an unordered pair $\{u, v\}$ in case of undirected graph, such that $u, v \in V$.
Unless otherwise specified, G will denote a directed graph, where $|V| = n$. Given
a graph G and a subset of vertices X, $G[X]$ denotes the subgraph of G induced
by X and $V(G)$ denotes the set of vertices present in the graph G. Given a
directed graph G, we denote the underlying undirected graph by $\hat{G}$. We follow
the standard model of computation to discuss the complexity measures of the
stated algorithms. In particular, we consider the computational model in which
an input appears on a read-only tape and the output is produced on a write-only
tape and we only consider the internal read-write tape in the measure of space
complexity. Throughout this paper, by $\widetilde{O}(s(n))$, we mean $O(s(n)(\log n)^{O(1)})$.

The notions of separator and separator family defined below are crucial in
this paper.

Definition 1. *A subset of vertices* S *of an undirected graph* G *is said to be a*
ρ-*separator (for any constant* ρ, $0 < \rho < 1$) *if removal of* S *disconnects* G *into*
two sub-graphs A *and* B *such that* $|A|, |B| \le \rho n$ *and the size of the separator is*
the number of vertices in S.

A subset of vertices $\overline{S}$ *of an undirected graph* G *with* n *vertices is said to*
be a $r(n)$-*separator family if the removal of* $\overline{S}$ *disconnects* G *into sub-graphs*
containing at most $r(n)$ *vertices.*

Now we restate the results and the main tools used in [4] to solve directed
planar reachability problem and these results are extensively used in this paper.
In [4], the authors construct a $\frac{8}{9}$-separator for a given undirected planar graph.

Theorem 3 ([4]). *(a) Given an undirected planar graph* G *with* n *vertices, there*
is an algorithm PlanarSeparator *that outputs a* $\frac{8}{9}$-*separator of* G *in polynomial*
time and $\widetilde{O}(\sqrt{n})$ *space.*
(b) For any $0 < \epsilon < 1/2$, *there is an algorithm* PlanarSeparatorFamily *that*
takes an undirected planar graph as input and outputs a $n^{1-\epsilon}$-*separator family*
of size $O(n^{\frac{1}{2}+\epsilon})$ *in polynomial time and* $\widetilde{O}(n^{\frac{1}{2}+\epsilon})$ *space.*

In [4], the above theorem was used to obtain a new algorithm for reachability in directed planar graph.

Theorem 4 ([4]). *For any constant* $0 < \epsilon < 1/2$, *there is an algorithm* DirectedPlanarReach *that, given a directed planar graph* G *and two vertices* s *and* t, *decides whether there is a path from* s *to* t. *This algorithm runs in time* $n^{O(1/\epsilon)}$ *and uses* $O(n^{1/2+\epsilon})$ *space, where* n *is the number of vertices of* G.

3 Red-Blue Path Problem

3.1 Deciding Red-Blue Path in Planar DAGs

Given a directed graph G with each edge colored either Red or Blue and two vertices s and t, a *red-blue path* denotes a path that alternate between Red and Blue edges and the RedBluePath problem decides whether there exists a directed red-blue path from s to t such that the first edge is Red and last edge is Blue. The RedBluePath problem is a generalization of the reachability problem in graphs, however this problem is NL-complete even when restricted to planar DAGs [6]. This makes it an interesting problem in the area of space bounded complexity as to the best of our knowledge, this is the only "reachabililty-like" problem in planar graphs that is hard for NL. We will now give a proof of Theorem 1.

Proof (of Theorem 1). Consider a planar DAG G. Let $\overline{S}$ be the $n^{(1-\epsilon)}$-separator family computed by PlanarSeparatorFamily on $\hat{G}$ and let $S = \overline{S} \cup \{s,t\}$. For the sake of convenience, we associate two numerical values to the edge colors – 0 to Red and 1 to Blue. We run the subroutine RedBluePathDetect (Algorithm 3) with the input $(G, s, t, n, 0, 1)$ and if the returned value is true, then there is a directed red-blue path from s to t such that the first edge is Red and last one is Blue. In Algorithm 2, we use the notation $(u, v) \in^{(init,temp)} \overline{E'}$ to decide whether there is a red-blue path from u to v that starts with an edge of color value *init* and ends with an edge of color value *temp*.

In Algorithm 1, we use general DFS type search to check the presence of a red-blue path between any two given vertices s' and t'. The only difference with DFS search is that here we explore edges such that color of the edges alternates between red and blue. If we start from a vertex s', then the for loop (Lines 3 – 8) explore the path starting from s' such that first edge of the path is of specified color. In the main algorithm (Algorithm 3), we use Algorithm 1 as a base case, i.e., when the input graph is small in size (is of size $n^{1/2}$). Otherwise, we first compute S and then run Algorithm 2 on the auxiliary graph $\overline{G} = (S, \overline{E})$. Algorithm 3 does not store the graph $\overline{G}$. Whenever it is queried with a pair of vertices to check if it forms an edge, it recursively runs Algorithm 3 on all the connected components of $G[V \setminus S]$ separately (Lines 10 – 14 of Algorithm 2) and produces an answer. Finally, we perform same DFS like search as in Algorithm 1 on $\overline{G}$ (Lines 1 – 9 of Algorithm 2).

In the base case, we use Algorithm 1 which takes linear space and polynomial time. Thus due to the restriction of the size of the graph in the base case, we

---

Input   : $G' = (V', E'), s', t', init, final$
Output   : "Yes" if there is a red-blue path from s' and t' starts with $init$ and
           ends with $final$

/* Use two sets- N_i, for $i = 0, 1$, to store all the vertices that have
  been explored with the color value i */

1 if $s' \notin N_{init}$ **then**
2   Add s' in N_{init};
3   **for** *each edge* $(s', v) \in E'$ *of color value init* **do**
4   **if** $v = t'$ *and* $init = final$ **then**
5   Return true;
6   **end**
7   Run `ColoredDFS`$(G', v, t', init + 1(mod\ 2), final)$
8   **end**
9 end

Algorithm 1. Algorithm `ColoredDFS`: One of the Building Blocks of `RedBluePath` Detect

---

Input   : $\overline{G'} = (\overline{V'}, \overline{E'}), G', s', t', init, final$
Output   : "Yes" if there is a red-blue path from s' and t' starts with $init$ and
           ends with $final$

/* Use two sets- R_i, for $i = 0, 1$, to store all the vertices that have
  been explored with the color value i */

1 if $s' \notin R_{init}$ **then**
2   Add s' in R_{init};
3   **for** *each* $(s', v) \in^{(init, temp)} \overline{E'}$ *for each* $temp \in \{0, 1\}$ **do**
4   **if** $v = t'$ *and* $temp = final$ **then**
5   Return true;
6   **end**
7   Run `ModifiedColoredDFS`$(\overline{G'}, G', v, t', temp + 1\ (mod\ 2), final)$
8   **end**
9 end

/* ''$(u, v) \in^{(init, temp)} \overline{E'}$?'' query will be solved using the following
  procedure   */

10 for *every* $a \in V$ **do**
      /* V be the set of vertices of G' */
      /* V_a = the set of vertices of $\hat{H}$'s connected component
        containing a, where $H = G[V \setminus \overline{V'}]$ */
11   **if** `RedBluePathDetect`$(G[V_a \cup \overline{V'}], u, v, n, init, temp)$ *is true* **then**
12   Return true for the query;
13   **end**
14 end
15 Return false for the query;
    /* End of the query procedure   */

Algorithm 2. Algorithm `ModifiedColoredDFS`: One of the Building Blocks of `RedBluePathDetect`

Input : $G', s', t', n, init, final$

Output : "Yes" if there is a red-blue path from s' and t' starts with $init$ and ends with $final$

1 **if** $n' \leq n^{\frac{1}{2}}$ **then**

2 $\quad|\quad$ Run ColoredDFS($G', s', t', init, final$);

3 **else**

$\quad|\quad$ /* let $r' = n'^{(1-\epsilon)}$ */

4 $\quad|\quad$ Run PlanarSeparatorFamily on $\hat{G}'$ to compute r'-separator family $\overline{S'}$;

5 $\quad|\quad$ Run ModifiedColoredDFS($\overline{G'} = (\overline{S'} \cup \{s', t'\}, \overline{E'}), G', s', t', init, final$);

6 **end**

Algorithm 3. Algorithm RedBluePathDetect: Algorithm for Red-Blue Path in planar DAG

have $\widetilde{O}(n^{1/2})$ space and polynomial time complexity. The sets N_0 and N_1 of algorithm ColoredDFS only store all the vertices of the input graph and we run ColoredDFS on a graph with $n^{1/2}$ vertices and it visits all the edges of the input graph at most once which results in the polynomial time requirement.

Let $\mathcal{S}$ and $\mathcal{T}$ denote its space and time complexity functions. Since $(1-\epsilon)^k \leq \frac{1}{2}$ for $k = O(\frac{1}{\epsilon})$, the depth of the recursion is $O(\frac{1}{\epsilon})$. Also, $|V_a \cup \overline{S'}| \leq 2n'^{(1-\epsilon)}$. This gives us the following recurrence relation:

$$\mathcal{S}(n') = \begin{cases} \tilde{O}(n'^{(\frac{1}{2}+\epsilon)}) + \mathcal{S}(2n'^{(1-\epsilon)}) & \text{if } n' > n^{\frac{1}{2}} \\ \tilde{O}(n^{\frac{1}{2}}) & \text{otherwise} \end{cases}$$

Thus, $\mathcal{S}(n) = O(\frac{1}{\epsilon})\tilde{O}(n^{\frac{1}{2}+\epsilon}) = \tilde{O}(n^{\frac{1}{2}+\epsilon})$.

For time analysis, we get the following recurrence relation:

$$\mathcal{T}(n') = \begin{cases} q(n)(p_1(n')\mathcal{T}(2n'^{(1-\epsilon)}) + p_2(n')) & \text{if } n' > n^{\frac{1}{2}} \\ q(n)\tilde{O}(n^{\frac{1}{2}}) & \text{otherwise} \end{cases}$$

As the recursion depth is bounded by $O(\frac{1}{\epsilon})$ (a constant), we have $\mathcal{T}(n) = p(n)^{O(\frac{1}{\epsilon})}$ for some polynomial $p(n)$.

Proof of correctness: We now give a brief idea about the correctness of this algorithm. In the base case, we use similar technique as DFS just by alternatively exploring Red and Blue edges and thus this process gives us a path where two consecutive edges are of different colors. Otherwise, we also do a DFS like search by alternatively viewing Red and Blue edges and we do this search on the graph $H = (\overline{S'} \cup \{s, t\}, \overline{E'})$. By this process, we decide on presence of a path in H from s to t such that two consecutive edges are of different colors in G and the edge coming out from s is Red and the edge going in at t is Blue. This is enough as each path P in G must be broken down into the parts $P_1, P_2, \cdots, P_k$ and each P_i must be a sequence of edges that starts and ends at some vertices of $\overline{S'} \cup \{s, t\}$

and also alternates in color. We find each such P_i, just by considering each connected component of $G(V' \setminus \overline{S'})$ and repeating the same steps recursively. $\square$

Due to [6], we know that the reachability problem in directed graphs reduces to RedBluePath in planar DAGs. For the class of graphs in which this reduction results the sub-quadratic increase in the number of vertices, we have an algorithm for reachability problem that takes sublinear space and polynomial time. As a special case of this we can state the following theorem.

Theorem 5. *Given a directed acyclic graph* $G = (V, E)$, *where* $|E| = \widetilde{O}(n)$, *with a drawing in a plane such that the number of edge crossings is* $\widetilde{O}(n)$ *and two vertices* s *and* t, *then for any constant* $0 < \epsilon < \frac{1}{2}$, *there is an algorithm that decides whether there is a path from* s *to* t *or not. This algorithm runs in polynomial time and uses* $O(n^{\frac{1}{2}+\epsilon})$ *space, where* n *is the number of vertices of* G.

Proof. We consider a reduction similar to the reduction from directed reachability problem to RedBluePath problem in planar DAG given in [6]. We do the following: (i) insert new vertices in between edges of G so that in the resulting graph each edge takes part in only one crossing and (ii) replace each crossing of the resulting graph with a *planarizing gadget* as in Fig. 1 and also replace each edge without any crossing with two edges as shown in Fig. 1. Denote the resulting graph as G_{planar} and the corresponding vertices of s and t as s' and t'. It is easy to see that there is a bijection between $s - t$ paths in G and $s' - t'$ red-blue paths in G_{planar} that starting with a Red edge and ending with a Blue edge.

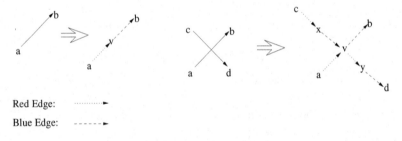

Red Edge: ···········▶

Blue Edge: ------▶

Fig. 1. Red-Blue Edge Gadget

If the drawing of the given graph G contains k edge crossings, then step (i) will introduce at most $2k$ many new vertices and say after this step the number of edges becomes m. Then step (ii) will introduce at most $(2m + 3k)$ many vertices. It is clear from the reduction itself that $m = \widetilde{O}(n)$ and thus the graph G_{planar} contains $\widetilde{O}(n)$ many vertices. Now by applying the algorithm RedBluePathDetect on G_{planar}, we get the desired result. $\square$

A large class of graphs will satisfy the conditions specified in Theorem 5. We now explicitly give an example of one such class of graphs. Before that, we give some definitions. *Crossing number* of a graph G, denoted as $cr(G)$, is the lowest number of edge crossings (or the crossing point of two edges) of a drawing of the graph G in a plane. A graph is said to be *k-planar* if it can be drawn on the plane in such a way that each edge has at most k crossing point (where it crosses a single edge). It is known from [9] that a k-planar graph with n vertices has at most $O(n\sqrt{k})$ many edges. Note that a k-planar graph has crossing number at most mk, where m is the number of edges. Now we can state the following corollary.

Corollary 1. *Given a directed acyclic graph, which is k-planar, where* $k = O(\log^c n)$*, for some constant c, with a drawing in a plane having minimum number of edge crossings and two vertices s and t, then for any constant* $0 < \epsilon < \frac{1}{2}$*, there is an algorithm that decides whether there is a path from s to t or not. This algorithm runs in polynomial time and uses* $O(n^{\frac{1}{2}+\epsilon})$ *space, where n is the number of vertices of the given graph.*

3.2 Deciding Even Path in Planar DAGs

Given directed graph G and two vertices s and t, EvenPath is the problem of deciding the presence of a (simple) directed path from s to t, that contains even number of edges. We can view this problem as a relaxation of RedBluePath problem as a path starting with Red edge and ending with Blue edge is always of even length. In this section, we establish a relation between EvenPath problem in planar DAG with detecting a odd length cycle in a directed planar graph with weight one (can also be viewed as an unweighted graph).

Lemma 1. *For directed planar graphs, for any constant* $0 < \epsilon < \frac{1}{2}$*, there is an algorithm that solves the problem of deciding the presence of odd length cycle in polynomial time and* $O(n^{\frac{1}{2}+\epsilon})$ *space, where n is the number of vertices of the given graph.*

The above lemma is true due to the fact that we can do BFS efficiently for undirected planar graph and it is enough to detect odd length cycle in each of the strong components of the undirected version of the given directed planar graph. For undirected graph, presence of odd length cycle can be detected using BFS algorithm and then put red and blue colors on the vertices such that vertices in the consecutive levels get the opposite colors. After coloring of vertices if there exists an monochromatic edge (edge where both vertices get the same color), then we can conclude that there is an odd length cycle in the graph otherwise there is no odd length cycle. But this is not the case for general directed graph. However, the following proposition will help us to detect odd length cycle in directed graph.

In the following proposition, we use $u \to v$ to denote a directed edge (u, v) and $x \xrightarrow{P} y$ to denote a directed path P from a vertex x to y.

Proposition 1. *A strongly connected directed graph contains an odd length cycle if and only if the underlying undirected graph contains an odd length cycle.*

Proof. The forward direction follows trivially. Now to prove the reverse direction, we will use the induction arguments on the length of the odd cycle in the undirected version of the graph. The base case is when the undirected version of the graph contains a 3-length cycle. If the undirected edges present in the undirected cycle also form directed cycle when we consider the corresponding edges in the directed graph, then there is nothing to prove. But if this is not the case, then the Fig. 2 will depict the possible scenarios. As the graph is strongly connected, so there must be a path P from t to s and if this path does not pass through the vertex x, then any one of the following two cycles $s \to t \xrightarrow{P} s$ or $s \to x \to t \xrightarrow{P} s$ must be of odd length. Now suppose P contains the vertex x and thus $P = P_1 P_2$, where P_1 is the path from t to x and P_2 is the path from x to s. It is easy to see that all the three cycles $s \to t \xrightarrow{P} s$, $x \to t \xrightarrow{P_1} x$ and $s \to x \xrightarrow{P_2} s$ cannot be of even length.

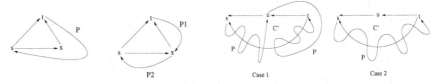

Fig. 2. For undirected cycle of length 3   **Fig. 3.** For undirected cycle of length $(k+2)$

Now by induction hypothesis, assume that if the undirected version has a cycle of k-length (k odd), then there exists an odd length cycle in the original directed graph.

Now lets prove this induction hypothesis for any undirected cycle of length $(k + 2)$. Consider the corresponding edges in the directed graph and without loss of generality assume that this is not a directed cycle. As $(k + 2)$ is odd, so there must be one position at which two consecutive edges are in the same direction. Now contract these two edges in both directed and undirected version of the graph and consider the resulting k-length cycle in the undirected graph. So according to the induction hypothesis, there must be one odd length cycle C in the resulting directed graph. Now if C does not contain the vertex u (where we contract the two edges), then expanding the contracted edges will not destroy that cycle and we get our desired odd length cycle in the directed version of the graph. But if this is not the case, then consider C after expanding those two contracted edges($t \to u \to s$), say the resulting portion is C'. If C' is a cycle, then there is nothing more to do. But if not, then consider the path P from s to t (there must be such path as the graph is strongly connected). Now there will be two possible cases: either P contains u or not. It is easy to see that for both the possible cases (case 1 and case 2 of Fig. 3 and in that figure every crossing of two paths denotes a vertex), all cycles generated by C' and P cannot be of even length. In case 1, if all the cycles generated by the paths $s \xrightarrow{P} u$ and $t \xrightarrow{C'} s$ and all the cycles generated by the paths $u \xrightarrow{P} t$ and $t \xrightarrow{C'} s$ are of even length, then as $t \xrightarrow{C'} s$ is of odd length, so the path $s \xrightarrow{P} u \xrightarrow{P} t$ must be of odd length. And then one of the following two cycles $s \xrightarrow{P} u \to s$ and $u \xrightarrow{P} t \to u$ is of odd length. Similarly in case 2, if all the cycles generated by $s \xrightarrow{P} t$ and $t \xrightarrow{C'} s$ are of odd length, then the path $s \xrightarrow{P} t$ is of odd length and so the cycle $s \xrightarrow{P} t \to u \to s$ is of odd length.

Proof (of Lemma 1). In a directed planar graph, any cycle cannot be part of two different strong component, so checking presence of odd cycle is same as checking presence of odd cycle in each of its strong components. Constructing strong components of a directed planar graph can be done by polynomial many times execution of `DirectedPlanarReach` algorithm (See Theorem 4), as a strong component will contain vertices x, y if and only if `DirectedPlanarReach` (G, x, y, n) and `DirectedPlanarReach` (G, y, x, n) both return "yes". And thus strong component construction step will take $\widetilde{O}(n^{\frac{1}{2}+\epsilon})$ space and polynomial time. After constructing strong components, it is enough to check presence of odd cycle in the underlying undirected graph (according to Proposition 1). So now on, without loss of generality, we can assume that the given graph G is strongly connected. Now execute `OddCycleUndirectedPlanar` $(\hat{G}, s, n)$ (Algorithm 4) after setting the color of s (any arbitrary vertex) to red. Here we adopt the well known technique used to find the presence of odd length cycle in a graph using BFS and coloring of vertices. In Algorithm 4, instead of storing color values for all the vertices, we only stores color values for the vertices present in the separator (Line 11) and we do the coloring recursively by considering the smaller connected components (Line 10). The algorithm is formally defined in Algorithm 4.

By doing the similar type of analysis as that of `RedBluePathDetect`, it can be shown that `OddCycleUndirectedPlanar` will take $O(n^{\frac{1}{2}+\epsilon})$ space and polynomial time and so over all space complexity of detecting odd length cycle in directed planar graph is $O(n^{\frac{1}{2}+\epsilon})$ and time complexity is polynomial in n.

Now we argue on the correctness of the algorithm `OddCycleUndirectedPlanar`. This algorithm will return "yes" in two cases. First case when there is a odd length cycle completely inside a small region ($n' \leq n^{\frac{1}{2}}$) and so there is nothing to prove for this case as it is an well known application of BFS algorithm. Now in the second case, a vertex v in the separator family will get two conflicting colors means that there exists at least one vertex u in the separator family such that there are two vertex disjoint odd as well as even length path from u to v and as a result, both of these paths together will form an odd length cycle.   □

Now we are ready to prove the main theorem of this subsection.

Input : $G' = (V', E'), s', n$, where G' is an undirected graph

Output : "Yes" if there is an odd length cycle

1 **if** $n' \leq n^{\frac{1}{2}}$ **then**

2 | Run BFS(G' , s') and color the vertices with red and blue such that vertices in the alternate layer get the different color starting with a vertex that is already colored;

3 | **if** *there is a conflict between stored color of a vertex and the new color of that vertex or there is an edge between same colored vertices* **then**

4 | | return "yes";

5 | **end**

6 **else**

| /* let $r' = n'^{(1-\epsilon)}$ */

7 | Run PlanarSeparatorFamily on $\hat{G}'$ to compute r'-separator family $\overline{S'}$;

8 | Set $S' := \overline{S'} \cup \{s'\}$;

9 | **for** *every* $x \in V'$ **do**

| | /* V_x = the set of vertices of $\hat{H}$'s connected component
| | containing x, where $H = G[V' \setminus S']$ */

10 | | Run OddCycleUndirectedPlanar($G[V_x \cup S'], s', n$);

11 | | Store color of the vertices of S' in an array of size $|S'|$;

12 | **end**

13 **end**

Algorithm 4. Algorithm OddCycleUndirectedPlanar: Checking Presence of Odd Cycle in an Undirected Planar Graph

Proof (of Theorem 2). Given a planar DAG G and two vertices s and t, first report a path from s to t, say P, which can easily be done by polynomially many invocation of the algorithm DirectedPlanarReach mentioned in Theorem 4 and thus requires polynomial time and $O(n^{\frac{1}{2}+\epsilon})$ space. If the path P is not of even length, then construct a directed graph G' which has the same vertices and edges as G except the edges in path P, instead we do the following: if there is an edge (u, v) in P, then we add an edge (v, u) in G'. Now we can observe that the new graph G' is a directed planar graph.

Claim. G has an even length path if and only if G' has an odd length cycle.

Proof. Suppose G' has an odd length cycle, then that cycle must contains the reverse edges of P in G. Denote the reverse of the path P by P_{rev}. Now lets assume that the odd cycle C' contains a portion of P_{rev} (See Fig. 4). Assume that the cycle C' enters into P_{rev} at x (can be t) and leaves P_{rev} at y (can be s). Then in the original graph G, the path $s \xrightarrow{P} y \xrightarrow{C'} x \xrightarrow{P} t$ is of even length. Now for the converse, lets assume that there exists an even length path P_1 from s to t in G. Both the paths P and P_1 may or may not share some edges and without loss of generality we can assume that they share some edges (See Fig. 5). Now if we consider all the cycles formed by P_{rev} and portions of P_1 in G', then it is easy to see that all the cycles cannot be of even length until length of

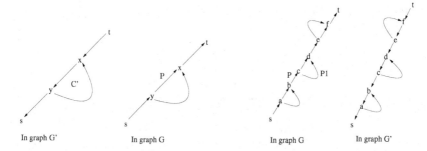

Fig. 4. When G' contains an odd length cycle

Fig. 5. When G contains an even length $s - t$ path

P and P_1 both are of same parity (either both odd or both even), but this is not the case.

Now we can check the presence of odd length cycle in the graph G' in polynomial time and $O(n^{\frac{1}{2}+\epsilon})$ space (by Lemma 1). □

Acknowledgement. The first author would like to thank Surender Baswana for some helpful discussions and comments. We also thank the anonymous reviewers for their helpful comments and suggestions.

References

1. Savitch, W.J.: Relationships between nondeterministic and deterministic tape complexities. J. Comput. Syst. Sci. 4, 177–192 (1970)
2. Wigderson, A.: The complexity of graph connectivity. In: Mathematical Foundations of Computer Science, pp. 112–132 (1992)
3. Barnes, G., Buss, J.F., Ruzzo, W.L., Schieber, B.: A sublinear space, polynomial time algorithm for directed s-t connectivity. In: Proceedings of the Seventh Annual Structure in Complexity Theory Conference, pp. 27–33 (1992)
4. Imai, T., Nakagawa, K., Pavan, A., Vinodchandran, N.V., Watanabe, O.: An $O(n^{1/2+\epsilon})$-Space and Polynomial-Time Algorithm for Directed Planar Reachability. In: 2013 IEEE Conference on Computational Complexity (CCC), pp. 277–286 (2013)
5. Chakraborty, D., Pavan, A., Tewari, R., Vinodchandran, N.V., Yang, L.: New time-space upperbounds for directed reachability in high-genus and h-minor-free graphs. Electronic Colloquium on Computational Complexity (ECCC) 21, 35 (2014)
6. Kulkarni, R.: On the power of isolation in planar graphs. TOCT 3(1), 2 (2011)
7. LaPaugh, A.S., Papadimitriou, C.H.: The even-path problem for graphs and digraphs. Networks 14(4), 507–513 (1984)
8. Nedev, Z.P.: Finding an Even Simple Path in a Directed Planar Graph. SIAM J. Comput. 29, 685–695 (1999)
9. Pach, J., Tóth, G.: Graphs drawn with few crossings per edge. Combinatorica 17(3), 427–439 (1997)

Non-repetitive Strings over Alphabet Lists

Neerja Mhaskar[1] and Michael Soltys[2,*]

[1] McMaster University
Dept. of Computing & Software
1280 Main Street West
Hamilton, Ontario L8S 4K1, Canada
pophlin@mcmaster.ca

[2] California State University Channel Islands
Dept. of Computer Science
One University Drive
Camarillo, CA 93012, USA
michael.soltys@csuci.edu

Abstract. A word is non-repetitive if it does not contain a subword of the form vv. Given a list of alphabets $L = L_1, L_2, \ldots, L_n$, we investigate the question of generating non-repetitive words $w = w_1 w_2 \ldots w_n$, such that the symbol w_i is a letter in the alphabet L_i. This problem has been studied by several authors (e.g., [GKM10], [Sha09]), and it is a natural extension of the original problem posed and solved by A. Thue. While we do not solve the problem in its full generality, we show that such strings exist over many classes of lists. We also suggest techniques for tackling the problem, ranging from online algorithms, to combinatorics over 0-1 matrices, and to proof complexity. Finally, we show some properties of the extension of the problem to abelian squares.

Keywords: Thue words, non-repetitive, square-free, abelian square.

1 Introduction

A *string* over a (finite) *alphabet* Σ is an ordered sequence of symbols from the alphabet: let $w = w_1 w_2 \ldots w_n$, where for each i, $w_i \in \Sigma$. In order to emphasize the array structure of w, we sometimes represent it as $w[1..n]$. We say that v is a *subword* of w if $v = w_i w_{i+1} \ldots w_j$, where $i \leq j$. If $i = j$, then v is a single symbol in w; if $i = 1$ and $j = n$, then $v = w$; if $i = 1$, then v is a *prefix* of w and if $j = n$, then v is a *suffix* of w. We can express that v is a subword more succinctly as follows: $v = w[i..j]$, and when the delimiters do not have to be expressed explicitly, we use the notation $v \leq w$. We say that v is a *subsequence* of w if $v = w_{i_1} w_{i_2} \ldots w_{i_k}$, for $i_1 < i_2 < \ldots < i_k$.

We now define the main concept in the paper, namely a string over an *alphabet list*. Let:

$$L = L_1, L_2, \ldots, L_n,$$

** Research supported in part by an NSERC Discovery Grant.

M.S. Rahman and E. Tomita (Eds.): WALCOM 2015, LNCS 8973, pp. 270–281, 2015.
© Springer International Publishing Switzerland 2015

be an ordered list of (finite) alphabets. We say that w is a string over the list L if $w = w_1w_2 \ldots w_n$ where for all i, $w_i \in L_i$. Note that we impose no conditions on the L_i's: they may be equal, disjoint, or have elements in common. The only condition on w is that the i-th symbol of w must be selected from the i-th alphabet, i.e., $w_i \in L_i$. Let Σ_k denote a fixed generic alphabet of k symbols and let $\Sigma_L = L_1 \cup L_2 \cup \cdots \cup L_n$.

Given a list L of finite alphabets, we can define the set of strings w over L with a regular expression R_L: $R_L := L_1 \cdot L_2 \cdot \ldots \cdot L_n$.

Let $L^+ := L(R_L)$ be the language of all the strings over the list L. For example, if $L_0 = \{\{a, b, c\}, \{c, d, e\}, \{a, 1, 2\}\}$, then

$$R_{L_0} := \{a, b, c\} \cdot \{c, d, e\} \cdot \{a, 1, 2\},$$

and $ac1 \in L_0^+$, but $2ca \notin L_0^+$. Also, in this case $|L_0^+| = 3^3 = 27$. We should point out that $\{a, b, c\}$ is often written as $(a + b + c)$, but we use the curly brackets since it is reminiscent of indeterminate strings, which is yet another way of looking at strings over alphabet lists. See, for example, [Abr87] or [SW09] for a treatment of indeterminates.

We say that w has a *repetition* (or a *square*) if there exists a v such that $vv \le w$. We say that w is *non-repetitive* (or *square-free*) if no such subword exists. An alphabet list L is *admissible* if L^+ contains a non-repetitive string. Let $\mathcal{L}$ represent a *class* of lists; the intention is for $\mathcal{L}$ to denote lists with a given property. For example, we are going to use $\mathcal{L}_{\Sigma_k}$ to denote the class of all lists $L = L_1, L_2, \ldots L_n$, where for each $i \in [n] = \{1, 2, \ldots, n\}$, $L_i = \Sigma_k$, and $\mathcal{L}_k$ will denote the class of all lists $L = L_1, L_2, \ldots, L_n$, where for each $i \in [n]$, $|L_i| = k$, that is, those lists consisting of alphabets of size k. Note that $\mathcal{L}_{\Sigma_k} \subseteq \mathcal{L}_k$. We say that a class of lists $\mathcal{L}$ is admissible if *every* list $L \in \mathcal{L}$ is admissible. For ease of reference, we include a table summarizing the notation for classes with different properties in a table at the end of the paper.

Since any string of length at least 4 over $\Sigma_2 = \{0, 1\}$ contains a square, it follows that $\mathcal{L}_2$ is not admissible. On the other hand, [Thu06] showed using substitutions that $\mathcal{L}_{\Sigma_3}$ is admissible. Using a probabilistic algorithm, [GKM10] showed that $\mathcal{L}_4$ is admissible; the algorithm works as follows: in its i-th iteration, it selects randomly a symbol from L_i, and continues if the string created thus far is square-free, and otherwise deletes the suffix consisting of the right side of the square it just created, and restarts from the appropriate position.

Our paper is motivated by the following question, already posed in [GKM10]: is the class $\mathcal{L}_3$ admissible? That is, given any list $L = L_1, L_2, \ldots, L_n$, where for all $i \in [n]$, $|L_i| = 3$, can we always find a non-repetitive string over such a list? We conjecture with [GKM10] that the answer to this question is affirmative, but we only show that certain (large) subclasses of $\mathcal{L}_3$ are admissible (Theorem 8). In Section 4 we propose different approaches for attacking this conjecture in its full generality.

2   Combinatorial Results

Consider the alphabet $\Sigma_3 = \{1, 2, 3\}$, and the following *substitution scheme*, i.e., *morphism*, due to A. Thue, as presented in [GKM10]:

$$S = \begin{cases} 1 \mapsto 12312 \\ 2 \mapsto 131232 \\ 3 \mapsto 1323132 \end{cases} \tag{1}$$

Given a string $w \in \Sigma_3^*$, we let $S(w)$ denote w with every symbol replaced by its corresponding substitution: $S(w) = S(w_1 w_2 \ldots w_n) = S(w_1) S(w_2) \ldots S(w_n)$.

Lemma 1. *If $w \in \Sigma_3^*$ is a square-free string, then so is $S(w)$.*

Thue's substitution (1) is not the only one; for example, [Lee57] proposes a different substitution[1]. [Ber95, Theorem 3.2], which is a translation of Thue's work on repetitions in words, gives a characterization of the properties of such substitutions (called therein *iterated morphism*). It requires the morphism to be square free for any w of length 3 over Σ_3. Our proof does not require this assumption.

Corollary 2 (A. Thue). $\mathcal{L}_{\Sigma_3}$ *is admissible.*

We are interested in the question whether $\mathcal{L}_3$ is admissible, i.e., whether every list $L = L_1, L_2, \ldots, L_n$, with $|L_i| = 3$, is admissible. Experimental data, with lists of length 20, seems to confirm it. Since we are not able to answer this question in its full generality, we examine different sub-classes of $\mathcal{L}_3$ for which it is true. The goal of this approach is to eventually show that $\mathcal{L}_3$ is admissible.

Recall that a System of Distinct Representatives (SDR) of a collection of sets $\{L_1, L_2, \ldots, L_n\}$ is a selection of n distinct elements $\{a_1, a_2, \ldots, a_n\}$, $a_i \in L_i$.

Claim 3. *If L has an SDR, then L is admissible.*

Proof. Simply let $w = a_1 a_2 \ldots a_n$ be the string consisting of the distinct representatives; as all symbols are distinct, w is necessarily square-free.   □

It is a celebrated result of P. Hall ([Hal87]) that a necessary and sufficient condition for a collection of sets to have an SDR is that they have the *union property*: for any sub-collection $\{L_{i_1}, \ldots, L_{i_k}\}$, $1 \le k \le n$, $|L_{i_1} \cup \cdots \cup L_{i_k}| \ge k$.

Corollary 4. *If L has the union property, then L is admissible.*

Given a list L, we say that the mapping $\Phi : L \longrightarrow \Sigma_3$, $\Phi = \langle \phi_i \rangle$, is *consistent* if for all i, $\phi_i : L_i \longrightarrow \Sigma_3$ is a bijection, and for all $i \neq j$, if $a \in L_i \cap L_j$, then $\phi_i(a) = \phi_j(a)$. In other words, Φ maps all the alphabets to the single alphabet Σ_3, in such a way that the same symbol is always mapped to the same unique symbol in $\Sigma_3 = \{1, 2, 3\}$.

---

[1] Leech's substitutions are longer than Thue's, and they are defined as follows (see [Tom10]): $1 \mapsto 1232132312321$; $2 \mapsto 2313213123132$; $3 \mapsto 3121321231213$.

Lemma 5. *If L has a consistent mapping, then L is admissible.*

Proof. Suppose that L has a consistent mapping $\Phi = \langle \phi_i \rangle$. By Corollary 2 we pick a non-repetitive $w = w_1 w_2 \ldots w_n$ of length n. Let

$$w' = \phi_1^{-1}(w_1)\phi_2^{-1}(w_2)\ldots\phi_n^{-1}(w_n),$$

then w' is a string over L, and it is also non-repetitive. If it were the case that $vv \leq w'$, then the subword vv of w' under Φ would be a square in w, which is a contradiction. □

Let CMP $= \{\langle L \rangle : L$ has a consistent mapping$\}$ be the "Consistent Mapping Problem," i.e., the language of lists $L = L_1, L_2, \ldots, L_n$ which have a consistent mapping. We show in Lemma 6 that this problem is **NP**-complete. It is clearly in **NP** as a given mapping can be verified efficiently for consistency.

Lemma 6. CMP *is* **NP***-hard.*

Proof. A graph $G = (V, E)$ is 3-colorable if there exists an assignment of three colors to its vertices such that no two vertices with the same color have an edge between them. The problem 3-color is **NP**-hard, and by [GJS76] it remains **NP**-hard even if the graph is restricted to be planar.

We show that CMP is **NP**-hard by reducing the 3-colorability of planar graphs to CMP. Given a planar graph $P = (V, E)$, we first find all its triangles, that is, all cliques of size 3. There are at most $\binom{n}{3} \approx O(n^3)$ such triangles, and note that two different triangles may have 0, 1, or 2 vertices in common. If the search yields no triangles in P, then by [Grö59] such a P is 3-colorable, and so we map P to a fixed list with a consistent mapping, say $L = L_1 = \{a, b, c\}$. (In fact, by [DKT11] it is known that triangle-free planar graphs can be colored in linear time.)

Otherwise, denote each triangle by its vertices, and let $T_1, T_2, \ldots, T_k$ be the list of all the triangles, each $T_i = \{v_1^i, v_2^i, v_3^i\}$; note that triangles may overlap. We say that an edge $e = (v_1, v_2)$ is *inside* a triangle if both v_1, v_2 are in some T_i. For every edge $e = (v_1, v_2)$ *not* inside a triangle, let $E = \{e, v_1, v_2\}$. Let $E_1, E_2, \ldots, E_\ell$ be all such triples, and the resulting list is:

$$L_P = T_1, T_2, \ldots, T_k, E_1, E_2, \ldots, E_\ell.$$

See example given in Figure 1.

We show that L_P has a consistent mapping if and only if P is 3-colorable.

Suppose that P is 3-colorable. Let the colors be labeled with $\Sigma_3 = \{1, 2, 3\}$; each vertex in P can be labeled with one of Σ_3 so that no edge has end-points labeled with the same color. This clearly induces a consistent mapping as each triangle $T_i = \{v_1^i, v_2^i, v_3^i\}$ gets 3 colors, and each $E = \{e, v_1, v_2\}$ gets two colors for v_1, v_2, and we give e the third remaining color.

Suppose, on the other hand, that L_P has a consistent mapping. This induces a 3-coloring in the obvious way: each vertex inside a triangle gets mapped to one of the three colors in Σ_3, and each vertex not in a triangle is either a singleton, in which case it can be colored arbitrarily, or the end-point of an edge not inside a triangle, in which case it gets labeled consistently with one of Σ_3. □

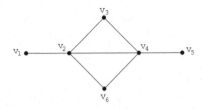

Fig. 1. In this case the list L_P is composed as follows: there are two triangles, $\{v_2, v_3, v_4\}, \{v_2, v_6, v_4\}$, and there are two edges not inside a triangle giving rise to $\{v_1, v_2, (v_1, v_2)\}, \{v_4, v_5, (v_4, v_5)\}$. Note that this planar graph is 3-colorable: $v_1 \mapsto 1$, $v_2 \mapsto 2$, $v_3 \mapsto 3$, $v_6 \mapsto 3$, $v_4 \mapsto 1$, and $v_5 \mapsto 2$. And the same assignment can also be interpreted as a consistent mapping of the list L_P.

We say that a collection of sets $\{L_1, L_2, \ldots, L_n\}$ is a *partition* if for all i, j, $L_i = L_j$ or $L_i \cap L_j = \emptyset$.

Corollary 7. *If L is a partition, then L is admissible.*

Proof. We show that when L is a partition, we can construct a consistent Φ, and so, by Lemma 5, L is admissible. For each i in $[n]$ in increasing order, if L_i is new i.e., there is no $j < i$, such that $L_i = L_j$, then let $\phi_i : L_i \longrightarrow \Sigma_3$ be any bijection. If, on the other hand, L_i is not new, there is a $j < i$, such that $L_i = L_j$, then let $\phi_i = \phi_j$. Clearly $\Phi = \langle \phi_i \rangle$ is a consistent mapping. $\square$

Note that by Lemma 5, the existence of a consistent mapping guarantees the existence of a square-free string. The inverse relation does not hold: a list L may not have a consistent mapping, and still be admissible. For example, consider $L = \{\{a, b, c\}, \{a, b, e\}, \{c, e, f\}\}$. Then, in order to have consistency, we must have $\phi_1(a) = \phi_2(a)$ and $\phi_1(b) = \phi_2(b)$. In turn, by bijectivity, this implies that $\phi_1(c) = \phi_2(e)$. Again, by consistency:

$$\phi_3(c) = \phi_1(c) = \phi_2(e) = \phi_3(e),$$

and so $\phi_3(c) = \phi_3(e)$, which violates bijectivity. Hence L does not have a consistent mapping, but $w = abc \in L^+$, and w is square-free.

Let $\mathcal{L}_{SDR}, \mathcal{L}_{Union}, \mathcal{L}_{Consist}$, and $\mathcal{L}_{Part}$, be classes consisting of lists with: an SDR, the union property, a consistent mapping, and the partition property, respectively. Summarizing the results in the above lemmas we obtain the following theorem.

Theorem 8. $\mathcal{L}_{SDR}, \mathcal{L}_{Union}, \mathcal{L}_{Consist}$, *and* $\mathcal{L}_{Part}$ *are all admissible.*

A natural way to construct a non-repetitive string over L is as follows: pick any $w_1 \in L_1$, and for $i + 1$, assuming that $w = w_1 w_2 \ldots w_i$ is non-repetitive, pick an $a \in L_{i+1}$, and if wa is non-repetitive, then let $w_{i+1} = a$. If, on the other hand, wa has a square vv, then vv must be a suffix (as w is non-repetitive by assumption). Delete the right copy of v from w, and restart.

The above paragraph describes the gist of the algorithm for computing a non-repetitive string over $\mathcal{L}_4$, presented in [GKM10]. The correctness of the algorithm relies on a beautiful probabilistic argument that we present partially in the proof of Lemma 11. For the full version of this result the reader is directed to the source [GKM10]. On the other hand, the correctness of the algorithm in [GKM10] also relies on Lemma 9 shown below, which was assumed but not shown [GKM10, line 7 of Algorithm 1, on page 2].

Incidentally, suppose that there is an $L \in \mathcal{L}_4$ with the following property: there exists an $L_i = \{a, b, c, d\}$ such that if w is a non-repetitive string in L^+, then $w_i = a$. That is, all non-repetitive strings in L^+ must select a from L_i. Then $\mathcal{L}_3$ would be inadmissible, since we could construct an inadmissible $L' \in \mathcal{L}_3$ as follows: $L'_i = \{b, c, d\}$, and for $j \neq i$, L'_j any 3-element subset of L_j.

Lemma 9. *If w is non-repetitive, then for any symbol a, either $w' = wa$ is still non-repetitive, or w' has a unique square (consisting of a suffix of w').*

Proof. Suppose that $w' = wa$ has a square; denote this square $v_\ell v_r$, where $v_\ell = v_r$, and $v_\ell v_r$ is a suffix of w'. Suppose that there is another square $v'_\ell v'_r$. We examine the following cases:

1. If $|v'_r| \leq \lfloor \frac{|v_r|}{2} \rfloor$, then $v'_\ell v'_r$ is a suffix of v_r, and hence $v'_\ell v'_r$ is also a suffix of v_ℓ, and hence w has a square — contradiction.

2. If $\lfloor \frac{|v_r|}{2} \rfloor < |v'_r| < |v_r|$, then let x be the (unique) suffix of v'_ℓ that corresponds to a prefix of v_r. Note that the case $|v'_r| = |v_r|$ is superfluous, as it means that $v'_r = v_r$, and since $|v'_r| < |v_r|$, $|x| > 0$. Since x is a suffix of v'_ℓ, it also must be a suffix of v'_r, and so x is also a suffix of v_r, and hence a suffix of v_ℓ. Thus, we must have xx straddling $v_\ell v_r$, and thus we have a square in w — contradiction.

3. The case $|v_r| < |v'_r| < |v_\ell v_r|$ is symmetric to the previous case, with the roles of v_r, v'_r and v_ℓ, v'_ℓ reversed.

4. Finally, $|v'_r| \geq |v_\ell v_r|$ means that $v_\ell v_r$ is also a subword of v'_ℓ, giving us a repetition $v'_\ell \leq w$, and hence a contradiction.

Thus, the only possible case is $v_\ell = v'_\ell, v_r = v'_r$, and this means that w' must have a unique repetition, if it has one at all. $\square$

An open question is how to de-randomize [GKM10, Algorithm 1]. The naïve way to de-randomize it is to employ an exhaustive search algorithm: given an L in $\mathcal{L}_4$, examine every $w \in L^+$ in lexicographic order until a non-repetitive is found, which by [GKM10, Theorem 1] must happen. In that sense, the correctness of the probabilistic algorithm implies the correctness of the deterministic exhaustive search algorithm. However, such an exhaustive search algorithm takes $4^{|L|}$ steps in the worst case; is it possible to de-randomize it to a deterministic polytime algorithm? Also, what is the expected running time of the probabilistic algorithm?

3   Abelian Squares

There are generalizations of the notion of a square in a string. For example, while a *square* in w is a subword $vv \leq w$, an *overlap* is a subword of the form $avava$, where a is a single symbol, and v is an arbitrary word (see [Sha09, pg. 37], and the excellent [Ram07]). The point is that the string $avava$ can be seen as two overlapping occurrences of the word ava. While there are no arbitrarily long square-free words over $\Sigma_2 = \{0, 1\}$, there are arbitrarily long overlap-free words over Σ_2 (see [Sha09, Theorem 2.5.1, pg. 38]).

An *abelian square* is a word of the form ww' where $|w| = |w'|$, and where w' is a permutation of w. That is, if $w = w_1 w_2 \ldots w_n$, then $w' = w_{\pi(1)} w_{\pi(2)} \cdots w_{\pi(n)}$, where $\pi : [n] \longrightarrow [n]$ is a bijection. A word w is abelian-square-free if there is no $vv' \leq w$ such that vv' is an abelian square. While there are arbitrarily long square-free words over Σ_3, the question was posed in [Sha09, Section 2.9, Problem 1(a), pg. 47] whether there are infinite abelian-square-free words (where aa is not counted as an abelian square, that is, abelian-square-of-size-at-least-2-free words). We show in Lemma 10 that there are no abelian-square-free words of size 8 or bigger; but allowing abelian squares of size 1 makes the problem more difficult. Here is a word of size 25, with no abelian-square-free but allowing abelian squares of size 1: $aaabaaacaaabbbaaacaa$.

Lemma 10. *If w is a word over Σ_3 such that $|w| \geq 8$, then w must have an abelian square.*

Proof. We show that if $w \in \Sigma_3^{\geq 8}$, i.e., w is a word over Σ_3 of size at least 8, then w necessarily has an abelian square.

Let $\tau : \Sigma_3 \longrightarrow \Sigma_3$ be a bijection, that is, τ is a permutation of $\{a, b, c\}$. (Note that this is not the same as the π above, which is a permutation of a string w.) It is easy to see that for each of the six possible τ's, w is an abelian square if and only if $\tau(w)$ is an abelian square. Therefore, if we show that for any w of the form $w = abx$, where $x \in \Sigma_3^*$, w has an abelian square, it will follow that every w has an abelian square. (If $w = aax, bbx, ccx$ then w has a square, which is also an abelian square, and for the six cases that arise from two distinct initial characters we apply a τ to reduce it to the $w = abx$ case.)

Consider Figure 2 which represents with a tree the prefixes of all the strings over Σ_3. Think of the labels on the nodes on any branch starting at the root (ε) as spelling out such a prefix. Note that all the branches starting with ab end in a $\times$-leaf, which denotes that adding any symbol in $\Sigma_3 = \{a, b, c\}$ would yield an abelian square. This proves the Lemma, as the other prefixes (starting with one of $\{ba, bc, ca, cb\}$) would also eventually yield an abelian square. $\square$

Adapting the method of [GKM10] we can also show that there are infinite abelian-square-free words over lists of size 4.

Lemma 11. *Let L be any list where for all i, $|L_i| = 4$. Then, there is an abelian-square-free word over L.*

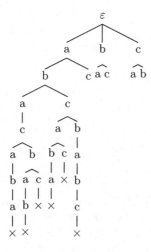

Fig. 2. No abelian squares of length greater than 8

Proof. Fix an ordering inside each L_i, and let $r = r_1, r_2, \ldots, r_m$ be a sequence over $\{1, 2, 3, 4\}$. We use r to build an abelian-square-free word as follows: starting with $w = \varepsilon$, in the i-th step, add to the right end of w the symbol in position r_i in $L_{|w|+1}$. If the resulting w' is abelian-square-free, continue. Otherwise, there is an abelian square (which, unlike in the case of regular squares, does not have to be unique — see Lemma 9). Let vv' be the longest abelian square so that $w' = xvv'$. Delete v' and restart the process. Let (D, s) be a log of the procedure, where D is a sequence of integers keeping track of the differences in size of the w's from one step to the next; let s be the final string after the entire r has been processed. Following the same technique as in [GKM10], we show that given (D, s) there is a unique r corresponding to it. By assuming that the total number of s's are less than a given n_0, we get a contradiction by letting r be sufficiently large, and bounding the number of logs with Catalan numbers [Sta99]. □

The authors have written a short Python program for checking abelian squares; you may find it on the second author's web page.

4   Future Directions

4.1   Online Algorithms and Games

In the online version of the problem, L is presented piecemeal, one alphabet at a time, and we select the next symbol without knowing the future, and once selected, we cannot change it later. More precisely, the L_i's are presented one at a time, starting with L_1, and when L_i is presented, we must select $w_i \in L_i$, without knowing $L_{i+1}, L_{i+2}, \ldots$, and without being able to change the selections already committed to in $L_1, L_2, \ldots, L_{i-1}$.

We present the online problem in a game-theoretic context. Given a class of lists $\mathcal{L}$, and a positive integer n, the players take turns, starting with the adversary. In the i-th round, the *adversary* presents a set L_i, and the *player* selects a $w_i \in L_i$; the first i rounds can be represented as:

$$G = L_1, w_1, L_2, w_2, \ldots, L_i, w_i.$$

The condition imposed on the adversary is that $L = L_1, L_2, \ldots, L_n$ must be a member of $\mathcal{L}$.

The player has a *winning strategy for* $\mathcal{L}$, if $\forall L_1 \exists w_1 \forall L_2 \exists w_2 \ldots \forall L_n \exists w_n$, such that $L = L_1, L_2, \ldots, L_n \in \mathcal{L}$ and $w = w_1 w_2 \ldots w_n$ is square-free. For example, the player does not have a winning strategy for $\mathcal{L}_1$ and $\mathcal{L}_2$; see Figure 3. On the other hand, the player has a winning strategy for $\mathcal{L}_{\Sigma_3}$: simply pre-compute a square-free w, and select w_i from L_i. However, this is not a bona fide online problem, as all future L_i's are known beforehand. In a true online problem we expect the adversary to have the ability to "adjust" the selection of the L_i's based on the history of the game.

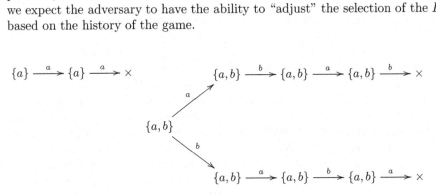

Fig. 3. Player loses if adversary is allowed subsets of size less than 3: the moves of the adversary are represented with subsets $\{a\}$ and $\{a, b\}$ and the moves of the player are represented with labeled arrows, where the label represents the selection from a subset

We present another class of lists for which the player has a winning strategy. Let $\text{size}_L(i) = |L_1 \cup \ldots \cup L_i|$. We say that L has the *growth* property if for all $1 \le i < n = |L|$, $\text{size}_L(i) < \text{size}_L(i + 1)$. We denote the class of lists with the growth property as $\mathcal{L}_{\text{Grow}}$.

Lemma 12. *The player has a winning strategy for $\mathcal{L}_{\text{Grow}}$.*

Proof. In the i-th iteration, select w_i that has not been selected previously; the existence of such a w_i is guaranteed by the growth property. □

The growth property places a rather strong restriction on L, as it allows the construction of square-free strings where all the symbols are different, and hence they are trivially square-free. Note that the growth property implies the union property discussed in Corollary 4. To see this note that the growth property implies the existence of an SDR (discussed in Claim 3), in the stronger sense of every L_i containing an a_i such that for all $j \neq i$, $a_i \notin L_j$.

It would be interesting to study the relationship between admissible $\mathcal{L}$ in the original sense, and those $\mathcal{L}$ for which the player has a winning strategy in the online game sense. Clearly, if there exists a winning strategy for $\mathcal{L}$, then $\mathcal{L}$ is admissible; what about the converse?

4.2   Boolean Matrices

Instead of considering alphabets, we consider sets of natural numbers, i.e., each $L_i \subseteq \mathbb{N}$, and $L = L_1, L_2, \ldots L_n$, and $\mathcal{L}$ is a class of lists as before. We say that $w \in L^+$ if $w = j_1, j_2, \ldots, j_n$, i.e., w is a sequence of numbers, such that for all $i \in [n]$, $j_i \in L_i$. The definition of repetitive (square) is analogous to the alphabet of symbols case.

Note that any $L = L_1, L_2, \ldots, L_n$ can be *normalized* to be $\hat{L}$, where each $\hat{L}_i$ is replaced with $\hat{L}_i \subseteq [3n]$. This can be accomplished by mapping all integers in $\cup L$, at most $3n$ many of them, in an order preserving way, to $[3n]$. Clearly, L is admissible iff $\hat{L}$ is admissible, and given a list L, it can be normalized in polynomial time. This allows us to restate the game theoretic approach given in the previous section with bounded quantification; this in turn places the problem in the polytime hierarchy, and hence in **PSPACE**. This is not surprising as many two-player zero-sum games are in this class (see [Pap94, Chapter 19]).

The integer restatement suggests an approach based on 0-1 matrices. Given a normalized list $L = \{L_1, L_2, \ldots, L_n\}$, we define the 0-1 $n \times 3n$ matrix A_L where row i of A_L is the incidence vector of L_i: $A_L(i, j) = 1 \iff j \in L_i$.

The attraction of this setting is that it may potentially allow us to use the machinery of combinatorial matrix theory to show that $\mathcal{L}_3$ is admissible.

It is easy to see that L is admissible iff there is a selection S that picks a single 1 in each row in such a way that there are no i consecutive rows equal to the next i consecutive rows. More precisely, L is admissible iff there does not exist i, j, such that $1 \leq i \leq j \leq \lfloor \frac{n}{2} \rfloor$, and such that the submatrix of A_L consisting of rows i through j is equal to the submatrix of A_L consisting of rows $j + 1$ through $2(j + 1) - i$.

Suppose that Σ_L is re-ordered bijectively by Γ, and let

$$L_\Gamma = \{\Gamma(L_1), \Gamma(L_2), \ldots, \Gamma(L_n)\}.$$

Then L is admissible iff L_Γ is admissible. Note that a bijective re-ordering of Σ_L is represented by a permutation of the columns of A_L. Thus, permuting the columns of A_L does not really change the problem; the same is not true of permuting the rows, which actually re-orders the list L, changing the constraints, and therefore changing the problem.

Consider the matrices $S = A_L A_L^t$ and $T = A_L^t A_L$. The element $[s_{ij}]$ record the number of elements common in the sets L_i and L_j, where $1 \leq i, j \leq n$. The diagonal elements $[s_{ii}]$ record the cardinality of the set L_i, which is 3. The element $[t_{ij}]$ record the number of times the numbers i and j, where $1 \leq i, j \leq 3n$, occur together in the sets of L. The diagonal elements of T display the total number of times each number in $[3n]$ appears in L. The properties of these matrices are studied to possibly use them in the construction of Φ(consistent mapping).

4.3   Proof Complexity

By restating the generalized Thue problem in the language of 0-1 matrices, as we did in Section 4.2, we can more easily formalize the relevant concepts in the language of first order logic, and use its machinery to attack the problem.

We are going to adopt the logical theory $\mathbf{V}^0$ as presented in [CN10], whose language is $\mathcal{L}_A^2 = [0, 1, +, \cdot, |\,|; =_1, =_2, \leq, \in]$ (see [CN10, Definition IV.2.2, pg. 76]). Without going into all the details, this language allows the indexing of a 0-1 string X; on the other hand, a 0-1 matrix A_L can be represented as a string X_L with the definition: $X_L(3n(i-1) + j) = A_L(i,j)$. Hence, $\mathcal{L}_A^2$ is eminently suitable for expressing properties of strings.

Define the following auxiliary predicates:

- Let Three(X_L) be a predicate which states that the matrix A_L corresponding to X_L has exactly three 1s per row.
- Let Sel(Y_L, X_L) be a predicate which states that Y_L is a selection of X_L, in the sense that Y_L corresponds to the 0-1 matrix which selects a single 1 in each row of A_L.
- Let SF(Y_L) be a predicate which states that Y_L is square-free (i.e., non-repetitive).

Lemma 13. *All three predicates* Three, SF, Sel *are* Σ_0^B.

Our conjecture can be stated as a Σ_1^B formula over $\mathcal{L}_A^2$ as follows:

$$\alpha(X_L) := \exists Y_L \leq |X_L|(\text{Three}(X_L) \wedge \text{Sel}(Y_L, X_L) \wedge \text{SF}(Y_L)).$$

Suppose we can prove that $\mathbf{V}^0 \vdash \alpha(X_L)$; then, we would be able to conclude that given any L, we can compute a non-repetitive string over L in $\mathbf{AC}^0$. Likewise, if $\mathbf{V}^1 \vdash \alpha(X_L)$, then we would be able to conclude that the non-repetitive string can be computed in polynomial time.

References

[Abr87] Abrahamson, K.R.: Generalized string matching. SIAM J. Comput. 16(6), 1039–1051 (1987)

[Ber95] Berstel, J.: Axel Thue's papers on repetitions in words: a translation. Technical report, Université du Québec a Montréal (1995)

[CN10] Cook, S.A., Nguyen, P.: Logical Foundations of Proof Complexity. Cambridge Univeristy Press (2010)

[DKT11] Dvořák, Z., Kawarabayashi, K.-I., Thomas, R.: Three-coloring triangle-free planar graphs in linear time. ACM Trans. Algorithms 7(4), 41:1–41:14 (2011)

[GJS76] Garey, M.R., Johnson, D.S., Stockmeyer, L.: Some simplified np-complete graph problems. Theoretical Computer Science 1(3), 237–267 (1976)

[GKM10] Grytczuk, J., Kozik, J., Micek, P.: A new approach to nonrepetitive sequences. arXiv:1103.3809 (December 2010)

[Grö59] Grötzsch, H.: Ein dreifarbensatz für dreikreisfreie netze auf der kugel 8, 109–120 (1959)

[Hal87] Hall, P.: On representatives of subsets. In: Gessel, I., Rota, G.-C. (eds.) Classic Papers in Combinatorics. Modern Birkhäuser Classics, pp. 58–62. Birkhäuser, Basel (1987)

[Lee57] Leech, J.: A problem on strings of beads. Mathematical Gazette, 277 (December 1957)

[Pap94] Papadimitriou, C.H.: Computational Complexity. Addison-Wesley (1994)

[Ram07] Rampersad, N.: Overlap-free words and generalizations. PhD thesis, Waterloo University (2007)

[Sha09] Shallit, J.: A second course in formal languages and automata theory. Cambridge Univeristy Press (2009)

[Sta99] Stanley, R.P.: Exercises on catalan and related numbers. Enumerative Combinatorics 2 (1999)

[SW09] Smyth, W.F., Wang, S.: An adaptive hybrid pattern-matching algorithm on indeterminate strings. Int. J. Found. Comput. Sci. 20(6), 985–1004 (2009)

[Thu06] Thue, A.: Über unendliche zeichenreichen. Skrifter: Matematisk-Naturvidenskapelig Klasse. Dybwad (1906)

[Tom10] Robinson Tompkins, C.: The morphisms with unstackable image words. CoRR, abs/1006.1273 (2010)

Summary of classes of lists

$\mathcal{L}$ denotes a class of lists		
$L = L_1, L_2, \ldots, L_n$ denotes a (finite) list of alphabets		
L_i denotes a finite alphabet		
Class name	**Description**	**Admissible**
$\mathcal{L}_{\Sigma_k}$	for all $i \in [n]$, $L_i = \Sigma_k$	for Σ_k, yes for $k \geq 3$; no for $k < 3$
$\mathcal{L}_k$	for all $i \in [n]$, $|L_i| = k$	yes for $k \geq 4$; no for $k \leq 2$; for $k = 3$?
$\mathcal{L}_{\mathrm{SDR}}$	L has an SDR	yes
$\mathcal{L}_{\mathrm{Union}}$	L has the union property	yes
$\mathcal{L}_{\mathrm{Consist}}$	L has a consistent mapping	yes
$\mathcal{L}_{\mathrm{Part}}$	L is a partition	yes
$\mathcal{L}_{\mathrm{Grow}}$	for all i, $\left| \cup_{j=1}^{i} L_j \right| < \left| \cup_{j=1}^{i+1} L_j \right|$	yes, even for online games

Dichotomy Theorems for Homomorphism Polynomials of Graph Classes

Christian Engels

Saarland University, Department of Computer Science, Saarbruecken, Germany
engels@cs.uni-saarland.de

Abstract. In this paper, we will show dichotomy theorems for the computation of polynomials corresponding to evaluation of graph homomorphisms in Valiant's model. We are given a fixed graph H and want to find all graphs, from some graph class, homomorphic to this H. These graphs will be encoded by a family of polynomials.

We give dichotomies for the polynomials for cycles, cliques, trees, outerplanar graphs, planar graphs and graphs of bounded genus.

1 Introduction

Graph homomorphisms are studied because they give important generalizations of many natural questions (k-coloring, acyclicity, binary CSP and many more cf. [16]). One of the first results, given by Hell and Nešetřil [15], was on the decision problem where they gave a dichotomy. The exact result was, that deciding if there exists a homomorphism from some graph G to a fixed undirected graph H is polynomial time computable if H is bipartite and NP-complete otherwise. A different side of graph homomorphisms was looked at by Chekuri and Rajaraman [4], Dalmau et al. [6], Freuder [10] and finally Grohe [14]. They studied the following: Given a restricted graph class $\mathcal{G}$, decide if there is a graph $G \in \mathcal{G}$ homomorphic to a given graph H. Instead of restricting the graph classes H as in the first problem, we restrict the graphs we map from. Later, focus shifted onto the counting versions of these two sides where we have to count the number of homomorphisms. Dyer and Greenhill [9] solved the first problem in the counting case and Dalmau and Jonsson [5] the second. The first problem was extended by Bulatov and Grohe [1] to graphs with multiple edges. They also notice some interesting connections to statistical physics and constraint satisfaction problems. A good introduction to the history of graph homomorphism was written by Grohe and Thurley [13] and research on these topics continues even today with two noticeable results being the works by Goldberg et al. [11] and Cai et al. [3].

However, the arithmetic circuit complexity was still open. The previous results could only show that the hard cases have no polynomial size circuits for counting the number of homomorphisms but it was unclear if these problems are VNP complete. The study of VNP complete problems and the arithmetic world was started in the seminal paper by Valiant [22]. In this world, we look at the complexity of computing a family of polynomials using a family of arithmetic

M.S. Rahman and E. Tomita (Eds.): WALCOM 2015, LNCS 8973, pp. 282–293, 2015.

circuits. Recently, a dichotomy for graph homomorphisms was shown by Rugy-Altherre [20]. Here a graph is encoded by a product of edge variables and sets of graphs as sums over these products. This is known as generating function and a detailed definition will be provided in Section 2. However, his result was for the first side of the graph homomorphism problem.

In this paper we look at the second side of the graph homomorphism problem to complete the picture for the arithmetic circuit world. While we could not get a general theorem as in [5], we show multiple hardness proofs for some classes. We will look at cycles, cliques, trees, outerplanar graphs, planar graphs and graphs of bounded genus.

Recently, homomorphism polynomials in a different form are even used for giving natural characterizations of VP independent of the circuit definition [8]. In this way our results can be interpreted as showing that some straightforward candidates originating from the counting world do not give a characterization of VP.

Section 2 gives a formal introduction to our model, related hard problems and states the problem precisely. We prove our dichotomies in Sections 3.1 to 3.6 where the constructions in Sections 3.4 to 3.6 build on each other. The construction in Section 3.3 will use a slightly different model as the other sections. We will give a brief introduction into concepts from graph genus in Section 3.6 but refer the reader to the textbook by Diestel [7].

2 Model and Definitions

Let us first give a brief introduction to the field of Valiant's classes. For further information the reader is referred to the textbook by Bürgisser [2]. In this theory, we are given an arithmetic circuit (a directed acyclic connected graph) with addition and multiplication gates over some field K. These gates are either connected to other gates or input gates from the set $K \cup X$ for some set of indeterminates X. At the end we have exactly one output gate. An arithmetic circuit computes a polynomial in $K[X]$ at the output gate in the obvious way.

As Valiant's model is non-uniform, a problem consists of families of polynomials. A p-family is a sequence of polynomials (f_n) over $K[X]$ where the number of variables is n and the degree is bounded by some polynomial in n. Additionally the family of polynomials (f_n) should be computed by a family of arithmetic circuits (C_n) where f_n is computed by C_n for all n. Valiant's Model focuses its study on p-families of polynomials.

We define $L(f)$ to be the number of gates for a minimal arithmetic circuit computing a given polynomial $f \in K[X]$. VP is the class of all p-families of polynomials where $L(f_n)$ is bounded polynomially in n. Let $q(n), r(n), s(n)$ be polynomially bounded functions. A p-family $(f_n) \in K[x_1, \ldots, x_{q(n)}]$ is in VNP if there exists a family $(g_n) \in K[x_1, \ldots, x_{r(n)}, y_1, \ldots, y_{s(n)}]$ in VP such that

$$f(x_1, \ldots, x_{q(n)}) = \sum_{\epsilon \in \{0,1\}^{s(n)}} g(x_1, \ldots, x_{r(n)}, \epsilon_1, \ldots, \epsilon_{s(n)}).$$

The classes VP and VNP are considered algebraic analogues to P and NP or more accurately #P. We can also define an algebraic version of AC_0, mentioned by Mahajan and Rao [17]. A p-family is in VAC_0 if there exists a family of arithmetic circuit of constant depth and polynomial size with unbounded fan-in that computes the family of polynomials.

The notion of a reduction in Valiant's model is given by p-projections. A p-family (f_n) is a p-projection of (g_n), written as $(f_n) \leq_p (g_n)$, if there exists a polynomially bounded function $q(n)$ such that for every n, $f(x_1, \ldots, x_n) = g(a_1, \ldots, a_{q(n)})$ for some $a_i \in K \cup \{x_1, \ldots, x_n\}$. Once we have a reduction, we get a notion of completeness in the usual way.

However, we use a different kind of reduction called a c-reduction. This is similar to a Turing reduction in the Boolean world. We define $L^g(f)$ as the number of gates for computing f where the arithmetic circuits is enhanced with an oracle gate for g. An oracle gate for the polynomial $g \in K[x_1, \ldots, x_{n'}]$ has as output $g(a_1, \ldots, a_{n'})$ where $a_1, \ldots, a_{n'}$ are the inputs to this gate. This allows us to evaluate g on $a_1, \ldots, a_{n'}$ in one step if we computed $a_1, \ldots, a_{n'}$ previously in our circuit.

We say f c-reduces to g, written $(f_n) \leq_c (g_n)$, if there exists a polynomial p such that $L^{g_{p(n)}}(f)$ is bounded by some polynomial. This reduction, however, is only useful for VNP and not for VAC_0 and VP. In this paper we will exclusively deal with c-reductions for our VNP completeness results.

2.1 Complete Problems

We continue with the basic framework of graph properties. In the following K will be an infinite field of characteristic not equal to two.

Definition 1. *Let X be a set of indeterminates. Let $\mathcal{E}$ be a graph property, that is, a class of graphs which contains with every graph also all of its isomorphic copies. Let $G = (V, E)$ be an edge weighted, undirected graph with a weight function $w : E \to K \cup X$. We extend the weight function by $w(E') := \prod_{e \in E'} w(e)$ to subsets $E' \subseteq E$.*

The generating function $GF(G, \mathcal{E})$ of the property $\mathcal{E}$ is defined as

$$GF(G, \mathcal{E}) := \sum_{E' \subseteq E} w(E')$$

where the sum is over all subsets E' such that the subgraph (V, E') of G has property $\mathcal{E}$.

The reader should notice that the subgraph still contains all vertices and just takes a subset of the edges.

In the following, let G be a graph and let $X = \{x_e \mid e \in E\}$. We label each edge e by the indeterminate x_e. We conclude by stating some basic VNP-complete problems. Proofs of these facts can be found in the textbook by Bürgisser [2].

Theorem 2 ([2]). *$GF(K_n, \mathcal{UHC}_n)$ is VNP-complete where $\mathcal{UHC}_n$ is the set of all hamiltonian cycles in K_n.*

Theorem 3 ([2]). *Let $\mathcal{CL}$ be the set of all cliques. Meaning, the set of all graphs, where one connected component is a complete graph and each of the remaining connected components consist of one vertex only. The family $\mathrm{GF}(K_n, \mathcal{CL})$ is VNP-complete.*

Theorem 4 ([2]). *Let $\mathcal{M}$ be the set of all graphs where all connected components have exactly two vertices. The family $\mathrm{GF}(K_n, \mathcal{M})$ is VNP-complete.*

This polynomial gives us all perfect matchings in a graph. It is well known that the original VNP-complete problem, the permanent, is equal to $\mathrm{GF}(K_{n,n}, \mathcal{M})$ for bipartite graphs which is a projection of $\mathrm{GF}(K_{n^2}, \mathcal{M})$.

2.2   The Problem and Related Definitions

We now formulate our problem. Let G, H be undirected graphs. We will generally switch freely between having the variable indexed by either edges (x_e) or vertices ($x_{i,j}$ for $i, j \in V$). We let x_j correspond to the self-loop at vertex j.
    A homomorphism from $G = (V, E)$ to $H = (V', E')$ is a mapping $f : V \to V'$ such that for all edges $\{u, v\} \in E$ there exist an edge $\{f(u), f(v)\} \in E'$. We can define the corresponding generating function as follows.

Definition 5. *Let $\mathcal{H}_H$ be the property of all connected graphs homomorphic to a fixed H. We denote by $\mathcal{F}^{H,n}$ the generating function $\mathcal{F}^{H,n} := \mathrm{GF}(K_n, \mathcal{H}_H)$.*

We can state now the first dichotomy theorem.

Theorem 6 ([20]). *If H has a loop or no edges, $\mathcal{F}^{H,n}$ is in VAC_0 and otherwise it is VNP-complete under c-reductions.*

Instead of looking at all graphs, we want to look at a restricted version. What happens if we do not want to find every graph homomorphic to a given H but every *cycle* homomorphic to a given H? We state our problem in the next definitions.

Definition 7. *Let $\mathcal{E}_n$ be a graph property. Then $\mathcal{F}_{\mathcal{E}_n}^{H,n}$ is the generating function for all graphs in $\mathcal{E}_n$ on n vertices homomorphic to a fixed graph H.*

Definition 8. *We define the following graph polynomials.*

- $\mathcal{F}_{cycle_n}^{H,n}$ *where $cycle_n$ is the property where one connected component is a cycle and the others are single vertices in a graph of size n.*
- $\mathcal{F}_{clique_n}^{H,n}$ *where $clique_n$ is the property where one connected component is a clique and the others are single vertices in a graph of size n.*
- $\mathcal{F}_{tress_n}^{H,n}$ *where $trees_n$ is the property where one connected component is a tree and the others are single vertices in a graph of size n.*
- $\mathcal{F}_{outerplanar_n}^{H,n}$ *where $outerplanar_n$ is the property where one connected component is a outerplanar graph and the others are single vertices in a graph of size n.*

- $\mathcal{F}^{H,n}_{planar_n}$ where $planar_n$ is the property where one connected component is a planar graph and the others are single vertices in a graph of size n.
- $\mathcal{F}^{H,n}_{genus(k),n}$ where genus(k),n is the property where one connected component has genus k and the others are single vertices in a graph of size n.

We will use the notation $\mathcal{F}_{cycle}$, $\mathcal{F}_{clique}$, $\mathcal{F}_{tree}$, $\mathcal{F}_{outerplanar}$, $\mathcal{F}_{planar}$ and $\mathcal{F}_{genus(k)}$ as a shorthand.

Let us now introduce the homogeneous degree of a polynomial.

Definition 9. Let $\bar{x} = x_{i_1}, \ldots, x_{i_l}$ be a subset of variables and (f_n) be a p-family. We can write f_n as

$$f_n = \sum_{\bar{i}} \alpha_{\bar{i}} \prod_{j=1}^{n} x_j^{i_j}.$$

The homogeneous component of f_n of degree k with variables $\bar{x}$ is

$$\mathrm{HOMC}_k^{\bar{x}}(f_n) = \sum_{\substack{i_1,\ldots,i_l \\ k=\sum_{j=1}^{l} i_j}} \alpha_{i_1,\ldots,i_l} x_{i_1}^{i_i} \ldots x_{i_l}^{i_j}.$$

Finally, we need a last lemma in our proofs. This lemma was stated explicit by Rugy-Altherre [20] and can also be found in [2]. It will give us a way to extract all polynomials of homogeneous degree k in some set of variables in c-reductions.

Lemma 10. Then for any sequence of integers (k_n) there exists a c-reduction from the homogeneous component to the polynomial itself:

$$\mathrm{HOMC}_{k_n}^{\bar{x}}(f_n) \leq_c (f_n).$$

The circuit for the reduction has size in $\mathcal{O}(n\delta_n)$ where δ_n is the degree of f_n.

The reader should note that using this theorem will blow up our circuit polynomially in size and can hence be used only a constant number of times in succession. However, we can use this lemma on subsets of vertices. We replace every variable x_i in the subset by $x_i y$ for a new variable y and take the homogeneous components of y. We will use this technique to *enforce* edges to be taken. Notice that enforcing n edges to be taken only increases the circuit size by a factor of n. Additionally, we can set edge variables to zero to *deny* our polynomial using these edges.

Let G be a graph that is homomorphic to a given H. We will, in general, ignore self-loops in G, i.e. assume G to never have any self-loops. If we have proven a theorem for all G without self loops, we can just take the homomorphism polynomial with self-loops, take the homogeneous component of degree zero of all self-loops and get the homomorphism polynomial without self-loops. As we will prove the dichotomy for these, the hardness will follow.

3   Dichotomies

3.1   Cycles

As a first graph class we look at cycles. The proof for the dichotomy will be relatively easy and gives us a nice example to get familiar with homomorphism polynomials and hardness proofs. Our proofs will, in general, reason first about the kind of monomials that exist for a given H and then try to extract or modify these via Lemma 10 to get a solution to a VNP-complete problem. This will yield the reduction. Our main dichotomy for cycles is the following theorem.

Theorem 11. *If H has at least one edge or has a self-loop, then $\mathcal{F}_{cycle}$ is VNP-complete under c-reductions. Else it is in VAC_0.*

The next simple fact shows us which cycles are homomorphic to a given graph H. Let n_0 be defined as n if n is even and $n - 1$ if n is odd.

Fact 1. *Given H a graph with at least one edge, all cycles of length n_0 are homomorphic to H.*

It is easy to see that by folding the graph in half we get one path which is trivially homomorphic to an edge. Our hardness proof will only be able to handle cycles of even length. Luckily this is enough to prove hardness.

Lemma 12. *Let $\mathcal{UHC}_{n_0,even}$ be the graph property of all cycles of length n_0. Then $\mathrm{GF}(K_{n_0}, \mathcal{UHC}_{n_0,even})$ is VNP-hard under c-reductions.*

Proof. If n is even, we can immediately use the hardness of $\mathrm{GF}(K_n, \mathcal{UHC}_n)$ (cf. Theorem 2). If n is odd, we use the following reduction. We have given all cycles of length $n - 1$ and want to get all cycles of length n. We evaluate the polynomial for K_{n+1} and get all cycles of length $n + 1$. We can contract one edge with the following argument. We enforce, via taking the homogeneous component of degree one of $x_{n+1,1}$, all cycles to use $x_{n+1,1}$. We then replace $x_{i,n+1}$ by $x_{i,1}$ for all i and set $x_{n+1,1}$ to one. This gives us all cycles of length n with a factor 2 for every monomial.

To see this let us look at the following argument. Let the edge $(n + 1, 1)$ be the edge we contract and let i, j be two arbitrary points picked in the graph. If we connect i, j with a path through every point we can complete this into a cycle two different ways. Either with the edges $(1, i), (n+1, j)$ or $(1, j), (n+1, i)$. Notice, that every different choice of i, j will construct a different cycle if we contract 1 and $n + 1$. This concludes our reduction to $\mathrm{GF}(K_n, \mathcal{UHC}_n)$. As our circuit can easily divide by two if the polynomial is over an infinite field (see. [21]).   □

Later proofs will also use the contracting idea from the previous lemma. A simple case distinction will give us the proof of the theorem.

Proof (of Theorem 11).
If H has at least one edge, we know from Lemma 1 that all even cycles are homomorphic to H and by this represented in our polynomial. If we take the

homogeneous components of degree n_0, we extract all even cycles of length n_0. This is VNP-hard via the previous Lemma (12).

If H has a self-loop, we can map all cycles to the one vertex in H. We can then extract the hamiltonian cycles of length n by using the homogeneous degree of n as all cycles are homogeneous to a self-loop.

If H has no edge, our polynomial is the zero polynomial as we cannot map any graph G containing an edge to H.

Using Valiant's Criterion, we can prove membership of $\mathcal{F}_{\text{cycle}}$ in VNP (cf.[2]).   □

3.2   Cliques

Here, we will not use cycles in the hardness proof but work directly with the clique polynomial defined by Bürgisser. The complete proof is an easy exercise. In contrast to the other results, we show that computing $\mathcal{F}_{\text{clique}}$ is easy for most choices of H.

Theorem 13. *If H has a self-loop then $\mathcal{F}_{clique}$ is VNP-complete under c-reductions. Otherwise $\mathcal{F}_{clique}$ is in VAC_0.*

3.3   Trees

As the new characterization of VP had a specific tree structure we want to look at the general problem. In previous sections our polynomial just contained the edges of the graph but for this section we need a slightly different model. If a monomial in our polynomial would select the edges E' we also select the vertices $\{u, v | \{u, v\} \in E'\}$ in our monomial. In essence, we will also select the vertices forming the edges, giving us polynomials with variables $X = \{x_e | e \in E\} \cup \{x_v | v \in V\}$. It will be clear later why we need this special form.

Theorem 14. *If H contains an edge, then $\mathcal{F}_{tree}$ is VNP-complete under c-reductions. Otherwise $\mathcal{F}_{tree}$ is in VAC_0.*

In the proof we construct a tree with vertices and edges as vertices in the graph. We connect every edge in the original graph to the vertices and then look at homogeneous components such that every vertex is covered but only $\frac{n}{2}$ edges chosen. For space reasons this proof is omitted. But we still want to mention an interesting fact of the proof. It does not use the fact that the graph class only contains trees. Instead we only use that it contains trees. Hence the theorem can be easily extended to other graph classes, provided we look at the homomorphism polynomials which contain edges and the vertices connected to these edges.

3.4   Outerplanar Graphs

Next we will show a dichotomy for outerplanar graphs. We start with the case of a triangle homomorphic to H.

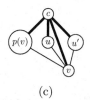

(a) Triangle graph

(b) Illustration of graph with buddy vertices

(c)

Fig. 1. Outerplanar Reductions

Lemma 15. *If a triangle is homomorphic to H then $\mathcal{F}_{outerplanar}$ is VNP hard under c-reductions.*

Proof. We will reduce to Hamiltonian Cycle by using a construction as in Figure 1a. This means, we pick an arbitrary vertex c and enforce all n outgoing edges from this vertex via homogeneous components. We further enforce the whole graph to have $n + n - 3$ edges. The graph given is obviously outerplanar but we still need to proof that no other graph fulfilling our criteria can be outerplanar.

We call the implied order of the graph, the order of the outer circle of vertices starting from the star and ending at it again without any edges crossing. As there are two such orderings let us fix an arbitrary one for every graph. Let us now look at a graph which has not an implied order of the outer vertices. This implies that there exists a vertex u which has degree 4. With our ordering every vertex (except c up to and including the later defined vertex v has a single parent. Furthermore, let v be the first vertex of degree 4 in this order and let $p(v)$ be the parent of v. Notice that by enforcing all n instead of just $n-2$ edges starting at the center, a parent $p(v) \neq c$ has to exist.

Let u, u' denote the other vertices adjacent to v different than $p(v)$ and c. As we enforced edges from c to every vertex, we can easily see the $K_{2,3}$ with v, c on the one side and $u, u', p(v)$ on the other side. Hence the graph cannot be outerplanar. This implies that every vertex except c and the two neighbouring vertices have degree at most 3. Enforcing the overall number of edges gives us at least degree 3 and hence implies equality.

Constructing all cycles in a K_{n-2} from this is now an easy task. We evaluate edges which we do not need anymore with the value one and connect the path in a similar way as in Lemma 12. This gives us again a weight of two for every monomial which is easily corrected. □

Theorem 16. *If H has an edge then $\mathcal{F}_{outerplanar}$ is VNP-complete under c-reductions and otherwise trivial.*

Proof. To make the graph homomorphic to a single edge we will modify it in the following way. For every vertex v, except c, we choose a *buddy vertex* v'. We enforce the edge between every vertex and his buddy vertex and set the edge between a buddy vertex and c to zero. Additionally, we set all vertices from v to any other non buddy vertex to zero and all edges from a buddy vertex to a different buddy vertex to be zero. In essence this splits every vertex into a left and right part (see Figure 1b). The hardness proof follows from Lemma 15 by contracting the edge between a vertex and his buddy vertex. Hence the combined degree of a vertex and his buddy vertex is at most three. Taking the homogeneous components increases the circuit size by a factor of n.

We know by [19] that checking if a graph is outerplanar is possible in linear time. With this we can use Valiant's Criterion to show the membership. □

3.5 Planar Graphs

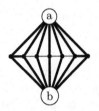

Fig. 2. Planar Gadget

Lemma 17. *All graphs isomorphic to Figure 2 with the thick edges fixed and $n + 2 + 2(n + 2)$ edges required are all permutations of the vertices $(1, \dots, n)$.*

Proof. Take an embedding in the plane of the graph without any crossings. If we show that every vertex has at most one edge going to the right, it follows that the set of vertices from left to right ordered is a permutation of the vertices.

Let us look at the following subgraph. Let v be a vertex with two right successors u, u' and a parent $p(v)$. By construction the parent always exists. We denote the top and bottom vertex by a and b in our graph. We can now build a $K_{3,3}$ minor in the following way. $S_1 = \{v, a, b\}$ and $S_2 = \{u, u', p'\}$. As a and b are connected to every vertex we only need to check that u is connected to u, u' and p which is by assumption. This proves that via edge deletion our graph would have a $K_{3,3}$ minor if the vertices would not give us a permutation. □

Theorem 18. *If H has an edge then $\mathcal{F}_{planar}$ is VNP-complete under c-reductions. Otherwise $\mathcal{F}_{planar}$ is in VAC_0.*

Proof. We again glue the second and second to last vertex in our planar gadget together in a similar manner as in the previous constructions to get all cycles from a path. Notice, how these are independent of the order and hence the same for all possible ordering.

However, this graph is not yet homomorphic to a single edge. To accomplish this, we will use a graph of size $2n$. We, as in the outerplanar case, enforce every vertex, except a and b, to have a buddy vertex u_v. Then we subdivide the edge (a, v) and (b, v) for every original, meaning none buddy, vertex v with a new vertex v'_a, v'_b respectively. This will give us for every part a square consisting of the vertices a, v, v'_a, u_v and the square b, v, v'_b, u_v.

Now it is easy to see that we can fold a to b which leaves us with a grid of height one. A grid can be easily folded to one edge. The size of the circuit is increased by a factor of at most $2n$.

As testing planarity is easy, we can use Valiant's Criterion to show membership. □

3.6   Genus k Graphs

Graph embeddings are one of the major relaxations of planarity. For this we find a surface of a specific type such that a graph can be embedded in this surface without any crossing edges. If we want to increase the *orientable genus* of a surface by one, we can glue a handle onto it which edges can use without crossing other edges. We call a graph a *genus k graph* if there exists a surface of orientable genus k such that G can be embedded in this surface and k is minimal. Notice, that a genus 0 graph is planar. While the topic of graph genus is vast, we will mostly use theorems as a blackbox and only reason about graphs of genus zero and one. For a detailed coverage of the topic, the reader is referred to [7].

With the planar result in place we can use the simple proof strategy. Construct a genus k graph where we append the planar construction. In this way the genus bound will ensure that our planar gadget gives us all permutation of vertices as long as the connection of these two graphs will not reduce the genus.

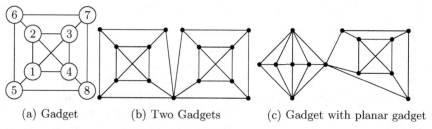

(a) Gadget   (b) Two Gadgets   (c) Gadget with planar gadget

Fig. 3. Genus k Reduction

Lemma 19. *The graph in Figure 3a has genus one.*

Proof. We can use the given embedding with one handle for the crossing in the middle to show an upper bound of one.

We again construct a $K_{3,3}$ with the sets $S_1 = \{2, 1, 6'\}, S_2 = \{3, 4, 7'\}$ where $6'$ is the vertex constructed from contracting the edge $(5, 6)$ and $7'$ from the edge $(7, 8)$. And hence the graph is not planar and has a lower bound for the genus of one.   □

The next theorem shows how we can glue graphs together to increase the genus in a predictable way.

Definition 20 ([18]). *G is a* vertex amalgam *of H_1, H_2 if G is obtained from disjoint graphs H_1 and H_2 where we identify one vertex form H_1 with one vertex from H_2.*

With this we restate a theorem from Miller [18] to compute the genus of a given graph.

Theorem 21 ([18]). *Let $\gamma(G)$ be the orientable genus of a graph G. Let G be constructed from vertex amalgams of graphs $G_1, \ldots, G_n$. Then $\gamma(G) = \sum_{i=1}^{n} \gamma(G_i)$.*

This now gives us immediately the result that a graph constructed as in Figure 3b with k gadgets has genus k.

Theorem 22. *If H has an edge then $\mathcal{F}_{genus(k)}$ is VNP-complete under c-reductions for any k. Otherwise $\mathcal{F}_{genus(k)}$ is in VAC_0.*

Proof. With Theorem 21, Lemma 17 and the construction in Figure 3 we are almost done. Because we enforced a genus k graph to occur, all graphs that are homomorphic to the planar gadget have genus zero and hence be planar.

The only thing left to do is to modify our graphs such that they are homomorphic to an edge without violating the properties. It is clear that we can fold our genus one gadgets together. If we then subdivide the edge $(1,3)$ and $(2,4)$ (which keeps our block property) we can first fold 7 to 5 and 3 to 1. Folding then again 6 to 8 and 2 to 4 we get a square with two dangling edges. The dangling edges can be folded onto the square and the square is homomorphic to one edge. This construction increases the size of the circuit at most by a factor of $14k + 2n$. As testing for a fixed genus is in NP, we can use Valiant's Criterion to show membership. $\square$

4   Conclusion

We have shown many dichotomy results for different graph classes but some classes are still open. We want to especially mention the case of our graph class being the class of trees. It is known that we can use Kirchoff's Theorem to find all spanning trees of a given graph. This, however, does not include monomials of total degree less than $n-1$ which our polynomials include. From the algebraic view, the knowledge ends here. In the counting view, where we solve the task of counting all trees in a graph, a bit more is known. Goldberg and Jerrum [12] showed that counting the number of subtrees that are distinct up to isomorphism is #P-complete. This, combined with our dichotomy for trees including the vertices, gives us a strong indication that the similar problem is VNP-hard in the algebraic world.

A different expansion of these results would be the case of bounded treewidth. As mentioned earlier, in the counting version the case of bounded treewidth is indeed the most general form and completely characterizes the easy and hard instances of counting graph homomorphisms. Additionally, recent advancements showed that graph homomorphisms of a specific type characterize VP. Can homomorphism from graph classes parameterized by treewidth, similar to the counting case, be used for a complete characterization of VP and VNP?

An interesting research direction would be the case of disconnected graph properties. Rugy-Altherre looked at the property that any graph is homomorphic to a given graph H. This includes disconnected graphs with connected components larger than one vertex. We instead only looked at restricted homomorphisms where one major connected component exists. It is unclear to the author if our proofs could be adapted to this case.

Acknowledgments. I want to thank my doctoral advisor M. Bläser for his guidance. I additionally want to thank R. Curticapean for many discussions on the counting versions of these problems and B. V. Raghavendra Rao for introducing me to this topic. I also want to thank the anonymous reviewers for their helpful comments.

References

1. Bulatov, A.A., Grohe, M.: The complexity of partition functions. Theor. Comput. Sci. 348(2-3), 148–186 (2005)
2. Bürgisser, P.: Completeness and reduction in algebraic complexity theory, vol. 7. Springer (2000)
3. Cai, J., Chen, X., Lu, P.: Graph homomorphisms with complex values: A dichotomy theorem. SIAM J. Comput. 42(3), 924–1029 (2013)
4. Chekuri, C., Rajaraman, A.: Conjunctive query containment revisited. Theor. Comput. Sci. 239(2), 211–229 (2000)
5. Dalmau, V., Jonsson, P.: The complexity of counting homomorphisms seen from the other side. Theor. Comput. Sci. 329(1-3), 315–323 (2004)
6. Dalmau, V., Kolaitis, P.G., Vardi, M.Y.: Constraint satisfaction, bounded treewidth, and finite-variable logics. In: Van Hentenryck, P. (ed.) CP 2002. LNCS, vol. 2470, pp. 310–326. Springer, Heidelberg (2002)
7. Diestel, R.: Graph Theory. Springer, GmbH & Company KG, Berlin and Heidelberg (2000)
8. Durand, A., Mahajan, M., Malod, G., de Rugy-Althere, N., Saurabh, N.: Homomorphism polynomials complete for VP. In: FSTTCS (to appear, 2014)
9. Dyer, M.E., Greenhill, C.S.: The complexity of counting graph homomorphisms (extended abstract). In: SODA, pp. 246–255 (2000)
10. Freuder, E.C.: Complexity of k-tree structured constraint satisfaction problems. In: AAAI, pp. 4–9 (1990)
11. Goldberg, L.A., Grohe, M., Jerrum, M., Thurley, M.: A complexity dichotomy for partition functions with mixed signs. SIAM J. Comput. 39(7), 3336–3402 (2010)
12. Goldberg, L.A., Jerrum, M.: Counting unlabelled subtrees of a tree is #p-complete. LMS J. Comput. Math. 3, 117–124 (2000)
13. Grohe, M., Thurley, M.: Counting homomorphisms and partition functions. Model Theoretic Methods in Finite Combinatorics 558, 243–292 (2011)
14. Grohe, M.: The complexity of homomorphism and constraint satisfaction problems seen from the other side. J. ACM 54(1) (2007)
15. Hell, P., Nešetřil, J.: On the complexity of h-coloring. Journal of Combinatorial Theory, Series B 48(1), 92–110 (1990)
16. Hell, P., Nešetřil, J.: Graphs and homomorphisms, vol. 28. Oxford University Press, Oxford (2004)
17. Mahajan, M., Rao, B.V.R.: Small space analogues of valiant's classes and the limitations of skew formulas. Computational Complexity 22(1), 1–38 (2013)
18. Miller, G.L.: An additivity theorem for the genus of a graph. J. Comb. Theory, Ser. B 43(1), 25–47 (1987)
19. Mitchell, S.L.: Linear algorithms to recognize outerplanar and maximal outerplanar graphs. Information Processing Letters 9(5), 229–232 (1979)
20. de Rugy-Altherre, N.: A dichotomy theorem for homomorphism polynomials. In: Rovan, B., Sassone, V., Widmayer, P. (eds.) MFCS 2012. LNCS, vol. 7464, pp. 308–322. Springer, Heidelberg (2012)
21. Strassen, V.: Vermeidung von divisionen. Journal für die reine und angewandte Mathematik 264, 184–202 (1973)
22. Valiant, L.G.: Completeness classes in algebra. In: STOC, pp. 249–261 (1979)

Common Unfolding of Regular Tetrahedron and Johnson-Zalgaller Solid

Yoshiaki Araki[1], Takashi Horiyama[2], and Ryuhei Uehara[3]

[1] Japan Tessellation Design Association, Japan
yoshiaki.araki@gmail.com
[2] Information Technology Center,
Saitama University, Japan
horiyama@al.ics.saitama-u.ac.jp
[3] School of Information Science,
Japan Advanced Institute of Science and Technology, Japan
uehara@jaist.ac.jp

Abstract. Common unfolding of a regular tetrahedron and a Johnson-Zalgaller solid is investigated. More precisely, we investigate the sets of all edge unfoldings of Johnson-Zalgaller solids. Among 92 Johnson-Zalgaller solids, some of edge unfolding of J17 and J84 admit to fold into a regular tetrahedron. On the other hand, there are no edge unfolding of the other Johnson-Zalgaller solids that admit to fold into a regular tetrahedron.

1 Introduction

In 1525 the painter and printmaker Albrecht Dürer published a book, translated as "The Painter's Manual," in which he explained the methods of perspective [13]. In the book, he includes a description of many polyhedra, which he presented as surface unfoldings, are now called "nets." An *edge unfolding* is defined by a development of the surface of a polyhedron to a plane, such that the surface becomes a flat polygon bounded by segments that derive from edges of the polyhedron. We would like an unfolding to possess three characteristics. (1) The unfolding is a single, simply connected piece. (2) The boundary of the unfolding is composed of (whole) edges of the polyhedron, that is, the unfolding is a union of polyhedron faces. (3) The unfolding does not self-overlap, that is, it is a simple polygon. We call a simple polygon that satisfies these conditions a *net* for the polyhedron.

Since then, nets for polyhedra have been widely investigated (rich background can be found in [5], and recent results can be found in [10]). For example, Alexandrov's theorem states that every metric with the global topology and local geometry required of a convex polyhedron is in fact the intrinsic metric of some convex polyhedron. Thus, if P is a net of a convex polyhedron Q, then the shape (as a convex polyhedron) is uniquely determined. Alexandrov's theorem was stated in 1942, and a constructive proof was given by Bobenko and Izmestiev in 2008 [4]. A pseudo-polynomial algorithm for Alexandrov's theorem, given by Kane et al. in 2009, runs in $O(n^{456.5} r^{1891}/\epsilon^{121})$ time, where r is the ratio of

M.S. Rahman and E. Tomita (Eds.): WALCOM 2015, LNCS 8973, pp. 294–305, 2015.
© Springer International Publishing Switzerland 2015

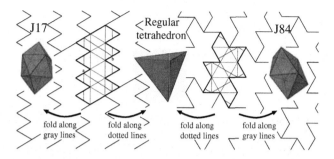

Fig. 1. (Left) an edge unfolding of the JZ solid J17, and (right) an edge unfolding of the JZ solid J84, which are also nets of a regular tetrahedron, respectively. These polygons are also p2 tilings.

the largest and smallest distances between vertices, and ϵ is the coordinate relative accuracy [9]. The exponents in the time bound of the result are remarkably huge.

Therefore, we have to restrict ourselves to smaller classes of polyhedra to investigate from the viewpoint of efficient algorithms. In this paper, we consider some classes of polyhedra that have common nets. In general, a polygon can be a net of two or more convex polyhedra. Such a polygon is called a *common net* of the polyhedra[1]. Recently, several polygons folding into two different polyhedra have been investigated (see [12] for comprehensive list). In this context, it is natural to ask whether there is a common net of two (or more) different Platonic solids. This question has arisen several times independently, and it is still open (see [5, Section 25.8.3]). In general nets, there is a polygon that can folds into a cube and an almost regular tetrahedron with small error $\epsilon < 2.89200 \times 10^{-1796}$ [12]. On the other hand, when we restrict ourselves to deal with only edge unfoldings, there are no edge unfolding of the Platonic solids except a regular tetrahedron that can fold into a regular tetrahedron [8]. This result is not trivial since a regular icosahedron and a regular dodecahedron have 43,380 edge unfoldings. In fact, it is confirmed that all the edge unfolding are nets (i.e., without self-overlapping) recently [6].

In this paper, we broaden the target of research from the set of five Platonic solids to the set of 92 Johnson-Zalgaller solids (JZ solids for short). A JZ solid is a strictly convex polyhedron, each face of which is a regular polygon, but which is not uniform, i.e., not a Platonic solid, Archimedean solid, prism, or antiprism (see, e.g., http://mathworld.wolfram.com/JohnsonSolid.html). Recently, the number of edge unfoldings of the JZ solids are counted [7], however, it has not been investigated how many nets (without self-overlapping) are there. On the other hand, the tilings of edge unfoldings of JZ solids are classified [2]. That is, they classified the class of the JZ solids whose edge unfoldings form tilings. Some tilings are well investigated in the context of nets; a polygon is a net of a regular tetrahedron if and only if it belongs to a special class of tilings [1].

---

[1] Note that an edge of an unfolding can passes through a flat face of the polyhedron.

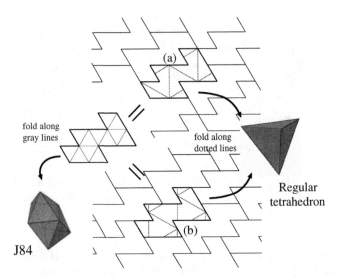

Fig. 2. An edge unfolding of the JZ solid J84. It has two different types of p2 tilings, and hence there are two different ways to fold into a regular tetrahedron.

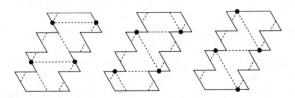

Fig. 3. An edge unfolding of the JZ solid J17 that can be folded into a regular tetrahedron in three different ways

In this paper, we concentrate on common nets of a regular tetrahedron and the JZ solids. More precisely, we classify the set of edge unfoldings of the JZ solids such that each of them is also folded into a regular tetrahedron. We first show that there exists edge unfoldings of some JZ solids that are also nets of a regular tetrahedra:

Theorem 1. *An edge unfolding of the JZ solid J17 and an edge unfolding of the JZ solid J84 fold into a regular tetrahedron.*

We will show that Fig. 1 certainly proves Theorem 1. Next we also compute all common nets that fold into both of a JZ solid and a regular tetrahedra:[2]

Theorem 2. *(1) Among 13,014 edge unfoldings of the JZ solid J17 [7], there are 87 nets that fold into a regular tetrahedron, which consist of 78 nets that have one way of folding into a regular tetrahedron, 8 nets that have two ways of*

---

[2] These numbers are counted on the "unlabeled" solids, and congruent unfoldings are not reduced. See [7] for further details.

Table 1. The JZ solids whose some edge unfoldings are nets of tetramonohedra

Name	J1	J8	J10	J12	J13	J14
Image						
# of □s	1	5	1	0	0	3
# of △s	4	4	12	6	10	6
L_{J_i}	$\sqrt{\frac{\sqrt3}{3}+1}$ $=1.255\cdots$	$\sqrt{\frac{5\sqrt3}{3}+1}$ $=1.971\cdots$	$\sqrt{\frac{\sqrt3}{3}+3}$ $=1.891\cdots$	$\sqrt{1.5}$ $=1.224\cdots$	$\sqrt{2.5}$ $=1.581\cdots$	$\sqrt{\sqrt3+\frac{3}{2}}$ $=1.797\cdots$

Name	J15	J16	J17	J49	J50	J51
Image						
# of □s	4	5	0	2	1	0
# of △s	8	10	16	6	10	14
L_{J_i}	$\sqrt{\frac{4\sqrt3}{3}+2}$ $=2.075\cdots$	$\sqrt{\frac{5\sqrt3}{3}+\frac{5}{2}}$ $=2.320\cdots$	2	$\sqrt{\frac{2\sqrt3}{3}+\frac{3}{2}}$ $=1.629\cdots$	$\sqrt{\frac{\sqrt3}{3}+\frac{5}{2}}$ $=1.754\cdots$	$\sqrt{3.5}$ $=1.870\cdots$

Name	J84	J86	J87	J88	J89	J90
Image						
# of □s	0	2	1	2	3	4
# of △s	12	12	16	16	18	20
L_{J_i}	$\sqrt{3}$ $=1.732\cdots$	$\sqrt{\frac{2\sqrt3}{3}+3}$ $=2.038\cdots$	$\sqrt{\frac{\sqrt3}{3}+4}$ $=2.139\cdots$	$\sqrt{\frac{2\sqrt3}{3}+4}$ $=2.270\cdots$	$\sqrt{\sqrt3+\frac{9}{2}}$ $=2.496\cdots$	$\sqrt{\frac{4\sqrt3}{3}+5}$ $=2.703\cdots$

folding into a regular tetrahedron, and 1 net that has three ways of folding into a regular tetrahedron. (2) Among 1,109 edge unfoldings of the JZ solid J84 [7], there are 37 nets that fold into a regular tetrahedron, which consist of 32 nets that have one way of folding into a regular tetrahedron, and 5 nets that have two ways of folding into a regular tetrahedron.

We note that some nets allow to fold into a regular tetrahedron in two or three different ways of folding. A typical example that has two ways of folding is shown in Fig. 2. We can tile the net of the JZ solid J84 in two different ways, hence we can fold a regular tetrahedron in two different ways according to the tilings. The unique net that has three ways of folding is shown in Fig. 3.

Among 92 JZ solids, Akiyama et al. found that 18 JZ solids have edge unfoldings that are also tilings [2]. We will show that all of them are also p2 tiling, which imply that they can be folded into tetramonohedra. As shown in Theorem 1, two of them can be folded into regular tetrahedra. On the other hand, the other 16 JZ solids do not have such edge unfoldings:

Theorem 3. *Except J17 and J84, there is no other JZ solid such that its edge unfolding is a net of a regular tetrahedron.*

Therefore, we classify the set of edge unfoldings of the JZ solids by the foldability of a regular tetrahedron.

2   Preliminaries

We first show some basic results about unfolding of a polyhedron.

Lemma 1 ([5, Sec. 22.1.3]). *All vertices of a polyhedron X are on the boundary of any unfolding of X.*

Let P be a polygon on the plane, and R be a set of four points (called *rotation centers*) on the boundary of P. Then P has a *tiling* called symmetry group p2 (*p2 tiling*, for short) if P fills the plane by the repetition of 2-fold rotations around the points in R. The filling should contain no gaps nor overlaps. The rotation defines an equivalence relation on the points in the plane. Two points p_1 and p_2 are mutually equivalent if p_1 can be moved to p_2 by the 2-fold rotations. More details of p2 tiling can be found, e.g., in [11]. Based on the notion of p2 tiling, any unfolding of a tetramonohedron[3] can be characterized as follows:

Theorem 4 ([1,3]). *P is an unfolding of a tetramonohedron if and only if (1) P has a p2 tiling, (2) four of the rotation centers consist in the triangular lattice formed by the triangular faces of the tetramonohedron, (3) the four rotation centers are the lattice points, and (4) no two of the four rotation centers belong to the same equivalent class on the tiling.*

We can obtain the characterization of the unfolding of a regular tetrahedron if each triangular face in Theorem 4 is a regular triangle. By Theorem 4, Theorem 1 is directly proved by Fig. 1. (Of course it is not difficult to check these nets in Fig. 1 by cutting and folding directly.)

In the classification in [2], they show only p1 tilings for the JZ solids J84, J86 and J89. However, they also have edge unfoldings that form p2 tilings as shown in Fig. 1 (J84) and Fig. 4 (J86 and J89), and hence they can fold into tetramonohedra.

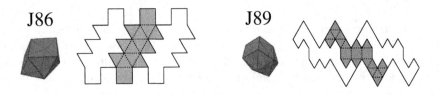

Fig. 4. p2 tilings by (left) an edge unfolding of JZ solid J86, and (right) an edge unfolding of JZ solid J89

Let L_{J_i} be the length of an edge of a regular tetrahedron T_{J_i} that has the same surface area of the JZ solid J_i. We assume that each face of J_i is a regular polygon that consists of edges of unit length. Thus it is easy to compute L_{J_i} from its surface area of J_i as shown in Table 1. If an edge unfolding P_{J_i} of the JZ

---

[3] A *tetramonohedron* is a tetrahedron that consists of four congruent triangular faces.

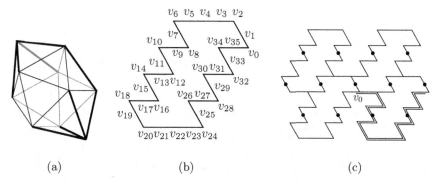

(a)   (b)   (c)

Fig. 5. (a) a spanning tree of the JZ solid J17, (b) its corresponding unfolding, and
(c) a p2 tiling around the rotation centers

solid J_i can be folded into a regular tetrahedron, the tetrahedron is congruent to
T_{J_i} since they have the same surface area. Moreover, by Theorem 4, P_{J_i} is a p2
tiling, and its four of the rotation centers form the regular triangular lattice filled
by regular triangles of edge length L_{J_i}. Let c_1 and c_2 be any pair of the rotation
centers of distance L_{J_i}. Then, by Lemma 1, c_1 and c_2 are on boundary of P_{J_i}
and P'_{J_i} for some polygons P_{J_i} and P'_{J_i}, respectively. By the same extension of
Theorem 25.3.1 in [5] used in [8, Lemma 8], we can assume that c_1 and c_2 are
on the corners or the middlepoints on some edges of regular faces of JZ solids J_i
without loss of generality. Summarizing them, we obtain the following lemma:

Lemma 2. *Assume that a polygon P_{J_i} is obtained by an edge unfolding of a JZ
solid J_i. If P_{J_i} can be folded into a regular tetrahedron T_{J_i}, P_{J_i} forms a p2 tiling
$\mathcal{T}$. Let c_1 and c_2 be any two rotation centers on $\mathcal{T}$ such that the distance between
c_1 and c_2 is L_{J_i}, equal to the length of an edge of T_{J_i}. Then, the vertices c_1 and
c_2 are on the corners or the middlepoints on edges of unit length in $\mathcal{T}$.*

3   The JZ Solids J17 and J84

In this section, we describe an algorithm to obtain Theorem 2. By applying the
technique in [6], we can enumerate a set of spanning trees of any polyhedron, where
a spanning tree is obtained as a set of edges. By traversing each spanning tree, we
can obtain its corresponding unfolding P_{J_i}. Since all edges of a JZ solid have the
same length, P_{J_i} can be represented by a cyclic list C_{J_i} of its interior angles a_j,
where vertices v_j of P_{J_i} correspond to the corners or the middlepoints on some
edges of the original JZ solid. Since a spanning tree has $n - 1$ edges, each edge
appears twice as the boundary of P_{J_i}, and each edge is broken into two halves, P_{J_i}
has $4(n-1)$ vertices. Fig. 5 illustrates (a) a spanning tree of the JZ solid J17, and (b)
its corresponding unfolding, which can be represented by $C_{J_i} = \{60, 180, 120, 180,$
$180, 180, 60, 180, 300, 180, 60, 180, 300, 180, 60, 180, 300, 180, 60, 180, 120, 180, 180,$
$180, 60, 180, 300, 180, 60, 180, 300, 180, 60, 180, 300, 180\}$.

Now, we use Theorem 4 and check if each edge unfolding is a p2 tiling or not. We can use the similar idea with the algorithm for gluing borders of a polyhedron (see [5, Chap. 25.2]): around each rotation center, check if the corresponding points make together 360°. If not, we dismiss this case, and otherwise, we obtain a gluing to form a regular tetrahedron.

We first consider the JZ solid J17. In this case, we can determine the length of each edge of the triangular lattice equals to 2, since each face of the (potential) regular tetrahedron consists of four unit tiles. We can check if each unfolding of the JZ solid J17 can be folded into a regular tetrahedron as follows:

1. For each pair of v_{j_1} and v_{j_2}, suppose they are rotation centers, and check if the distance between them is 2.
2. Obtain a path $v_{j_1'}-v_{j_1}-v_{j_1''}$ which is glued to $v_{j_1''}-v_{j_1}-v_{j_1'}$ by a 2-fold rotation around v_{j_1}. So do a path $v_{j_2'}-v_{j_2}-v_{j_2''}$ for v_{j_2}.
3. Replace interior angles of $a_{j_1'}, \ldots, a_{j_1}, \ldots, a_{j_1''}$ in C_{J_i} with angle a_{j_1}', where $a_{j_1}' = a_{j_1'} + a_{j_1''}$ if $j_1' \neq j_1''$ and $a_{j_1}' = a_{j_1} + 180$ if $j_1' = j_1'' = j_1$. So do $a_{j_2'}, \ldots, a_{j_2''}$. Let C_{J_i}' be the resulting cyclic list.
4. For each pair of v_{j_3} and v_{j_4}, suppose they are rotation centers, and check if a path $v_{j_3}-v_{j_4}$ in C_{J_i}' is glued to the remaining path $v_{j_4}-v_{j_3}$.
5. Check if v_{j_3} and v_{j_4} are the lattice points of the regular triangular lattice defined by v_{j_1}, v_{j_2}, and check if no two of $v_{j_1}, v_{j_2}, v_{j_3}$ and v_{j_4} belong to the same equivalent class.

In Step 1, since every face of the JZ solid J17 is a triangle, a_j is always a multiple of 60. The relative position of v_j from v_0 can be represented as a linear combination of two unit vectors $\boldsymbol{u}$ and $\boldsymbol{v}$ that make a 60° angle. Thus, we check if vector $\overrightarrow{v_{j_1}v_{j_2}}$ is one of $\pm 2\boldsymbol{u}, \pm 2\boldsymbol{v}, \pm 2(\boldsymbol{u} - \boldsymbol{v})$ in this step.

In Step 2, two vertices $v_{j_1'}$ and $v_{j_1''}$ are obtained as v_{j_1-k} and v_{j_1+k} with an integer k satisfying $a_{j_1-k} + a_{j_1+k} < 360$ and $a_{j_1-k'} + a_{j_1+k'} = 360$ for all $0 \leq k' < k$. In Fig. 5, v_1 and v_7 are supposed to be rotation centers, and paths $v_0-v_1-v_2$ and $v_4-v_7-v_{10}$ are glued to $v_2-v_1-v_0$ and $v_{10}-v_7-v_4$, respectively. By rotating P_{J_i} around v_{j_1} and v_{j_2} repeatedly, we obtain a horizontally infinite sequence of P_{J_i} as shown in Fig. 5(c), whose upper and lower borders are the repetition of the path denoted in double line. The list of the interior angles along the double line is obtained as C_{J_i}' in Step 3. In Fig. 5(c), C_{J_i}' is $\{180, 180, 240, 180, 300, 180, 60, 180, 300, 180, 60, 180, 120, 180, 180, 180, 60, 180, 300, 180, 60, 180, 300, 180, 60, 180, 300, 180\}$.

In Step 4, we check if $a_{j_3} = a_{j_4} = 180$ holds and $a_{j_3-k} + a_{j_3+k} = 360$ for other gluing of vertices v_{j_3-k} and v_{j_3+k} in C_{J_i}'. If P_{J_i} passes all checks in Steps 1–4, P_{J_i} has a p2 tiling with rotation centers $v_{j_1}, v_{j_2}, v_{j_3}$ and v_{j_4}. In Step 5, we check if the four points meet Theorem 4(2)–(4) and if each triangular face is a regular triangle. As in Step 1, this check can be done from the positions of vertices represented as a linear combination of $\boldsymbol{u}$ and $\boldsymbol{v}$.

For the JZ solid J84, we can check in the same way by letting the length of the triangular lattice equal to $\sqrt{3}$, and thus, in Step 1, we check if vector $\overrightarrow{v_{j_1}v_{j_2}}$ is one of $\pm(\boldsymbol{u} + \boldsymbol{v}), \pm(2\boldsymbol{u} - \boldsymbol{v}), \pm(2\boldsymbol{v} - \boldsymbol{u})$. The complete catalogue of 87 and 37 nets of the JZ solids J17 and J84, respectively, that fold into a regular tetrahedron is given in `http://www.al.ics.saitama-u.ac.jp/horiyama/res earch/unfolding/common/`.

4   The Other JZ Solids

In this section, we prove Theorem 3. Combining the results in [2] and the tilings in Fig. 1 and Fig. 4, the set $\mathcal{J}$ of JZ solids whose edge unfoldings can be p2 tiling is $\mathcal{J} = \{$J1, J8, J10, J12, J13, J14, J15, J16, J17, J49, J50, J51, J84, J86, J87, J88, J89, J90$\}$. In other words, some edge unfoldings of the JZ solids in $\mathcal{J}$ can be folded into tetramonohedra. Among them, J17 and J84 allow to fold into regular tetrahedra from their edge unfoldings as shown in Fig. 1. We will show that the other JZ solids do not. Hereafter, we only consider the JZ solids in $\mathcal{J}$. Then each face is either a unit square or a unit triangle. We call each of them a *unit tile* to simplify. We consider the rotation centers form the regular triangular lattice of size L_{J_i}. Let c_1 and c_2 be any pair of the rotation centers of distance L_{J_i}. We use the fact that the distance between c_1 and c_2 is equal to L_{J_i}, and show that any combination of unit tiles cannot achieve the length.

Intuitively, two points c_1 and c_2 are joined by a sequence of edges of unit length that are supported by unit tiles in $\mathcal{T}$. Thus, by Lemma 2, we can observe that there exists a linkage $L_{J_i} = (p_0, q_1, p_1, q_2, p_2, \ldots, q_k, p_k)$ such that (1) c_1 is on either p_0 or q_1, (2) c_2 is on either q_k or p_k, (3) the length of $p_i p_{i+1}$ is 1, (4) the length of $p_{i-1} q_i$ and $q_i p_i$ is $1/2$ (in other words, q_i is the center point of $p_{i-1} p_i$), (5) each angle at q_i ($1 \leq i \leq k$) is $180°$, (6) each angle at p_i ($1 \leq i \leq k - 1$) is in $\{60°, 90°, 120°, 150°, 180°, 210°, 240°, 270°, 300°\}$, and (7) the linkage is not self-crossing. (See [5] for the definition of the notion of *linkage*.) Without loss of generality, we suppose that L_{J_i} has the minimum length among the linkages satisfying the conditions from (1) to (7). By the minimality, we also assume that (8) $p_i \neq p_j$ for each $i \neq j$, and (9) if $|i - j| > 1$, the distance between p_i and p_j is not 1 (otherwise, we obtain a shorter linkage).

Therefore, by Theorem 4, for sufficiently large k, if all possible pairs c_1 and c_2 on the linkages satisfying the conditions from (1) to (9) do not achieve the required distance L_{J_i}, any edge unfolding of the JZ solid J_i cannot be folded into a corresponding regular tetrahedron T_{J_i}. We show an upper bound of k:

Theorem 5. *Let $\mathcal{J}$ be the set $\{$J1, J8, J10, J12, J13, J14, J15, J16, J17, J49, J50, J51, J84, J86, J87, J88, J89, J90$\}$ of the JZ solids that have some edge unfoldings which are also p2 tilings. For some $J_i \in \mathcal{J}$, suppose that the linkage $L_{J_i} = (p_0, q_1, p_1, q_2, p_2, \ldots, q_k, p_k)$ defined above exist. Then $k \leq 10$.*

Proof. By simple calculation, L_{J_i} takes the maximum value $\sqrt{4\sqrt{3}/3 + 5} = 2.703\cdots$ for J90 in $\mathcal{J}$. Thus the length of the line segment $c_1 c_2$ is at most $2.703\cdots$.

Now we assume that the line segment $c_1 c_2$ passes through a sequence $C_1 C_2 \cdots C_h$ of unit tiles in this order. That is, the line segment $c_1 c_2$ has nonempty intersection with each of C_i in this order. If $c_1 c_2$ passes an edge shared by two

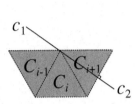

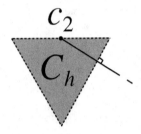

Fig. 6. The shortest way to pierce three consecutive unit tiles

Fig. 7. The shortest intersection with the last unit tile C_h

unit tiles, we take arbitrary one of two in the sequence. We consider the minimum length of the part of c_1c_2 that intersects three consecutive unit tiles $C_{i-1}C_iC_{i+1}$ in the sequence. Since they are unit triangles and/or squares, three unit tiles make greater than or equal to $180°$ at a vertex. Therefore the minimum length is achieved by the three consecutive triangles arranged in Fig. 6, and in this case, the length is greater than or equal to $\sqrt{3}/2 = 0.866\cdots$. Thus, if c_1c_2 passes through nine unit tiles, the intersection has length at least $3\sqrt{3}/2 = 2.598\cdots$. On the other hand, the last point c_2 is on the vertex or a midpoint of an edge of the last unit tile C_h. Then the intersection of c_1c_2 and C_h has at least $\sqrt{3}/4 = 0.433\cdots$ (Fig. 7). Since $3\sqrt{3}/2 + \sqrt{3}/4 = 3.03\cdots > 2.703$, c_1c_2 passes through at most 9 unit tiles.

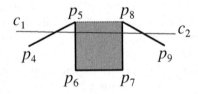

Fig. 8. Unit square can contribute three edges to the linkage

Now we turn to the linkage $L_{J_i} = (p_0, q_1, p_1, q_2, p_2, \ldots, q_k, p_k)$ supported by the unit tiles $C_1C_2\cdots C_h$ with $h \leq 9$. We consider the number of edges of a unit tile that contributes to L_{J_i}. Locally, the worst case is that a unit square that contributes three edges to L_{J_i} (Fig. 8). However, in this case, the length of the intersection of the square and c_1c_2 has length at least 1. Therefore, further analysis for the remaining length at most $2.703\cdots - 1 = 1.703\cdots$, and we can confirm that this case does not give the worst value of k. In the same reason, if c_1c_2 passes through an entire edge of length 1, it does not give the worst value of k. Next considerable case is that a unit tile C_i contributes two edges to L_{J_i}

independent from C_{i-1} and C_{i+1}. That is, c_1c_2 passes through two vertices of C_i. Then C_i is not a unit triangle since we can replace two edges by the third edge and obtain a shorter linkage. Thus C_i is a unit square, and c_1c_2 passes through the diagonal of C_i since two edges are not shared by C_{i-1} and C_{i+1}. Then the intersection of c_1c_2 and C_i has length $\sqrt{2} = 1.414\cdots$, and hence this case does not give the worst value of k again. Therefore, in the worst case, each unit tile contributes exactly two edges to L_{J_i}, and each edge is shared by two consecutive unit tiles in the sequence $C_1C_2\cdots C_h$, where $h \le 9$. Therefore, the linkage consists of at most 10 unit length edges, that is, $k \le 10$. □

Now, for $k \le 10$, if all possible pairs c_1 and c_2 on the linkages satisfying the conditions from (1) to (9) do not realize any distance L_{J_i} in Table 1, any edge unfolding of the JZ solid J_i cannot be folded into a corresponding regular tetrahedron T_{J_i}. However, the number of possible configurations of the linkage is still huge. To reduce the number, we use the following theorem:

Theorem 6. *Let $\mathcal{J}$ be the set $\{J1, J8, J10, J12, J13, J14, J15, J16, J17, J49, J50, J51, J84, J86, J87, J88, J89, J90\}$ of the JZ solids that have some edge unfoldings which are also p2 tilings. For some $J_i \in \mathcal{J}$, suppose that the linkage $L_{J_i} = (p_0, q_1, p_1, q_2, p_2, \ldots, q_k, p_k)$ defined above exist. Let I be the set of integers and $I_{+1/2}$ be the set defined by $I \cup \{i + 1/2 \mid i \in I\}$. Let $\mathbf{u_1} = (1, 0)$, $\mathbf{u_2} = (\sqrt{3}/2, 1/2)$, $\mathbf{u_3} = (1/2, \sqrt{3}/2)$, $\mathbf{u_4} = (0, 1)$, $\mathbf{u_5} = (-1/2, \sqrt{3}/2)$, and $\mathbf{u_6} = (-\sqrt{3}/2, 1/2)$ be six unit length vectors (Fig. 9). Then there are four integers $k_2, \ldots, k_5$ in I and two numbers k_1, k_6 in $I_{+1/2}$ such that $\sum_{i=1}^{6} |k_i| \le 10$ and $c_2 = c_1 + \sum_{i=1}^{6} k_i \mathbf{u_i}$.*

Proof. When we regard each edge in the linkage as a unit vector, since vectors are commutative, we can swap two edges without changing the coordinate of c_2 (see Fig. 10; one can find the same idea in, e.g., [5, Section 5.1.1]). Thus we have the theorem. □

Corollary 1. *For the two points c_1 and c_2 with $c_2 = c_1 + \sum_{i=1}^{6} k_i \mathbf{u_i}$ in Theorem 6, There are four numbers h_1, h_2, h_3, h_4 in $I_{+1/2}$ such that $c_2 = c_1 + \sum_{i=1}^{4} h_i \mathbf{u_i}$.*

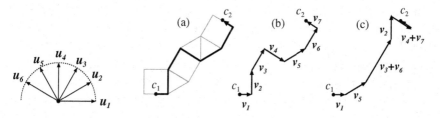

Fig. 9. Six unit vectors

Fig. 10. Linkage as the set of unit vectors: (a) given tiling and linkage, (b) corresponding vectors, and (c) reorganized vectors

Proof. Since $u_6 = u_4 - u_2$ and $u_5 = u_3 - u_1$, we can remove two vectors from the equation. Precisely, we have $\sum_{i=1}^{6} k_i u_i = (k_1 - k_5)u_1 + (k_2 - k_6)u_2 + (k_3 + k_5)u_3 + (k_4 + k_4)u_4$.   □

Now we prove the main theorem in this section:

Proof. (of Theorem 3) First we consider two points c_1 and c_2 given in Corollary 1: $c_2 = c_1 + \sum_{i=1}^{4} h_i u_i$ for some four numbers h_1, h_2, h_3, h_4 in $I_{+1/2}$. Then $|c_1 c_2|^2 = L_{J_i}^2 = h_1^2 + h_4^2 + h_1 h_3 + h_2 h_4 + 13(h_2^2 + h_3^2)/4 + 2\sqrt{3}(h_1 h_2 + h_2 h_3 + h_3 h_4)$.

Now we fix some JZ solid Ji for some i. Let m_i and n_i be the number of triangles and squares in Ji, respectively. Then we have $L_{J_i}^2 = m_i/4 + \sqrt{3}n_i/3$.

By the condition that $h_1, h_2, h_3, h_4 \in I_{+1/2}$, we can observe that

$$m_i = 4h_1^2 + 4h_4^2 + 4h_1 h_3 + 4h_2 h_4 + 13h_2^2 + 13h_3^2 \tag{1}$$

$$n_i = 6(h_1 h_2 + h_2 h_3 + h_3 h_4) \tag{2}$$

From the second equation, we can observe that n_i is a multiple of 3. Thus the JZ solids J1, J8, J10, J15, J16, J49, J50, J86, J87, J88, and J90 have no edge unfolding that is a net of a regular tetrahedron.

For the remaining JZ solids J12 ($n = 0, m = 6$), J13 ($n = 0, m = 10$), J14 ($n = 3, m = 6$), J51 ($n = 0, m = 14$), and J89 ($n = 3, m = 18$), we check them by a brute force. More precisely, we generate all possible $k_1, k_2, \ldots, k_6 \in [-10..10] \cap I_{+1/2}$ with $\sum_{i=1}^{6} |k_i| \le 10$, and compute $h_1 = k_1 - k_6, h_2 = k_2 - k_5, h_3 = k_3 + k_6, h_4 = k_4 + k_5$, and n and m by the above equations. Then no 6-tuple $(k_1, k_2, \ldots, k_6)$ generates any pair of $(n = 0, m = 6)$, $(n = 0, m = 10)$, $(n = 3, m = 6)$, $(n = 0, m = 14)$, and $(n = 3, m = 18)$ [4].

Therefore, in $\mathcal{J}$, only J17 and J84 have feasible solutions $L_{J_{17}} = 2$ and $L_{J_{84}} = \sqrt{3}$ in the distances.   □

5   Convex Polyhedra with Regular Polygonal Faces

According to the classification in [2], there are 23 polyhedra with regular polygonal faces whose edge unfoldings allow tilings. Among them, 18 JZ solids have been discussed in Section 4, and four Platonic solids were discussed in [8]. The remaining one is hexagonal antiprism that consists of two regular hexagons and 12 unit triangles. By splitting each regular hexagon into six unit triangles, which is called *coplanar deltahedron*, we can show the following theorem using the same argument above:

Theorem 7. *The hexagonal antiprism has no edge unfolding that can fold into a regular tetrahedron.*

---

[4] From the viewpoint of the programming, we introduce integer variables $k_1' = 2k_1$, $k_2' = 2k_2$, $\ldots$, and $k_6' = 2k_6$, and compute $4m$ and $4n$. Then all computation can be done on integers. Hence we can avoid computational errors, and the program runs in a second.

Thus we can conclude as follows:

Corollary 8. *Among convex polyhedra with regular polygonal faces, including the Platonic solids, the Archimedean solids, and the JZ solids, regular prisms, and regular anti-prisms, only the JZ solids J17 and J84 (and regular tetrahedron) admit to fold into regular tetrahedra from their edge unfoldings.*

6   Concluding Remarks

In this paper, we show that the JZ solids J17 and J84 are exceptionally in the sense that their edge unfoldings admit to fold into regular tetrahedra. Especially, some edge unfoldings can fold into a regular tetrahedron in two or three different ways. In this research, the characterization of nets by tiling (Theorem 4) plays an important role. In general, even the decision problem that asks if a polyhedron can be folded from a given polygon is quite difficult problem [5, Chapter 25]. More general framework to solve the problem is future work.

References

1. Akiyama, J.: Tile-Makers and Semi-Tile-Makers. The Mathematical Association of Amerika, Monthly 114, 602–609 (2007)
2. Akiyama, J., Kuwata, T., Langerman, S., Okawa, K., Sato, I., Shephard, G.C.: Determination of All Tessellation Polyhedra with Regular Polygonal Faces. In: Akiyama, J., Bo, J., Kano, M., Tan, X. (eds.) CGGA 2010. LNCS, vol. 7033, pp. 1–11. Springer, Heidelberg (2011)
3. Akiyama, J., Nara, C.: Developments of Polyhedra Using Oblique Coordinates. J. Indonesia. Math. Soc. 13(1), 99–114 (2007)
4. Bobenko, A.I., Izmestiev, I.: Alexandrov's theorem, weighted Delaunay triangulations, and mixed volumes. arXiv:math.DG/0609447 (February 2008)
5. Demaine, E.D., O'Rourke, J.: Geometric Folding Algorithms: Linkages, Origami, Polyhedra. Cambridge University Press (2007)
6. Horiyama, T., Shoji, W.: Edge unfoldings of Platonic solids never overlap. In: Proc. CCCG 2011, pp. 65–70 (2011)
7. Horiyama, T., Shoji, W.: The Number of Different Unfoldings of Polyhedra. In: Cai, L., Cheng, S.-W., Lam, T.-W. (eds.) ISAAC 2013. LNCS, vol. 8283, pp. 623–633. Springer, Heidelberg (2013)
8. Horiyama, T., Uehara, R.: Nonexistence of Common Edge Developments of Regular Tetrahedron and Other Platonic Solids. In: Proc. China-Japan Joint Conference on Computational Geometry, Graphs and Applications (CGGA 2010), pp. 56–57 (2010)
9. Kane, D., Price, G.N., Demaine, E.D.: A pseudopolynomial algorithm for Alexandrov's Theorem. In: Dehne, F., Gavrilova, M., Sack, J.-R., Tóth, C.D. (eds.) WADS 2009. LNCS, vol. 5664, pp. 435–446. Springer, Heidelberg (2009)
10. O'Rourke, J.: How to Fold It: The Mathematics of Linkage, Origami and Polyhedra. Cambridge University Press (2011)
11. Schattschneider, D.: The plane symmetry groups: their recognition and notation. American Mathematical Monthly 85, 439–450 (1978)
12. Shirakawa, T., Horiyama, T., Uehara, R.: Construct of Common Development of Regular Tetrahedron and Cube. In: Proc. EuroCG 2011, pp. 47–50 (2011)
13. Strauss, W.S., Dürer, A.: The Painter's Manual. Abaris Books (1977)

Threshold Circuits for Global Patterns in 2-Dimensional Maps

Kei Uchizawa[1], Daiki Yashima[2], and Xiao Zhou[2]

[1] Faculty of Engineering, Yamagata University,
Jonan 4-3-16, Yonezawa-shi Yamagata, 992-8510, Japan
[2] Graduate School of Information Sciences, Tohoku University,
Aramaki-aza Aoba 6-6-05, Aoba-ku, Sendai, 980-8579, Japan
uchizawa@yz.yamagata-u.ac.jp
yashima.daiki@ec.ecei.ac.jp
zhou@ecei.tohoku.ac.jp

Abstract. In this paper, we consider a biologically-inspired Boolean function, called P_D^n, which models a task for detecting specific global spatial arrangements of local visual patterns on a 2-dimensional map. We prove that P_D^n is computable by a threshold circuit of size $O(\sqrt{n}\log n)$, which is improvement on the previous upper bound $O(n)$. We also show that the size of our circuit is almost optimal up to logarithmic factor: we show that any threshold circuit computing P_D^n needs size $\Omega(\sqrt{n}/\log n)$.

1 Introduction

A threshold circuit is a combinatorial circuit consisting of logic gates computing linear threshold functions, and is one of the most well-studied computational models in circuit complexity theory. Through the decades, threshold circuits receive much attention in the literature. It is known that threshold circuits have surprising computational power: polynomial-size and even constant-depth threshold circuits are able to compute a variety of Boolean functions including basic arithmetic operations such as ADDITION, ITERATED ADDITION, MULTIPLICATION, DIVISION, SORTING, etc. ([3, 5–8]).

On the other hand, there is another aspect of threshold circuit: a threshold circuit is a theoretical model of a neural network in the brain ([3, 4, 8]). It is known that a threshold gate, the basic element of a threshold circuit, captures a basic input-output characteristic of a biological neuron. As information processing of a neural network is carried out by neurons emitting electrical signals, computation of a threshold circuit is carried out by gates outputting Boolean values. In the line of research, we expect to make a step for understanding how a neural network realize an information processing (e.g., sensory processing) with high speed and low energy consumption.

However, classical circuit complexity theory offers little insight into the question; one of the reasons, as pointed out in the paper [1], is because focus of circuit complexity theory lies on a different set of computational problems such

M.S. Rahman and E. Tomita (Eds.): WALCOM 2015, LNCS 8973, pp. 306–316, 2015.
© Springer International Publishing Switzerland 2015

as the arithmetic operations mentioned above. Motivated by this reason, Legenstein and Maass [1] proposed several Boolean functions that model particular tasks for visual information processing. Among such functions, we focus on a Boolean function P_D^n modeling the following task (See Section 2.2 for the precise definition). Suppose local feature detectors are arranged in 2-dimensional map that reflects spatial relationship in the outside world, where each of the local feature detectors is able to detect the presence of a salient local feature, such as a center which emits higher intensity than its surrounding, line segment in a certain direction, or even more complex local visual patterns like an eye or a nose. Then we wish to detect a global pattern arrangements consisting of the local features such as the letter "T," or human face as an ultimate goal. P_D^n is a simplified variation of the pattern detection problem. It was shown in [1] that P_D^n is computable by such a threshold circuit of size (i.e., the number of gates) $O(n)$ and depth $O(\log n)$ that is suitable for VLSI implementation: their circuit consists of $O(n)$ Boolean gate of fan-in 2 and fan-out 2 together with threshold gates of fan-in $O(\log n)$, and moreover, its total wire length, one of the most influential complexity measures for the size of VLSI, is $O(n)$.

While the circuit given by Legenstein and Maass is optimized to and quite suitable for VLSI implementation, the minimum size of threshold circuits computing P_D^n is of independent interest. We can prove, as main result of this paper, that P_D^n is computable by a threshold circuit of size $O(\sqrt{n}\log n)$ and depth $O(\sqrt{n})$. Thus, by considering unbounded fan-in, fan-out and depth, we improve on the size of threshold circuits computing P_D^n. We obtain the result by constructing the desired circuit, and hence our proof exhibit the explicit structure of the circuit. As a complement to the result, we also show that any threshold circuit computing P_D^n requires size $\Omega(\sqrt{n}/\log n)$. Thus, our construction is optimal up to a polylogarithmic factor.

The rest of the paper is organized as follows. In Section 2, we define some terms on threshold circuits, and give formal definition of P_D^n. In Section 3, we show that P_D^n is computable by a threshold circuit of $\sqrt{n}\log n$ gates. In Section 4, we provide the $\sqrt{n}/\log n$ lower bound. In Section 5, we conclude with some remarks.

2   Definitions

2.1   Threshold Circuits

A *threshold gate* with an arbitrary number z of inputs computes a linear threshold function with z inputs: for every input $\boldsymbol{x} = (x_1, x_2, \ldots, x_z) \in \{0,1\}^z$, the output $g(\boldsymbol{x})$ of a threshold gate g with integer weights $w_1, w_2, \ldots, w_z$ and threshold t is defined as

$$g(\boldsymbol{x}) = \text{sign}\left(\sum_{i=1}^{z} w_i x_i - t\right)$$

where, for any number η, $\text{sign}(\eta) = 1$ if $\eta \geq 0$, $\text{sign}(\eta) = 0$, otherwise.

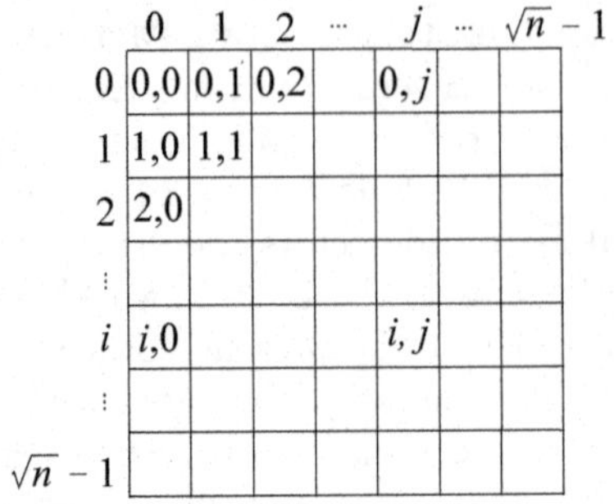

	0	1	2	$\cdots$	j	$\cdots$	$\sqrt{n}-1$
0	0,0	0,1	0,2		0,j		
1	1,0	1,1					
2	2,0						
$\vdots$							
i	i,0				i,j		
$\vdots$							
$\sqrt{n}-1$							

Fig. 1. Arrangement of indices of $\boldsymbol{x} = (x_{0,0}, x_{0,1}, \ldots, x_{\sqrt{n}-1,\sqrt{n}-1}) \in \{0,1\}^n$ and $\boldsymbol{y} = (y_{0,0}, y_{0,1}, \ldots, y_{\sqrt{n}-1,\sqrt{n}-1}) \in \{0,1\}^n$ on the 2-dimensional grid

A *threshold circuit* C is a combinatorial circuit of threshold gates, and is defined by a directed acyclic graph. Let n be the number of input variables to C. Then each node of in-degree 0 in C corresponds to one of the n input variables $x_1, x_2, \ldots, x_n$, and the other nodes correspond to threshold gates. The *size s* of a threshold circuit is defined to be the number of threshold gates in the circuit.

Let C be a threshold circuit with n input variables $x_1, x_2, \ldots, x_n$, and have size s. Let $g_1, g_2, \ldots, g_s$ be the gates in C, where $g_1, g_2, \ldots, g_s$ are topologically ordered with respect to the underlying directed acyclic graph of C. We regard the output of g_s as the *output* $C(\boldsymbol{x})$ *of* C, and call the gate g_s the *top gate* of C. The *depth d* of C is defined to be the number of gates on the longest path to the top gate g. A threshold circuit C *computes* a Boolean function $f : \{0,1\}^n \rightarrow \{0,1\}$ if $C(\boldsymbol{x}) = f(\boldsymbol{x})$ for every input $\boldsymbol{x} = (x_1, x_2, \ldots, x_n) \in \{0,1\}^n$.

2.2 Function P_D^n

Let n be a positive integer. In the rest of the paper, we assume that n is a square number. Let $\boldsymbol{x} = \{x_{0,0}, x_{0,1}, \ldots, x_{\sqrt{n}-1,\sqrt{n}-1}\} \in \{0,1\}^n$ be input variables that are arranged in a 2-dimensional square grid and represent detectors for a particular local feature; and, similarly, let $\boldsymbol{y} = \{y_{0,0}, y_{0,1}, \ldots, y_{\sqrt{n}-1,\sqrt{n}-1}\} \in \{0,1\}^n$ be input variables that are arranged in the square grid and represent detectors for another particular local feature. (See Fig. 1.)

The function P_D^n represents a simple task relating global patterns concerning the two local features, and is defined as

$$P_D^n(\boldsymbol{x}, \boldsymbol{y}) = \begin{cases} 1 & \text{if } \exists i, j, k, l : x_{i,j} = y_{k,l} = 1 \text{ such that } i > k \text{ and } j < l; \\ 0 & \text{otherwise} \end{cases}$$

for every pair of $\boldsymbol{x} = (x_{0,0}, x_{0,1}, \ldots, x_{\sqrt{n}-1,\sqrt{n}-1})$. Thus, more intuitively, $P_D^n(\boldsymbol{x}, \boldsymbol{y}) = 1$ if and only if there exists a pair of locations $x_{i,j}$ and $y_{k,l}$ such that (i) $x_{i,j} = y_{k,l} = 1$, and (ii) $x_{i,j}$ is below and to the left of $y_{k,l}$ (See Fig. 2).

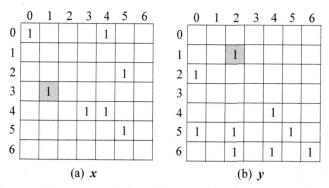

(a) $\boldsymbol{x}$ (b) $\boldsymbol{y}$

Fig. 2. An input assignment $(\boldsymbol{x}, \boldsymbol{y})$ for P_D^7, where (a) depicts $\boldsymbol{x}$ and (b) does $\boldsymbol{y}$. Note that we omit zeros in $\boldsymbol{x}$ and $\boldsymbol{y}$ for simplicity. In this case, $P_D^6(\boldsymbol{x}, \boldsymbol{y}) = 1$, since $x_{3,1} = 1$ and $y_{1,2} = 1$ (the corresponding locations are shaded).

3   Construction of Circuit

In this section we give an upper bound on the size of threshold circuits computing P_D^n, as in the following theorem.

Theorem 1. P_D^n *is computable by a threshold circuit of size* $O(\sqrt{n}\log n)$ *and depth* $O(\sqrt{n})$.

We prove the theorem by construction. Before proceeding to giving the explicit construction, we define some terms, and obtain a useful lemma. Throughout the proof, we denote an arbitrary pair of inputs to P_D^n by

$$\boldsymbol{x} = (x_{0,0}, x_{0,1}, \ldots, x_{\sqrt{n}-1,\sqrt{n}-1}) \in \{0,1\}^n$$

and

$$\boldsymbol{y} = (y_{0,0}, y_{0,1}, \ldots, y_{\sqrt{n}-1,\sqrt{n}-1}) \in \{0,1\}^n.$$

[Terms and idea]
For each j, $0 \le j \le \sqrt{n} - 1$, we consider the jth column of $\boldsymbol{x}$ (i.e., $x_{0,j}, x_{1,j}, \ldots, x_{\sqrt{n}-1,j}$) and define $\alpha_j^*(\boldsymbol{x})$ as the maximum index i satisfying $x_{i,j} = 1$:

$$\alpha_j^*(\boldsymbol{x}) = \max\{i \mid x_{i,j} = 1\};$$

if there is no index i satisfying $x_{i,j} = 1$, we define $\alpha_j^*(\boldsymbol{x}) = 0$. Similarly, for each j, $0 \le j \le \sqrt{n} - 1$, we below define β_j^*. Consider the lth column of $\boldsymbol{y}$ (i.e., $y_{0,l}, y_{1,l}, \ldots, y_{\sqrt{n}-1,l}$), and let β_j be the integer obtained by the minimum index k satisfying $y_{k,l} = 1$:

$$\beta_l(\boldsymbol{y}) = \sqrt{n} - 1 - \min\{k \mid y_{k,l} = 1\};$$

if there is no index k satisfying $y_{k,l} = 1$, we define $\beta_l(\boldsymbol{y}) = 0$.

Using $\beta_1(\boldsymbol{y}), \beta_2(\boldsymbol{y}), \ldots, \beta_{\sqrt{n}-1}(\boldsymbol{y})$, we now inductively define β_l^* for l, $0 \leq l \leq \sqrt{n}-1$ as follows:

$$\beta_{\sqrt{n}-1}^*(\boldsymbol{y}) = \beta_{\sqrt{n}-1}(\boldsymbol{y});$$

and, for each l, $\sqrt{n}-2 \geq l \geq 0$,

$$\beta_l^*(\boldsymbol{y}) = \max\left(\beta_l(\boldsymbol{y}), \beta_{l+1}^*(\boldsymbol{y})\right).$$

Clearly, $\beta_l^*(\boldsymbol{y})$ is the maximum over $\beta_l(\boldsymbol{y}), \beta_{l+1}(\boldsymbol{y}), \ldots, \beta_{\sqrt{n}-1}(\boldsymbol{y})$.

The following claim shows that we can determine $P_D^n(\boldsymbol{x}, \boldsymbol{y})$ by the values of $\alpha_j^*(\boldsymbol{x})$ and $\beta_l^*(\boldsymbol{y})$, $0 \leq j, l \leq \sqrt{n}-1$.

Lemma 1. $P_D^n(\boldsymbol{x}, \boldsymbol{y}) = 1$ *if and only if there exists a pair of indices j and l such that*

$$j = l - 1 \tag{1}$$

and

$$\sqrt{n} \leq \alpha_j^*(\boldsymbol{x}) + \beta_l^*(\boldsymbol{y}). \tag{2}$$

Proof. ($\Leftarrow$) Suppose there exists a pair of indices j and l satisfying (1) and (2). Let

$$i = \alpha_j^*(\boldsymbol{x}) \quad \text{and} \quad k = \sqrt{n} - 1 - \beta_l^*(\boldsymbol{y}). \tag{3}$$

By the definition of α_j^* and β_l^*,

$$x_{i,j} = 1 \tag{4}$$

and there exists an index l', $l \leq l' \leq \sqrt{n}-1$, satisfying

$$y_{k,l'} = 1. \tag{5}$$

Consider then the indices i, j, k, l'. Since (2) and (3) imply that $\sqrt{n} \leq i + \sqrt{n} - 1 - k$, it holds that

$$i > k. \tag{6}$$

Furthermore, we have by (1) that $j = l - 1$, and hence

$$j < l \leq l'. \tag{7}$$

Thus, by (4)–(7), $P_D^n(\boldsymbol{x}, \boldsymbol{y}) = 1$ holds.

($\Rightarrow$) Suppose $P_D^n(\boldsymbol{x}, \boldsymbol{y}) = 1$, that is, there exist indices i, j, k and l such that $x_{i,j} = 1$ and $y_{k,l} = 1$ satisfying

$$i > k \tag{8}$$

and

$$j < l. \tag{9}$$

Let $l' = j+1$. Then $j = l'-1$. Since (9) holds, we have $l' \leq l$. Then, by definition, we have $i \leq \alpha_j^*(\boldsymbol{x})$ and $\sqrt{n} - 1 - k \leq \beta_{l'}^*(\boldsymbol{y})$, and hence it holds that

$$\sqrt{n} + (i - k - 1) \leq \alpha_j^*(\boldsymbol{x}) + \beta_{l'}^*(\boldsymbol{y}).$$

Thus (8) implies that $\sqrt{n} \leq \alpha_j^*(\boldsymbol{x}) + \beta_{l'}^*(\boldsymbol{y})$, as required.   □

Based on Lemma 3, we below construct the desired threshold circuit C. Let $\tau = \lceil \log \sqrt{n} \rceil$ for simplicity.

[Construction of C]
Firstly, for each j, $0 \leq j \leq \sqrt{n} - 1$, we construct a set of τ threshold gates $g_{j,0}^*, g_{j,1}^*, \ldots, g_{j,\tau}^*$ so that the outputs of τ gates represent $\alpha_j^*(\boldsymbol{x})$ in binary system; we employ a circuit construction used in [1]. For each pair of i, $0 \leq i \leq \sqrt{n} - 1$, and t, $0 \leq t \leq \tau$, let

$$p_{i,t} = \left\lfloor \frac{i}{2^t} \right\rfloor$$

and

$$u_{i,t} = (-1)^{1+p_{i,t}} \cdot 2^i. \tag{10}$$

Clearly, if the $(t + 1)$st bit of the binary representation of i is one, then $p_{i,t}$ is odd, and hence $u_{i,t} = 2^i$; and otherwise, $p_{i,t}$ is even, and hence $u_{i,t} = -2^i$. Let j, $0 \leq j \leq \sqrt{n} - 1$, be an arbitrarily fixed index. For each t, $0 \leq t \leq \tau - 1$, the gate $g_{j,t}^*$ has threshold one, and receives every input in the jth column: $g_{j,t}^*$ receives $x_{i,j}$ with weight $u_{i,t}$ for every i, $0 \leq i \leq \sqrt{n} - 1$. Thus, for every $\boldsymbol{x} \in \{0,1\}^n$,

$$g_{j,t}^*(\boldsymbol{x}) = \text{sign} \left(-1 + \sum_{i=0}^{\sqrt{n}-1} u_{i,t} x_{i,j} \right). \tag{11}$$

Equations (10) and (11) imply that the output of $g_{j,t}^*$ is determined by $x_{i',j}$ satisfying

$$i' = \max \left\{ i \mid x_{i,j} = 1 \right\},$$

and thus equals to the $(t + 1)$st bit of the binary representation of α_j^*. Consequently,

$$\alpha_j^*(\boldsymbol{x}) = \sum_{t=0}^{\tau-1} 2^t \cdot g_{j,t}^*(\boldsymbol{x}). \tag{12}$$

Secondly, for each l, $0 \leq l \leq \sqrt{n} - 1$, we similarly construct a set of τ threshold gates $h_{l,0}, h_{l,1}, \ldots, h_{l,\tau-1}$ so that the outputs of the τ gates represent $\beta_l(\boldsymbol{y})$ in binary system. Let

$$q_{k,t} = \left\lfloor \frac{\sqrt{n} - 1 - k}{2^t} \right\rfloor.$$

For each t, $0 \leq t \leq \tau - 1$, the gate $h_{l,t}$ has threshold one, and receives every input in the lth column: $h_{l,t}$ receives $y_{k,l}$ with weight

$$v_{k,t} = (-1)^{1+q_{k,t}} \cdot 2^{\sqrt{n}-1-k}, \tag{13}$$

for every k, $0 \leq k \leq \sqrt{n} - 1$. Thus, for every $\boldsymbol{y} \in \{0,1\}^n$,

$$h_{l,t}(\boldsymbol{x}) = \text{sign}\left(-1 + \sum_{k=0}^{\sqrt{n}-1} v_{k,t} y_{k,l}\right). \tag{14}$$

Then the output of $h_{k,t}$ is determined by $y_{k',l}$ satisfying

$$k' = \min\{k \mid y_{k,l} = 1\},$$

and equals to the $(t+1)$st bit of the binary representation of β_j. Therefore, Eqs. (13) and (14) imply that

$$\beta_l(\boldsymbol{y}) = \sum_{t=0}^{\tau-1} 2^t \cdot h_{l,t}(\boldsymbol{y}). \tag{15}$$

Using $h_{l,0}, h_{l,1}, \ldots, h_{l,\tau-1}$, we construct gates $h_{l,0}^*, h_{l,1}^*, \ldots, h_{l,\tau-1}^*$ that represent $\beta_l^*(\boldsymbol{y})$; the construction is inductive on l from $\sqrt{n} - 1$ to 0. For the case where $l = \sqrt{n} - 1$, we do not create any new gate, and simply identify $h_{\sqrt{n}-1,t}^*$ with $h_{\sqrt{n}-1,t}$ for every t, $0 \leq t \leq \tau - 1$, since we have $\beta_{\sqrt{n}-1}^*(\boldsymbol{y}) = \beta_{\sqrt{n}-1}(\boldsymbol{y})$. For each l, $\sqrt{n} - 2 \geq l \geq 0$, we introduce two gates a^l, b^l and 2τ gates $a_0^l, a_1^l, \ldots, a_\tau^l$, $b_0^l, b_1^l, \ldots, b_\tau^l$ whose outputs will be used for inputs to $h_{l,0}^*, h_{l,1}^*, \ldots, h_{l,\tau}^*$. The gates a^l and b^l determine whether or not $\beta_l(\boldsymbol{y})$ is larger than $\beta_{l+1}^*(\boldsymbol{y})$, and are defined as

$$a^l(\boldsymbol{y}) = \text{sign}\left(-1 + \sum_{t=0}^{\tau-1} 2^t h_{l,t}(\boldsymbol{y}) - \sum_{t=0}^{\tau-1} 2^t h_{l+1,t}^*(\boldsymbol{y})\right)$$

and

$$b^l(\boldsymbol{y}) = \text{sign}\left(-\sum_{t=0}^{\tau-1} 2^t h_{l,t}(\boldsymbol{y}) + \sum_{t=0}^{\tau-1} 2^t h_{l+1,t}^*(\boldsymbol{y})\right)$$

By Eq. (15), a^l outputs one if and only if $\beta_l(\boldsymbol{y}) > \beta_{l+1}^*(\boldsymbol{y})$; while b^l outputs one if and only if $\beta_l(\boldsymbol{y}) \leq \beta_{l+1}^*(\boldsymbol{y})$. Thus, exactly one of a_l and b_l outputs one for any input. Then, for each t, $0, \leq t \leq \tau - 1$, the gate a_t^l computes AND of the outputs of $h_{l,t}$ and a^l:

$$a_t^l(\boldsymbol{y}) = \text{sign}(h_{l,t}(\boldsymbol{y}) + a^l(\boldsymbol{y}) - 2).$$

Clearly, the output of a_t^l equals to that of $h_{l,t}$ if $a^l(\boldsymbol{y}) = 1$ (i.e. $\beta_l(\boldsymbol{y}) > \beta_{l+1}(\boldsymbol{y})$); and equals to zero, otherwise. Similarly, the gate b_t^l computes AND of the outputs of $h_{l+1,t}^*$ and b^l:

$$b_t^l(\boldsymbol{y}) = \text{sign}(h_{l+1,t}^*(\boldsymbol{y}) + b^l(\boldsymbol{y}) - 2);$$

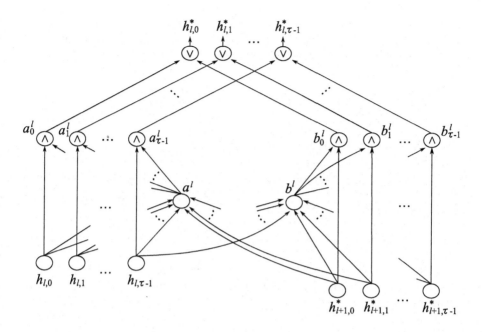

Fig. 3. Overview of the circuit for $h_{l,0}^*, h_{l,1}^*, \ldots, h_{l,\tau-1}^*$, where the gates labeled with "$\wedge$" compute AND of two inputs, and the gates labeled with "$\vee$" computes OR of two inputs

the output of b_t^l equals to that of $h_{l+1,t}^*$ if $b^l(\boldsymbol{y}) = 1$ (i.e. $\beta_l(\boldsymbol{y}) \le \beta_{l+1}(\boldsymbol{y})$); and equals to zero, otherwise. For each t, $0 \le t \le \tau - 1$, we obtain the gate $h_{l,t}^*$ simply computing OR of the outputs of a_t^l and b_t^l;

$$h_{l,t}^*(\boldsymbol{y}) = \text{sign}(a_t^l(\boldsymbol{y}) + b_t^l(\boldsymbol{y}) - 1).$$

Clearly, we have

$$\beta_j^*(\boldsymbol{y}) = \sum_{t=0}^{\tau-1} 2^t \cdot h_{j,t}^*(\boldsymbol{y}). \tag{16}$$

See Fig. 3 which depicts the circuit for computing $h_{l,0}^*, h_{l,1}^*, \ldots, h_{l,\tau-1}^*$.

Lastly, we construct $\sqrt{n} - 1$ gates $r_0, r_1, \ldots, r_{\sqrt{n}-2}$ such that, for each j, $0 \le j \le \sqrt{n} - 1$, the gate r_j determines if there exists an index l satisfying Eqs. (1) and (2). Eqs. (12) and (16) imply that we can obtain such r_j, as follows: r_j has threshold $-\sqrt{n}$ and receives the outputs of $g_{j,t}^*$ and $h_{l,t}^*$, where $j = l - 1$, with weight 2^t for every t, $0 \le t \le \tau - 1$. More formally,

$$r_j(\boldsymbol{x}, \boldsymbol{y}) = \text{sign}\left(-\sqrt{n} + \sum_{t=0}^{\tau-1} 2^t \cdot g_{j,t}^*(\boldsymbol{x}) + \sum_{t=0}^{\tau-1} 2^t \cdot h_{j+1,t}^*(\boldsymbol{y})\right).$$

Consequently, Lemma 3 implies that $P_D^n(\boldsymbol{x}, \boldsymbol{y}) = 1$ if and only if there exists an index j, $0 \le j \le \sqrt{n} - 1$, such that $r_j(\boldsymbol{x}, \boldsymbol{y}) = 1$. Therefore, our construction of C is completed by adding the top gate s computing OR of $r_0, r_1, \ldots, r_{\sqrt{n}-2}$:

$$s(\boldsymbol{x}, \boldsymbol{y}) = \text{sign}\left(-1 + \sum_{j=0}^{\sqrt{n}-2} r_j(\boldsymbol{x}, \boldsymbol{y})\right).$$

We now evaluate the size and depth of C. For every j, $0 \le j \le \sqrt{n} - 1$, we have τ gates $g_{j,0}^*, g_{j,1}^*, \ldots, g_{j,\tau-1}^*$. For every l, $0 \le l \le \sqrt{n} - 1$, we have τ gates $h_{j,0}, h_{j,1}, \ldots, h_{j,\tau-1}$. In addition, for every l, $0 \le l \le \sqrt{n}-1$, we have $2\tau+2$ gates $a^l, b^l, a_0^l, a_1^l, \ldots, a_{\tau-1}^l, b_0^l, b_1^l, \ldots, b_{\tau-1}^l$ together with τ gates $h_{l,0}^*, h_{l,1}^*, \ldots h_{l,\tau-1}^*$. Besides, we have $r_0, r_1, \ldots, r_{\sqrt{n}-2}$ and s. Consequently, the size of C is

$$\tau\sqrt{n} + \tau\sqrt{n} + (2\tau + 2)(\sqrt{n} - 1) + \tau(\sqrt{n} - 1) + \sqrt{n} - 1 + 1 = O(\sqrt{n}\log n).$$

Moreover, we require at most three layers to obtain $h_{l,0}^*, h_{l,1}^*, \ldots h_{l,\tau-1}^*$ for each l, $0 \le l \le \sqrt{n} - 1$, followed by the two layers containing $r_0, r_1, \ldots, r_{\sqrt{n}-2}$ and the top gate s. Thus, the depth is $O(\sqrt{n})$.

4 Lower Bound

In this section, we show that the circuit given in Theorem 1 is optimal up to logarithmic factor, as in the following theorem.

Theorem 2. *Let C be an arbitrary threshold circuit computing P_D^n. Then C has size $\Omega(\sqrt{n}/\log n)$.*

Let C be an arbitrary threshold circuit computing P_D^n. We prove the theorem by reducing the disjointness function $DISJ^n$ to our function, where $DISJ^n$ is defined as follows: For every pair of $\boldsymbol{x} = (x_1, x_2, \ldots, x_n) \in \{0,1\}^n$ and $\boldsymbol{y} = (y_1, y_2, \ldots, y_n) \in \{0,1\}^n$,

$$DISJ^n(\boldsymbol{x}, \boldsymbol{y}) = \begin{cases} 1 \text{ if } \forall i : x_i \ne y_i; \\ 0 \text{ otherwise.} \end{cases}$$

It is known that any threshold circuit computing $DISJ^n$ has almost linear size in n.

Lemma 2 ([2]). *Let C be an arbitrary threshold circuit computing $DISJ^n$. Then C has size $\Omega(n/\log n)$.*

Thus, it suffices to show that we can construct a circuit C' computing $DISJ^{O(\sqrt{n})}$ from C so that C' has same size as that of C. We obtain the desired circuit C' by just fixing some of the input variables of C to zeros, as follows.

Let X and Y be sets of the input variables to P_D^n:

$$X = \{x_{0,0}, x_{0,1}, \ldots, x_{\sqrt{n}-1, \sqrt{n}-1}\}$$

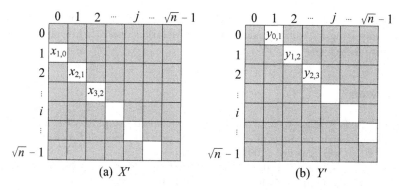

Fig. 4. In (a) (and (b), respectively), the unshaded locations indicate X' (and Y'), while shaded locations indicate $X\backslash X'$ (and $Y\backslash Y'$)

and

$$Y = \{y_{0,0}, y_{0,1}, \ldots, y_{\sqrt{n}-1,\sqrt{n}-1}\}.$$

We define sets X' and Y' of the input variables as

$$X' = \{x_{j+1,j} \mid 0 \leq j \leq \sqrt{n} - 2\} \subseteq X$$

and

$$Y' = \{y_{k,k+1} \mid 0 \leq k \leq \sqrt{n} - 2\} \subseteq Y.$$

(See Fig. 4.) We then fix every input in $(X\backslash X') \cup (Y\backslash Y')$ of C to zero. We denote by C^* the resulting circuit. Clearly, C^* computes P_D^n over $X' \cup Y'$. By the definition of P_D^n, C^* outputs one if and only if there exist indices j and k such that $j + 1 > k$ and $j < k + 1$:

$$k - 1 < j < k + 1,$$

and hence $j = k$. Consequently, C^* outputs one if and only if there exists an index j, $0 \leq j \leq \sqrt{n} - 2$, such that $x_{j+1,j} = y_{j,j+1} = 1$; thus C^* computes the complement of $DISJ^{\sqrt{n}-1}$.

We complete the construction of C' by replacing the top gate g of C^* by a new gate g^* computing its complement. Suppose g has threshold t and weight $w_1, w_2, \ldots, w_z$ for a number z of inputs consisting of $x_{j+1,j}$ and $y_{j,j+1}$, $0 \leq j \leq \sqrt{n} - 2$, together with the outputs of the gates in C^*. We then replace g by g^* with threshold $-2t + 1$ and weight $-2w_1, -2w_2, \ldots, -2w_z$.

5   Conclusion

In this paper, we consider a Boolean function P_D^n that models a simple task for information processing on 2-dimensional square grid. We show that P_D^n is computable by a threshold circuit of size $O(\sqrt{n} \log n)$, while any threshold circuit computing P_D^n requires size $\Omega(\sqrt{n}/\log n)$. Note that our circuit has smaller size than one given in [1], but has larger depth $\Theta(\sqrt{n})$ than the ones in [1]. It is

K. Uchizawa, D. Yashima, and X. Zhou

interesting if we can find a relationship between size and depth of threshold circuits computing P_D^n; smaller size may require larger depth, while smaller depth may require larger size.

References

1. Legenstein, R.A., Maass, W.: Foundations for a circuit complexity theory of sensory processing. In: Proceedings of Advances in Neural Information Processing Systems (NIPS 2000), pp. 259–265 (2001)
2. Nisan, N.: The communication complexity of threshold gates. In: Proceeding of Combinatorics, Paul Erdös is Eighty, pp. 301–315 (1993)
3. Parberry, I.: Circuit Complexity and Neural Networks. MIT Press, Cambridge (1994)
4. Sima, J., Orponen, P.: General-purpose computation with neural networks: A survey of complexity theoretic results. Neural Computation 15, 2727–2778 (2003)
5. Siu, K.Y., Bruck, J.: On the power of threshold circuits with small weights. SIAM Journal on Discrete Mathematics 4(3), 423–435 (1991)
6. Siu, K.-Y., Bruck, J., Kailath, T., Hofmeister, T.: Depth efficient neural networks for division and related problems. IEEE Transactions on Information Theory 39(3), 946–956 (1993)
7. Siu, K.Y., Roychowdhury, V.: On optimal depth threshold circuits for multiplication and related problems. SIAM Journal on Discrete Mathematics 7(2), 284–292 (1994)
8. Siu, K.Y., Roychowdhury, V., Kailath, T.: Discrete Neural Computation; A Theoretical Foundation. Prentice-Hall, Inc., Upper Saddle River (1995)

Superset Generation on Decision Diagrams

Takahisa Toda[1,*], Shogo Takeuchi[2], Koji Tsuda[2,3,4], and Shin-ichi Minato[2,5]

[1] Graduate School of Information Systems,
The University of Electro-Communications, Chofu, Japan
[2] ERATO MINATO Discrete Structure Manipulation System Project,
Japan Science and Technology Agency, at Hokkaido University, Sapporo, Japan
[3] Graduate School of Frontier Sciences, The University of Tokyo, Kashiwa, Japan
[4] Computational Biology Research Center, National Institute of Advanced Industrial
Science and Technology (AIST), Tokyo, Japan
[5] Graduate School of Information Science and Technology,
Hokkaido University, Sapporo, Japan

Abstract. Generating all supersets from a given set family is important, because it is closely related to identifying cause-effect relationship. This paper presents an efficient method for superset generation by using the compressed data structures BDDs and ZDDs effectively. We analyze the size of a BDD that represents all supersets. As a by-product, we obtain a non-trivial upper bound for the size of a BDD that represents a monotone Boolean function in a fixed variable ordering.

Keywords: superset, binary decision diagram, zero-suppressed binary decision diagram, cause-effect relationship, Boolean completion.

1 Introduction

Generating all supersets from a given family of sets is an important task, because it is closely related to the problem of identifying cause-effect relationship [1]. Suppose we observed that some combinations of items caused an effect P. The problem is to compute all combinations of items that can cause the effect P from the obvervation. For example, consider potential factors that may cause a disease as items and combinations of such factors that are actually identified in an occurrence of the disease as observed itemsets. If there is no prior knowledge on P, then we would have to conduct exhaustive search over all combinations of items, which results in exponential explosion. However, such a situation is rare and one can usually assume some regularity. This paper assumes monotonicity: if a combination of items causes an effect P, then any combination extending it causes the same effect. For a simple example, consider that if a combination of foods causes a stomachache, then it is very likely that any combination that contains those foods also causes a stomachache. Monotonicity allows us to identify all potential causes that can be derived from observation. Monotonicity frequently appears in practical situations. It is mentioned [2] that the property

---

* The first author is supported by JSPS KAKENHI Grant Number 26870011.

M.S. Rahman and E. Tomita (Eds.): WALCOM 2015, LNCS 8973, pp. 317–322, 2015.

of monotonicity for data has potential to significantly improve the reliability of computer-aided diagnosis. Monotonicity can also be seen in the context of product design [3], linkage problem of records in different databases [4], etc.

Our problem can be formally described as follows. We have a set of items V and a set family $\mathcal{S}$ over V. Each set in $\mathcal{S}$ means a combination of items that caused some effect. We want to compute all subsets of V that contain at least one set in $\mathcal{S}$. In this paper, we say that U is a *superset* of T if $T \subseteq U$, and we denote $\mathcal{S}^\uparrow := \{U \subseteq V : \exists T \in \mathcal{S}, \ T \subseteq U\}$.

Since the size of $\mathcal{S}^\uparrow$ can grow exponentially, one may feel impractical to compute $\mathcal{S}^\uparrow$. However, the difficulty can often be reduced largely by considering $\mathcal{S}^\uparrow$ as a Boolean formula and implicitly computing it on the compressed data structure, called *binary decision diagram* (*BDD* for short). A BDD is a graphical representation for Boolean functions, and it has been widely used in the design and verification of VLSI circuits (see [5, pp.257-258]). An advantage of BDDs is that BDDs tend to achieve a high compression efficiency in practice, and furthermore various basic operations to manipulate Boolean functions such as logical conjunction, disjunction and negation, etc are available on BDDs. Once the BDD for $\mathcal{S}^\uparrow$ is built, one can efficiently execute computation such as optimization and enumeration [5, pp.209-212].

In this paper, we present an efficient method for computing the BDD that represents all supersets for a given family of sets. Although BDDs are well suited for families of supersets, this does not apply to arbitrary set families, because the compression rule (i.e., the node elimination rule) of BDD is not so effective if a set family does not satisfy monotonicity. We thus separately use the two different data structures BDDs and ZDDs, where ZDDs are a yet another data structure specialized for set families. Our approach is that we first convert an input set family into a ZDD representation, and then construct the BDD for supersets from the ZDD. We analyze the size of an output BDD because it greatly affects the performance of whole computation. To do this, we introduce the notion of *separator* of a ZDD, which is related to the notion of cutwidth of a CNF or a hypergraph. We show that the size of an output BDD is linear in $|V|$ and exponential only in the maximum size of a separator in the ZDD for $\mathcal{S}$. This result implies that it is quite important in our method to find a good order of V so that the size of a separator is as small as possible. As a by-product, we obtain a non-trivial upper bound for the size of a BDD that represents a monotone Boolean function in a fixed variable ordering.

2   Decision Diagrams

A *zero-suppressed binary decision diagram* (*ZDD* for short) is a graphical representation for set families. Exactly one node has indegree 0, which is called the *root*. Each nonterminal node f has a label and exactly two children, which are indicated by the three fields V (f), LO (f), HI (f) associated with f. Each node has an element in a ground set V as its label. The children indicated by LO (f) and HI (f) are called the LO *child* and HI *child* of f, respectively. The arc to a LO child is called a LO *arc* and illustrated by a dashed arrow, while the arc to a HI child is called a HI *arc* and illustrated by a solid arrow. There are only two terminal nodes, denoted by $\top$ and $\bot$. ZDDs satisfy the following two conditions. They must be *ordered*: if a node u points to a nonterminal node v, then $V(u) < V(v)$. They must be *reduced*: no further application of the following reduction operation rules is possible.

1. If there is a nonterminal node u whose HI arc points to $\bot$, then redirect all the incoming arcs of u to the LO child, and then eliminate u.
2. If there are two nonterminal nodes u and v such that the subgraphs rooted by them are equivalent, then merge them.

It is known (see for example [5]) that if the order in V is fixed, then set families on V correspond in a one-to-one way to ZDDs whose labels are taken from V.

A *binary decision diagram* (*BDD* for short) is a similar graphical representation to ZDDs , but BDDs represent Boolean functions. BDDs have two different features from ZDDs. Firstly, BDDs have the following reduction rules.

1. If there is a nonterminal node u whose arcs both point to v, then redirect all the incoming arcs of u to v, and then eliminate u.
2. If there are two nonterminal nodes u and v such that the subgraphs rooted by them are equivalent, then merge them.

Secondly, given a BDD, each path from the root to $\top$ corresponds to a $(0, 1)$-assignment to Boolean variables and the value of a Boolean function. That is, assign 1 to x_k if the HI arc of a node with label k is selected; otherwise, assign 0 to x_k, and for such an assignment, the function value becomes 1. For efficiency, BDD nodes are usually implemented using a hash table, called a *uniquetable*, so that for any triple (k, lo, hi) of a node label and two BDD nodes, there is a unique BDD node f with $V(f) = k$, $LO(f) = lo$, and $HI(f) = hi$. Given a triple (k, lo, hi), the function getbddnode returns an associated node in the uniquetable if exists; otherwise, create a new node f such that $V(f) = k$, $LO(f) = lo$, and $HI(f) = hi$; register f to the uniquetable and return f.

3   Algorithm

Suppose that we are given a set family $\mathcal{S} := \{S_1, \ldots, S_m\}$ over a ground set $V := \{1, \ldots, n\}$. We present a method of computing all supersets for $\mathcal{S}$ as a Boolean function, thereby as a BDD.

Recall that the family of all supersets for $\mathcal{S}$ is denoted by $\mathcal{S}^\uparrow$. Let us observe that $\mathcal{S}^\uparrow$ can be identified with the following DNF:

$$\psi_\mathcal{S}(x_1, \ldots, x_n) := \bigvee_{1 \leq i \leq m} \bigwedge_{j \in S_i} x_j.$$

For convenience, we call a vector $v \in \{0, 1\}^n$ with $\psi_\mathcal{S}(v) = 1$ a *solution* of $\psi_\mathcal{S}$, where $\psi_\mathcal{S}$ is seen as a Boolean function. If no confusion, we identify v with the set representation $\{i \in V : v_i = 1\}$, where v_i denotes the i-th component of v. Now, we show that all solutions of ψ_S form $S^\uparrow$. Suppose that v is a solution of ψ_S. There is a term in ψ_S that is satisfied by v. Let $\bigwedge_{j \in S_k} x_j$ be such a term. It is clear that $S_k \subseteq \{i \in V : v_i = 1\}$. The converse direction would be now clear.

We are now ready to describe our algorithm for computing the BDD $B(\psi_S)$ that represents all supersets for S. Our algorithm is as follows.

1. Given a set family S over V, compute the ZDD $Z(S)$ that represents S.
2. Compute $B(\psi_S)$ from $Z(S)$.

In the former part, we use the bottom-up construction method of ZDDs after sorting S (see [6] for details). This computation requires $O(\sum \{|U| : U \in S\})$ time and extra space in the worst case except for the cost for sorting S. Thus, the latter part is a computation bottleneck. In the latter part, we execute the function SUP defined in Algorithm 1. Since the correctness of Algorithm 1 would be clear, we omit a proof.

Algorithm 1. Given a ZDD for a set family S, compute the BDD that represents all supersets for S. The function bddor computes the logical disjunction.

---

  function SUP(g)
    if $g = \top_{\mathrm{ZDD}}$ **then**
      return $\top_{\mathrm{BDD}}$;
    else if $g = \bot_{\mathrm{ZDD}}$ **then**
      return $\bot_{\mathrm{BDD}}$;
    end if
    $l \leftarrow$ SUP (LO (g)); $t \leftarrow$ SUP (HI (g)); $h \leftarrow$ bddor (l, t);
    return getbddnode (V (g), l, h);
  end function

---

4   Analysis

We analyze the size of an output BDD in Algorithm 1. Huang and Darwiche [7] used the notions of *cutwidth* for CNFs, and proved that given a CNF with a fixed variable ordering, the size of the BDD represented by the CNF is linear in the number of variables and exponential only in the cutwidth. Inspired by their appoarch, we introduce a related notion for ZDDs and analyze a BDD size (see Fig. 1).

Definition 1. *The i-th separator in a ZDD is the set of nonterminal nodes g such that g is a tail of some arc (j, k) with $j \leq i$ and $i < k$, where j and k denote node labels. The maximum size of a separator in a ZDD f is denoted by $s(f)$.*

Theorem 1. *Let S be a set family over a ground set V in a fixed order. Let Z be the ZDD that represents S. The BDD size for all supersets of S is bounded above by $|V| \cdot 2^{s(Z)}$.*

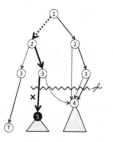 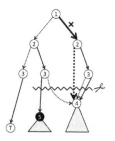 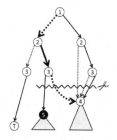

Fig. 1. The paths in a ZDD reaching ⑤ and ④ under the instanciation with $x_1 = 0, x_2 = 1, x_3 = 0$, where the 3-th separator is indicated just below a zigzag line

Proof. Let B be the output BDD, and let $f(x_1, \ldots, x_n)$ be the Boolean function of the DNF that is represented by Z, where $n := |V|$. Observe that nodes in B exactly correspond to subfunctions of f, where a subfunction means a function obtained from f by instanciating some consecutive variables $x_1, \ldots, x_i$ from the first. Indeed, given a node in B, the subgraph rooted by the node is also a BDD, and thus it represents a Boolean function. Clearly, this gives a surjective mapping from BDD nodes onto subfunctions of f. To see it is injective, consider that if two subgraphs represent an identical (i.e. logically equivalent) subfunction, then these graphs must be equivalent and thus merged because of the node sharing rule. This means that if two different instanciations of $x_1, \ldots, x_i$ yield an identical subfunction, then the paths going down B along these instanciations must merge at the node corresponding to the subfunction.

In order to bound the size of B, we consider the number of different subfunctions after instanciations of the i-th or less variables. Without loss of generality, we exclude constant subfunctions, i.e., the functions that always return true or false values, respectively. Given an instanciation, the terms of the DNF that is represented by Z are partitioned into two groups: unsatisfiable terms under the instanciation and the other terms. This partition induces a 2-coloring of the i-th separator. For example, let us see the ZDD illustrated in Fig. 1. Recall that each path from the root to ⊤ represents a term. Consider the instanciation with $x_1 = 0, x_2 = 1, x_3 = 0$. The two nodes in the 3-th separator is then colored in the following way. All paths starting with ① --→ ② → ③ → become unsatisfiable because the corresponding terms have x_3 in common. Thus let us color the node ⑤ in the 3-th separator black. For the node ④ in the 3-th separator, although all paths starting with ① → also become unsatisfiable, the path ① --→ ② → ③ --→ can reach ④ without becoming unsatisfiable, thus let us color ④ white. In general, we color nodes in the i-th separator white if they have a reachable path without becoming unsatisfiable in the current instanciation. This coloring contains a complete information to decide if a DNF is satisfied. Thus, if another instanciation of the i-th or less variables happens to yield the same coloring in the i-th separator, then the two instantiations must have the same subfunction in common. Since the number of all possible colorings is at most $2^{s(Z)}$, the number of different subfunctions after instanciations of the i-th or less variables is bounded above by $2^{s(Z)}$. Therefore, we conclude that the size of an output BDD is bounded above by $|V|2^{s(Z)}$. ☐

Corollary 1. *Let ψ be a DNF of n variables without negative literals, and let s be the maximum size of a separator in the ZDD that represents ψ as a set family. The BDD that represents ψ has size at most $n \cdot 2^s$.*

The following theorem assumes that the function getbddnode runs in constant time and the function bddor runs in the product of two input sizes, which both can be achieved in average time if one implements using a hashtable.

Theorem 2. *Algorithm 1 can be implemented to run in $O(|Z| \cdot |V|^2 \cdot 2^{2s(Z)})$ time, where V is a ground set and $|Z|$ denotes the number of nodes in an input ZDD Z.*

Proof. Suppose that each ZDD node has an auxiliary field AUX. For each recursive call with an input ZDD g, set the output BDD to AUX (g). It is then sufficient to compute SUP (g) exactly once for each ZDD node g in Z. Thus, the number of recursive calls is $2|Z|$, since it is bounded above by the number of arcs in Z. Clearly the maximum sizes of a separator in LO (g) and HI (g) are at most the maximum size of a separator in Z. From Theorem 1, it follows that the sizes of SUP (LO (g)) and SUP (HI (g)) are both at most $|V| \cdot 2^{s(Z)}$. Since the logical disjunction operation requires the time proportional to the product of input sizes, the total time is proportional to $|Z| \cdot (|V| \cdot 2^{s(Z)})^2$. ☐

References

1. Crama, Y., Hammer, P., Ibaraki, T.: Cause-effect relationships and partially defined Boolean functions. Annals of Operations Research 16(1), 299–325 (1988)
2. Kovalerchuk, B., Triantaphyllou, E., Ruiz, J.F., Torvik, V.I., Vityaev, E.: The reliability issue of computer-aided breast cancer diagnosis. Computers and Biomedical Research 33(4), 296–313 (2000)
3. Kovalerchuk, B., Vityaev, E., Triantaphyllou, E.: How can AI procedures become more effective for manufacturings? In: Proc. of the Artificial Intelligence and Manufacturing Research Planning Workshop, Albuquerque, New, Mexico, pp. 103–111 (June 1996)
4. Judson, D.: Statistical rule induction in the presence of prior information: The bayesian record linkage problem. In: Triantaphyllou, E., Felici, G. (eds.) Data Mining and Knowledge Discovery Approaches Based on Rule Induction Techniques. Massive Computing, vol. 6, pp. 655–694. Springer US (2006)
5. Knuth, D.E.: The Art of Computer Programming Volume 4a. Addison-Wesley Professional, New Jersey (2011)
6. Toda, T.: Fast compression of large-scale hypergraphs for solving combinatorial problems. In: Fürnkranz, J., Hüllermeier, E., Higuchi, T. (eds.) DS 2013. LNCS, vol. 8140, pp. 281–293. Springer, Heidelberg (2013)
7. Huang, J., Darwiche, A.: Using DPLL for efficient OBDD construction. In: Hoos, H.H., Mitchell, D.G. (eds.) SAT 2004. LNCS, vol. 3542, pp. 157–172. Springer, Heidelberg (2005)

On Triangle Cover Contact Graphs

Md. Iqbal Hossain, Shaheena Sultana, Nazmun Nessa Moon, Tahsina Hashem,
and Md. Saidur Rahman

Department of Computer Science and Engineering,
Bangladesh University of Engineering and Technology (BUET),
Dhaka-1000, Bangladesh
{mdiqbalhossain,saidurrahman}@cse.buet.ac.bd
{zareefas.sultana,tahsinahas̀hem}@gmail.com, moon_ruet@yahoo.com

Abstract. Let $S = \{p_1, p_2, \ldots, p_n\}$ be a set of pairwise disjoint geometric objects of some type in a $2D$ plane and let $C = \{c_1, c_2, \ldots, c_n\}$ be a set of closed objects of some type in the same plane with the property that each element in C covers exactly one element in S and any two elements in C are interior-disjoint. We call an element in S a *seed* and an element in C a *cover*. A *cover contact graph (CCG)* has a vertex for each element of C and an edge between two vertices whenever the corresponding cover elements touch. It is known how to construct, for any given point seed set, a disk or triangle cover whose contact graph is 1- or 2-connected but the problem of deciding whether a k-connected CCG can be constructed or not for $k > 2$ is still unsolved. A *triangle cover contact graph (TCCG)* is a cover contact graph whose cover elements are triangles. In this paper, we give an algorithm to construct a 4-connected $TCCG$ for a given set of point seeds. We also show that any outerplanar graph has a realization as a $TCCG$ on a given set of collinear point seeds. Note that, under this restriction, only trees and cycles are known to be realizable as CCG.

1 Introduction

Let $S = \{p_1, p_2, \ldots, p_n\}$ be a set of pairwise disjoint geometric objects of some type in a $2D$ plane and let $C = \{c_1, c_2, \ldots, c_n\}$ be a set of closed objects of some type in the same plane with the property that each element in C covers exactly one element in S and any two elements in C can intersect only on their boundaries. We call an element in S a *seed* and an element in C a *cover*. The seeds may be points, disks or triangles and covering elements may be disks or triangles. The *cover contact graph (CCG)* consists of a set of vertices and a set of edges where each vertex corresponds to a cover and each edge corresponds to a connection between two covers if they touch at their boundaries. In other words, two vertices of a cover contact graph are adjacent if the corresponding cover elements touch at their boundaries. Note that the vertices of the cover contact graph are in one-to-one correspondence to both seeds and covering objects. In a cover contact graph, if disks are used as covers then it is called a *disk cover contact graph* and if triangles are used as covers then it is called a *triangle*

M.S. Rahman and E. Tomita (Eds.): WALCOM 2015, LNCS 8973, pp. 323–328, 2015.

cover contact graph (TCCG). Figure 1(b) depicts the disk cover contact graph induced by the disk covers in Fig. 1(a), whereas Fig. 1(d) depicts the triangle cover contact graph induced by the triangle covers in Fig. 1(c).

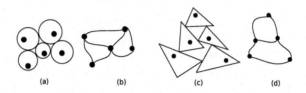

(a)   (b)   (c)   (d)

Fig. 1. Illustration for CCG and $TCCG$; (a) a disk cover, (b) a CCG, (c) a triangle cover and (d) a TCCG

A *coin graph* is a graph formed by a set of disks, no two of which have overlapping interiors, by making a vertex for each circle and an edge for each pair of circles that touches. Koebe's theorem [5,7] states that every planar graph can be represented as a coin graph. There are several works [8,9,4] in the geometric-optimization community where the problem is how to cover geometric objects such as points by other geometric objects such as convex shapes, disks. The main goal is to minimize the radius of a set of k disks to cover n input points. Applications of such covering problems are found in geometric optimization problems such as facility location problems [8,9]. Abellanas *et al.* [1] worked on a "coin placement problem," which is NP-complete. They tried to cover n points using n disks (each having different radius) by placing each disk in the center position at one of the points so that no two disks overlap. Further Abellanas *et al.* [2] considered another related problem. They showed that a given set of points in the plane, it is also NP-complete to decide whether there are disjoint disks centered at the points such that the contact graph of the disks is connected.

Recently, Atienza *et al.* [3] introduced the concept of cover contact graphs where the seeds are not necessarily the center of the disks. They gave an $O(n \log n)$ time algorithm to decide whether a given set of point seeds can be covered with homothetic triangles or disks such that the resulting cover contact graph is 1- or 2-connected. The k-connectivity problem is still unsolved for $k > 2$. Atienza *et al.* [3] also considered the problem from another direction which they called "realization problem." In a realization problem we are given a graph G of n vertices and a set S of n seeds and we are asked whether there is any covering so that the resulting cover contact graph is G. They gave some necessary conditions and then showed that it is NP-hard to decide whether a given graph can be realized as a disk cover contact graph if the correspondence between vertices and point seeds is given. They also showed that every tree and cycle have realizations as CCGs on a given set of collinear point seeds.

In this paper, we consider a set of arbitrary seeds in the plane where the seeds are points and the covers are triangles. In Section 2 we show that every set of six or more seeds admits a cover with 4-connected $TCCG$ and such a cover can be constructed in $O(n \log n)$ time. Section 3 gives an algorithm that realizes a given outerplanar graph as a triangle cover contact graph $(TCCG)$ for a given set of seeds on a line. Finally, Section 4 concludes the paper by suggesting some future works.

2   Connectivity in $TCCG$

In this section we consider the connectivity problem in $TCCG$ and develop an $O(n \log n)$ time algorithm to show that a set of given seeds can always be covered with triangles such that the resulting cover contact graph is 4-connected. We refer [6,10] for the graph theoretic terminologies used in this paper.

Consider a given set S of seeds such that no two seeds in S lie on a vertical line or on a horizontal line. It is trivial to show that S admits a path $TCCG$ as follows. We first sort the seeds according to x-coordinate value. We then cover the seeds with triangles such that every triangle touches the previous and next triangle (except the first and the last triangle). Thus the contact graph of the triangles forms a path. However constructions of $TCCG$s with higher connectivities look gradually difficult. We next deal with the construction of a 4-connected $TCCG$ and prove the following theorem.

Theorem 1. *Let S be a set of six or more seeds such that no two seeds are neither on a vertical line nor on a horizontal line. Then S admits a 4-connected $TCCG$, and such a $TCCG$ can be found in $O(n \log n)$ time.*

We give a constructive proof of Theorem 1. The outline of our construction is as follows. We first sort the seeds $p_1, p_2, \ldots, p_n$ according to their left-to-right order. We next select the topmost and the bottommost seeds p_t and p_b, respectively, among the seeds except p_1 and p_n. We enclose $S - \{p_1, p_n, p_t, p_b\}$ by a trapezoid as illustrated in Fig. 2(a). We then cover p_1 and p_n seeds by the triangles T_1 and T_n such that they touch at a point. We next cover p_t and p_b seeds by the triangles T_t and T_b such that they touch T_1 and T_n as illustrated in Fig. 2(b). Finally, each of the remaining uncovered seeds is covered by a triangle such that it touches T_t and T_b, the left and the right triangles as illustrated in Fig. 2(c). We prove that resulting graph is a 4-connected $TCCG$ by showing that there exist four internally vertex-disjoint paths between every pair of vertices.

Before giving a formal proof of Theorem 1 we need some definitions. For a point a in a $2D$ plane, we denote by x_a and y_a the x-coordinate and the y-coordinate of a, respectively. We specify the position of a in the plane by (x_a, y_a). We denote the straight-line segment that passes through the points a and b by L_{ab}. We denote a triangle connecting three points a_1, a_2, a_3 by $T(a_1, a_2, a_3)$ and a trapezoid connecting four points a_1, a_2, a_3, a_4 by $trap(a_1, a_2, a_3, a_4)$. We denote a path in a simple graph by the ordered sequence of vertices on the path. We are now ready to prove Theorem 1.

Proof of Theorem 1: We first construct a triangle cover of each seed in S as follows. Let p_l and p_r be the leftmost and the rightmost seeds in S, respectively.

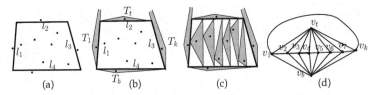

Fig. 2. (a) A set S of n seeds and a trapezoid enclosing $n-4$ seeds, (b) illustration for the triangles T_1, T_k, T_t and T_b, (c) S is covered by n triangles and (d) a 4-connected $TCCG$

Let p_t and p_b be the topmost and the bottommost seeds in $S - \{p_l, p_r\}$, respectively. Let $p_1 = p_l, p_2, \ldots, p_k = p_r$ be the seeds in $S - \{p_t, p_b\}$ according to their left-to-right ordering. Then $k = n - 2$. Let $S' = \{p_2, p_3, \ldots, p_{k-1}\}$. We denote by (x_{lm}, y_{lm}) the crossing point of two straight-lines l and m. We now enclose S' by a trapezoid $trap(l_1, l_2, l_3, l_4)$ so that $trap(l_1, l_2, l_3, l_4)$ satisfies the properties: (a) l_2 and l_4 lines are horizontal; (b) l_1 and l_3 lines are not parallel; (c) l_1 divides the seeds into $\{p_1\}$ and $S - \{p_1\}$; (d) l_2 divides the seeds into $\{p_t\}$ and $S - \{p_t\}$; (e) l_3 divides the seeds into $\{p_k\}$ and $S - \{p_k\}$; (f) l_4 divides the seeds into $\{p_b\}$ and $S - \{p_b\}$; (g) $x_{l_1 l_2}$ and $x_{l_1 l_4}$ are smaller than x_{p_2}; and (h) $x_{l_2 l_3}$ and $x_{l_3 l_4}$ are greater than $x_{p_{k-1}}$. Figure 2(a) illustrates an example of the trapezoid. One can easily see that such a trapezoid can be found because p_t, p_b, p_1 and p_k are the topmost, the bottommost, the leftmost and the rightmost seeds, respectively.

We first cover the four seeds p_1, p_k, p_t and p_b by the triangles T_1, T_k, T_t and T_b, respectively, as follows. Let the equation of l_2 be $y = y_t - c_1 = y_{l_2}$ and equation of l_4 be $y = y_b + c_2 = y_{l_4}$, where c_1 and c_2 are two small constants. (Precisely, $c_1 < y_t - y'_t$ and $c_2 < y'_b - y_b$; where y_t and y_b are y-coordinates of p_t and p_b, respectively, and y'_t and y'_b are y-coordinates of the topmost point and the bottommost point among the points in S', respectively.)

Since l_1 and l_3 are not parallel, without loss of generality assume that l_1 and l_3 cross at $(x_{l_1 l_3}, y_{l_1 l_3})$, where $y_{l_1 l_3} > y_{l_2}$. We now cover the seed p_1 by the triangle $T_1((x_{p_1} - c, y_{p_1}), (x_{l_1 l_4}, y_{l_1 l_4}), (x_{l_1 l_3}, y_{l_1 l_3}))$, where c is any small positive constant. Then we cover the seed p_k by the triangle $T_k((x_{p_k} + c, y_{p_k}), (x_{l_3 l_4}, y_{l_3 l_4}), (x_{l_1 l_3}, y_{l_1 l_3}))$. Next we cover the seed p_t by the triangle $T_t((x_{p_t}, y_{p_t} + c), (x_{l_1 l_2}, y_{l_1 l_2}), (x_{l_2 l_3}, y_{l_2 l_3}))$. Triangle $T_b((x_{p_b}, y_{p_b} - c), (x_{l_3 l_4}, y_{l_3 l_4}), (x_{l_1 l_4}, y_{l_1 l_4}))$ covers the seed p_b. Clearly, T_1 and T_k touch both the triangles T_t and T_b, and T_1 touches T_k.

We next cover the seeds in S' by triangles as follows. First we cover the seed p_2 by the triangle $T_2((A, y_{p_2}), (x_{p_2} + \frac{x_{p_3} - x_{p_2}}{2}, y_{l_4}), (x_{p_2} + \frac{x_{p_3} - x_{p_2}}{2}, y_{l_2}))$, where (A, y_{p_2}) lies on l_1. Clearly, T_2 touches T_1, T_t and T_b. We then cover each p_i for $2 < i < k - 1$ by the triangle $T_i((x_{p_{i-1}} + \frac{x_{p_i} - x_{p_{i-1}}}{2}, y_{p_i}), (x_{p_i} + \frac{x_{p_{i+1}} - x_{p_i}}{2}, y_{l_4}), (x_{p_i} + \frac{x_{p_{i+1}} - x_{p_i}}{2}, y_{l_2}))$. Note that the point $(x_{p_{i-1}} + \frac{x_{p_i} - x_{p_{i-1}}}{2}, y_{p_i})$ lies on the boundary of T_{i-1}. This implies that T_i touches T_{i-1}, T_t and T_b. We finally cover the seed p_{k-1} by $T_{k-1}((x_{p_{k-2}} + \frac{x_{p_{k-1}} - x_{p_{k-2}}}{2}, y_{p_{k-1}}), (x_{p_{k-1}} + \frac{x_{p_k} - x_{p_{k-1}}}{2}, y_{l_4}), (x_{l_2 l_3}, y_{l_2 l_3}))$. Clearly, T_{k-1} touches T_{k-2}, T_t, T_k and T_b.

Let $G = (V, E)$ be the resulting $TCCG$ of S with the vertex set $V = \{v_1, v_2, \ldots, v_k, v_t, v_b\}$ such that each v_i, $1 \le i \le k$, corresponds to the triangle T_i, v_t corresponds to the triangle T_t and v_b corresponds to the triangle T_b. We now show that G is 4-connected. Since T_1 touches T_2, T_k, T_t and T_b, we have $(v_1, v') \in E$, where $v' \in \{v_2, v_k, v_t, v_b\}$. Since T_t touches T_i, $1 \le i \le k$, we have $(v_t, v') \in E$, where $v' \in \{v_1, v_2, v_3, \ldots, v_{k-1}, v_k\}$. Since T_b touches T_i, $1 \le i \le k$, we have $(v_b, v') \in E$, where $v' \in \{v_1, v_2, v_3, \ldots, v_{k-1}, v_k\}$. Since T_k touches T_1, T_{k-1}, T_t and T_b, we have $(v_k, v') \in E$, where $v' \in \{v_1, v_{k-1}, v_t, v_b\}$. Since T_i, $2 \le i \le k-1$, touches T_{i-1} and T_{i+1}, we have the path $(v_1, v_2, \ldots, v_k)$. Figure 2(d) shows the $TCCG$ of the seed set in Fig. 2(a).

Since S has six or more vertices, S' has at least two vertices and hence every vertex of G has degree at least four. Thus to prove our claim that G is 4-connected, it is now sufficient to show that four internally vertex-disjoint paths exist between each of the pairs $\{v_i, v_j\}$, $\{v_i, v_t\}$, $\{v_i, v_b\}$ and $\{v_t, v_b\}$ of vertices, where $1 \le i < j \le k$, (see Fig. 2(d)). The paths between the vertices v_i and v_j are path $(v_i, v_{i+1}, \ldots, v_j)$, path (v_i, v_t, v_j), path (v_i, v_b, v_j) and path $(v_i, v_{i-1}, \ldots, v_1, v_k, v_{k-1}, \ldots, v_j)$. The paths between the vertices v_1 and v_t are path (v_1, v_t), path (v_1, v_2, v_t), path (v_1, v_k, v_t) and path (v_1, v_b, v_3, v_t). The paths between the vertices v_k and v_t are path (v_k, v_t), path (v_k, v_{k-1}, v_t), path (v_k, v_1, v_t) and path (v_k, v_b, v_{k-2}, v_t). The paths between the vertices v_i and v_t are path (v_i, v_t), path (v_i, v_{i-1}, v_t), path (v_i, v_{i+1}, v_t) and path (v_i, v_b, v_k, v_t) if $i \ne k-1$ or path (v_i, v_b, v_1, v_t) if $i \ne 2$. Similarly there are at least four internally vertex-disjoint paths exist between vertices v_i and v_b. It is not difficult to see that k paths exist between v_t and v_b through v_i. Thus four internally vertex-disjoint paths exist between each pair of vertices.

We now analyze the time complexity of our construction. For computing the triangle covers of n seeds we need $O(n)$ time. Since we are sorting the seeds, 4-connected $TCCG$ can be found in $O(n \log n)$ time. □

Our construction for a 4-connected $TCCG$ given in the proof of Theorem 1 can be used for any set of six or more seeds; if two seeds are on a vertical line or on a horizontal line then rotate the plane such that no two points remain on a vertical line or on a horizontal line.

3   Realizability of Outerplanar Graphs

In this section we show that an outerplanar graph has a realization as a triangle cover contact graph (TCCG) on a given set of seeds on a line as in the following theorem.

Theorem 2. *Let G be a connected outerplanar graph of n vertices. Let S be a set of n seeds aligned on a straight line. Then G is realizable on S as a $TCCG$ in $O(n \log n)$ time.*

In fact, we prove a claim stronger than that in Theorem 2 as in Theorem 3 on weighted version of the problem. A *weighted triangle cover contact graph (WTCCG)* G is a weighted CCG of n covers and $\sum_{i=1}^{n} w_i$ seeds where each vertex v_i of G has a weight w_i and the cover c_i corresponds to v_i covers w_i seeds of $\sum_{i=1}^{n} w_i$ ($w_i \in \mathbb{Z}$) seeds. We now give Theorem 3 on $WTCCG$.

Theorem 3. *Let w_i be the weight of a vertex v_i of a weighted connected outer-planar graph G of n vertices, and let S be a set of $\sum_{i=1}^{n} w_i$ seeds on a line. Then G is realizable on S as a WTCCG in $O(n \log n)$ time.*

The proof of Theorem 3 is involved. We only give our idea in this short version. Let G be a weighted connected outerplanar graph. We contract each biconnected components and bridges of G into a single vertex. Doing the operation recursively, the resulting graph G' is either a biconnected graph or a single vertex, and also a weighted graph where the weight of each vertex is the sum of the weights of the vertices of the corresponding uncontracted biconnected component or bridge of G. We then construct $WTCCG$ of G' and expanding the $WTCCG$ of G' we obtain a $WTCCG$ of G.

4 Conclusion

In this paper we have shown that any set of six or more point seeds admits covers with 4-connected $TCCG$ and such covers can be found in $O(n \log n)$ time. We also have shown that every connected outerplanar graph and weighted outerplanar graph can be realized as a $TCCG$ of a set of point seeds on a straight line. Our future work is to investigate which larger classes of graphs are realizable as $TCCG$.

References

1. Abellanas, M., Bereg, S., Hurtado, F., Olaverri, A.G., Rappaport, D., Tejel, J.: Moving coins. Comput. Geom. 34(1), 35–48 (2006)
2. Abellanas, M., Castro, N., Hernández, G., Márquez, A., Moreno-Jiménez, C.: Gear System Graphs (2006) (manuscript)
3. Atienza, N., Castro, N., Cortés, C., Garrido, M.A., Grima, C.I., Hernández, G., Márquez, A., Moreno, A., Nöllenburg, M., Portillo, J.R., Reyes, P., Valenzuela, J., Trinidad Villar, M., Wolff, A.: Cover contact graphs. Journal of Computational Geometry 3(1), 102–131 (2012)
4. Durocher, S., Mehrabi, S., Skala, M., Wahid, M.A.: The cover contact graph of discs touching a line. In: CCCG, pp. 59–64 (2012)
5. Koebe, P.: Kontaktprobleme der konformen abbildung. Ber. Sächs. Akad. Wiss. Leipzig, Math.-Phys. Klasse 88(1-3), 141–164 (1936)
6. Nishizeki, T., Rahman, M.S.: Planar Graph Drawing. Lecture Notes Series on Computing. World Scientific, Singapore (2004)
7. Pach, J., Agarwal, P.K.: Combinatorial Geometry. John Wiley and Sons, New York (1995)
8. Robert, J.M., Toussaint, G.T. (eds.): Computational geometry and facility location. In: Proc. Int. Conf. Oper. Res. Manage. Sci., vol. 68 (1990)
9. Welzl, E.: Smallest enclosing disks (balls and ellipsoids). In: Maurer, H.A. (ed.) New Results and New Trends in Computer Science. LNCS, vol. 555, pp. 359–370. Springer, Heidelberg (1991)
10. West, D.B.: Introduction to Graph Theory, 2nd edn. Prentice-Hall (2001)

Logspace and FPT Algorithms for Graph Isomorphism for Subclasses of Bounded Tree-Width Graphs

Bireswar Das*, Murali Krishna Enduri**, and I. Vinod Reddy

IIT Gandhinagar, India
{bireswar,endurimuralikrishna,reddy_vinod}@iitgn.ac.in

Abstract. We give a deterministic logspace algorithm for the graph isomorphism problem for graphs with bounded tree-depth. We also show that the graph isomorphism problem is fixed parameter tractable for a related parameterized graph class where the graph parameter is the length of the longest cycle.

1 Introduction

The Graph Isomorphism (GI) problem is to determine whether two given graphs G_1 and G_2 are isomorphic. This problem is important in complexity theory due to its unknown complexity status. Despite its unresolved complexity status, efficient algorithms for GI are known for restricted classes of graphs. Lindell studied the space complexity of tree isomorphism and showed that the problem is in logspace [11]. Reingold's logspace algorithm for undirected reachability [15] made it possible to design many of the recent logspace algorithms for GI for restricted classes of graphs [6,5]. The space complexity of GI for bounded treewidth remains an open problem. The GI problem has been studied and was shown to be in logspace for some subclasses of bounded treewidth [1,2]. In this paper we give a deterministic logspace algorithm for GI for bounded tree-depth graphs which is an interesting subclass of bounded treewidth graphs. Our isomorphism algorithm uses another algorithm as a subroutine that computes the tree-depth decomposition of bounded tree-depth graphs in logspace.

Bouland et al. [3] gave an fpt algorithm for GI parameterized by tree-depth of the graph. Following their work we give an fpt algorithm for the GI problem where the parameter is the length of the longest cycle. Fixed parameter tractable algorithms for GI are known for many interesting parameterization [16,9,10,3]. A graph whose longest cycle length is k has treewidth at most k. Thus, GI parameterized by longest cycle length becomes a natural subproblem of GI parameterized by treewidth. Our main motivation was to study the parameterized complexity of the GI problem where the parameter is treewidth. In a recent paper Lokshtanov et al. [12] gave an fpt algorithm for the GI problem parameterized by

---

* DIMACS postdoctoral fellow, supported by IUSSTF fellowship.
** Supported by Tata Consultancy Services (TCS) research fellowship.

M.S. Rahman and E. Tomita (Eds.): WALCOM 2015, LNCS 8973, pp. 329–334, 2015.
© Springer International Publishing Switzerland 2015

treewidth thus subsuming our result. Nevertheless, our algorithm is very different from [12] and may be useful to give a different fpt algorithm or new logspace algorithm for bounded treewidth graphs.

2   Preliminaries

In our fpt algorithm for GI parameterized by length of longest cycle , we use the idea of *block tree* representation of connected graphs. A maximal biconnected subgraph of a graph is called a *block*. The *block graph* representation of a graph G is a bipartite graph whose bipartition consists of the set of articulation points and the set of blocks. There is an edge between an articulation point v and a block B if and only if $v \in B$. The block graph of a connected graph is a tree [7].

One interesting graph parameter is tree-depth introduced by [14] Nešetřil et al, which measures how close a graph is to star graphs. The tree-depth decomposition of a graph is defined as follows: Let T be a rooted tree. The *closure* of T denoted $\mathsf{clos}(T)$ is the graph obtained by adding edges from each vertex v to all vertices which lie on a path from the root to vertex v. A *tree-depth decomposition* of a graph G is a tree T over $V(G)$ such that G is a subgraph of $\mathsf{clos}(T)$. The *depth* of the tree-depth decomposition T is the height of T. The *tree-depth* of a graph G, denoted $\mathsf{td}(G)$ is the minimum depth among all possible tree-depth decompositions of G. The tree-depth of a graph can be equivalently defined as follows:

Definition 1. *Let G be a graph with connected components $G_1, \cdots, G_p$. Then the tree-depth $\mathsf{td}(G)$ of G is 1 if $|V(G)| = 1$. If $|V(G)| > 1$ and $p = 1$ then $\mathsf{td}(G) := 1 + \min_{v \in V(G)} \mathsf{td}(G - v)$ else $\mathsf{td}(G) := \max_{i=1\cdots p} \mathsf{td}(G_i)$.*

We state a lemma due to Nešetřil et al. that we use for our fpt algorithm.

Lemma 1 ([13]). *Let G be a biconnected graph with length of longest cycle not more than k. Then $1 + \lceil \log_2 k \rceil \leq \mathsf{td}(G) \leq 1 + (k-2)^2$.*

3   Logspace Algorithm for Bounded Tree-depth Graphs

In this section we present a logspace algorithms to recognize and compute tree-depth decomposition of bounded tree-depth graphs.

Lemma 2. *Let d be a fixed constant. There is a logspace algorithm $\mathcal{A}_R$ that on input a connected graph G decides if $\mathsf{td}(G) \leq d$.*

Proof. The Turing machine for $\mathcal{A}_R$ uses one of its tape as a stack to store vertices appearing in the root to leaf paths in the tree-depth decomposition. If the tree-depth is indeed at most d then we do not have to store more than d vertices. First the algorithm attempts a vertex $v_1 \in V(G)$ as the root of the decomposition and puts it on the stack and tries to verify if $\mathsf{td}(G - v) \leq d - 1$. Let $G_1, \cdots, G_p$ be the connected components of $G - v_1$. The algorithm inductively verifies that

$\mathsf{td}(G_i) \leq d - 1$ for each i. If for some i, $\mathsf{td}(G_i) > d - 1$ then the algorithm removes v_1 from the stack tries a different vertex. The connected components are identified by the smallest vertex according to the input order appearing in them with the help of Reingold's algorithm for undirected reachability [15].

In the general inductive step there will be vertices $v_1, \cdots, v_m$ on the stack and the goal is to verify if a connected component G'' of $G' - v_m$ has tree-depth at most $d - m$ where G' is the connected component of $G - \{v_1, \cdots, v_{m-1}\}$ containing v_m. To check if $\mathsf{td}(G'') \leq d - m$, the algorithm tries different vertices v_{m+1} of G'' as the root of its decomposition just as above. Clearly, $\mathsf{td}(G'') \leq d - m$ iff one of the choices for v_{m+1} works. Once the algorithm verifies that $\mathsf{td}(G'') \leq d - m$, it removes v_{m+1} from the stack and goes to the next connected component or to the previous level of recursion depending on whether there are next connected components or not.   □

Next we give a logspace algorithm for computing tree-depth decomposition based on the previous algorithm.

Lemma 3. *There is a logspace algorithm $\mathcal{A}_D$ that on input a graph G with tree-depth at most d outputs a tree-depth decomposition of G.*

Proof. Notice that in $\mathcal{A}_R$ (Lemma 2) many of the vertices on the stack does not work as the roots of the decompositions of the connected components. These vertices gets replaced by other vertices in the backtracking process. There are also some vertices that does work. The key idea is to isolate all the vertices that are part of the total successful computation and output them along with their positions on the stack. Since the set of vertices along with their positions on the stack gives the complete information about all the root to leaf paths in the decomposition, it can be used to compute the decomposition easily. To find out if a vertex is a part of the total successful computation (i.e., it never gets replaced because of *failure*) the recursive calls has to be made multiple times.

If the tree-depth is indeed at most d, the algorithm $\mathcal{A}_R$ finally says "yes" when there is one vertex v_1 on the stack and no more connected component to be tested. At this point $(v_1, 1)$ could be safely output. The next task is to find the decompositions of the connected components of $G - v_1$. To do this we start the computation of $\mathcal{A}_R$ again with v_1 on the stack. Just before $\mathcal{A}_R$ decides that $\mathsf{td}(G') \leq d - 1$ where G' is a connected component of $G - v_1$, there will be a vertex v_2 in the second level of the stack and the algorithm can safely output $(v_2, 2)$. The computation proceeds in a similar way until $\mathcal{A}_D$ finds the complete decomposition. Since $\mathcal{A}_R$ is logspace this algorithm is also logspace.   □

Theorem 1. *Let d be a fixed constant. Then there is a logspace algorithm $\mathcal{A}_{ISO}$ that given a pair of graphs G, H with tree-depth at most d decides if $G \cong H$.*

Proof. First using algorithm $\mathcal{A}_D$ from Lemma 3 we find a tree-depth decomposition T_G of G. The idea is to use T_G to find an isomorphic decomposition[1] of H and thus establishing $G \cong H$. Similar idea has been used in [4]. We pick a vertex v_1 in H to check if it can act as the root of an isomorphic decomposition of H.

---

[1] Two decompositions are isomorphic if there is an isomorphism between the two trees which is also an isomorphism of the original graphs.

Suppose $G_1, \cdots, G_p$ are the connected components of $G - u_1$ where u_1 is the root of T_G. Let $H_1, \cdots, H_p$ be the connected components of $H - v_1$. The goal is to check if the connected components in both the graphs can be isomorphically matched in such a way that it gives an isomorphic tree-depth decomposition of the whole graph H. In general for G we will have a partial path $u_1, \cdots, u_m$ where u_1 is the root of T_G and u_i is the parent of u_{i+1}. Similarly for H we will have a set $\{v_1, \cdots, v_m\}$ of vertices that are being tested as candidates for the image of $u_1, \cdots, u_m$ in the isomorphic decomposition. Let G'' and H'' be two connected components of $G' - u_m$ and $H' - v_m$ respectively, where G' is the connected component of $G - \{u_1, \cdots, u_{m-1}\}$ containing u_m and H' is the connected component of $H - \{v_1, \cdots, v_{m-1}\}$ containing v_m. Let u_{m+1} be the root of G'' in T_G. We pick a vertex v_{m+1} in H'' to check if it can act as the image of u_{m+1}. To do this the algorithm checks if 1) $\{v_{m+1}, v_i\} \in E(H)$ iff $\{u_{m+1}, u_i\} \in E(G)$ for each $i = 1, \cdots, m$ and 2) all the connected components of $G'' - u_{m+1}$ and $H'' - v_{m+1}$ can be inductively matched isomorphically. If v_{m+1} does not work, the algorithm tries the next vertex in H'' in the input order.

The logspace implementation is similar to $\mathcal{A}_R$. Let $G_1'', \cdots, G_p''$ and $H_1'', \cdots, H_p''$ be the connected components of $G'' - u_{m+1}$ and $H'' - v_{m+1}$ respectively. The algorithm starts matching the G_i'''s one after another according to the order given by the decomposition[2] with $H_1'', \cdots, H_p''$. To match G_i'' the algorithm first counts the number of isomorphically matching copies of G_i'' among $G_1'', \cdots, G_{i-1}''$. To keep track of this number the algorithm maintains a counter in each level. Suppose the number is x. Then G_i'' is the $(x+1)^{st}$ copy and it has to be matched with $x + 1$ different copies among $H_1'', \cdots, H_p''$. If the corresponding number of copies in both the graphs in each level is found to be equal then the two graphs are isomorphic. $\qquad \square$

4   An FPT Algorithm

First we show that to design an fpt algorithm for a graph class it is enough to design an fpt algorithm for colored biconnected graphs in that class.

Lemma 4. *Let $\mathcal{G}$ be a parameterized graph class with the property that if $G \in \mathcal{G}$ then the biconnected components of G are also in $\mathcal{G}$. Let $\mathcal{A}$ be an algorithm to test isomorphism of colored biconnected graphs in $\mathcal{G}$ that runs in time $T(n, k)$ where n is the input size and k is the parameter. Then there is an isomorphism algorithm $\mathcal{B}$ in $\mathcal{G}$ that runs in time $O(n^2 T(n, k))$.*

Proof. Without loss of generality we assume that G and H are connected but not biconnected. First compute the block trees T_G and T_H for G and H. *Block nodes* and *articulation nodes* are vertices in a block tree that correspond to blocks and articulation points in the original graph. The leaves of a block tree are block nodes. Also notice that the block nodes and the articulation nodes alternate in any path in a block tree. Thus a block tree has a unique center. We assume that T_G and T_H are rooted at their centers.

---

[2] W.l.g., we assume that $G_1'', \cdots, G_p''$ is the listing according to the order.

The algorithm starts from the leaves of the block trees and proceeds in stages where each stage corresponds to the levels of the trees. In each stage the algorithm finds out if the subgraphs D_1 and D_2 induced in G and H by pairs of subtrees T_1 and T_2 of T_G and T_H are isomorphic. If they are isomorphic, we call T_1 and T_2 an *isomorphically matching pair* (imp). For each pair of subtrees in a level the algorithm maintains a table to remember if they are imp.

In the base case T_1 and T_2 correspond to block nodes. Any block corresponding to these blocks is biconnected with at most one articulation point. We run $\mathcal{A}$ on pairs of such blocks by coloring the articulation points with color "red" to decide if they are imps. For each such pair T_1 and T_2 the result is stored in the table.

The stages corresponding to articulation nodes are easier to handle. We just have to check if the subtrees $T_{11}, T_{12}, \cdots, T_{1p}$ and $T_{21}, T_{22}, \cdots, T_{2p}$ of T_1 and T_2 can be pairwise matched isomorphically. We use the imps (T_{1i}, T_{2j}) already computed in the previous stage to decide if (T_1, T_2) is imp.

Next we describe the stages corresponding to non-leaf block nodes. Let B_1 and B_2 be the blocks corresponding to roots r_1 and r_2 of T_1 and T_2. The algorithm tests the isomorphism of D_1 and D_2 by checking the isomorphism of B_1 and B_2 by coloring the articulation points in B_1 and B_2. Let a_1 and a_2 be the articulation points in B_1 and B_2 corresponding to the parents of r_1 and r_2. We color a_1 and a_2 by red color[3]. The other articulation points will corresponds to subtrees hanging below from the blocks B_1 and B_2. We color these articulation points by the isomorphism type of the subtree hanging below. We run $\mathcal{A}$ on the colored blocks B_1, B_2 to decide if (T_1, T_2) is imp. Since $\mathcal{A}$ is used $O(n^2)$ many times, the run-time of the algorithm is $O(n^2 T(n, k))$.   □

Apart from Lemma 4, one of the main ingredients is an fpt algorithm for graph isomorphism for colored graphs parameterized by the tree-depth. The algorithm is similar to the algorithm by Bouland et al.[3], except that we have to consider colors. But this is not difficult. Bouland et al. defines an isomorphism ordering for their algorithm. Colors of the vertices can be handled by modifying their definition in the obvious way. For details see [3].

Next we combine Bouland's algorithm and Lemma 4 to obtain the following theorem.

Theorem 2. *Let G and H be two graphs with length of longest cycle not more than k. Then we can check whether $G \cong H$ in time $f(k)n^{O(1)}$ where f is a fixed function.*

Proof. We use Bouland's algorithm (colored version) with tree-depth parameter $d = 1 + (k-2)^2$ in place of $\mathcal{A}$ in Lemma 4. Notice that by Lemma 1 the tree-depth of the biconnected components in the Lemma 4 is at most $1 + (k - 2)^2$.   □

5   Conclusion

The recognition problem for graphs with bounded tree-depth is in AC^0 [8]. It would be interesting to know if our logspace upper bound for GI can be improved

---

[3] We do not use the color red to color any other articulation points.

to AC^0 or NC^1. The question of whether there is a logspace algorithm for bounded treewidth graphs seems to be a much harder problem and probably needs new techniques. Although the result on fpt algorithm for GI is subsumed by [12], our algorithm is very different and it would be interesting to see if this method can be extended to give a different fpt algorithm for GI parameterized by treewidth.

References

1. Arvind, V., Das, B., Köbler, J.: A logspace algorithm for partial 2-tree canonization. In: Hirsch, E.A., Razborov, A.A., Semenov, A., Slissenko, A. (eds.) CSR 2008. LNCS, vol. 5010, pp. 40–51. Springer, Heidelberg (2008)
2. Arvind, V., Das, B., Köbler, J., Kuhnert, S.: The isomorphism problem for k-trees is complete for logspace. Information and Computation 217, 1–11 (2012)
3. Bouland, A., Dawar, A., Kopczyński, E.: On tractable parameterizations of graph isomorphism. In: Thilikos, D.M., Woeginger, G.J. (eds.) IPEC 2012. LNCS, vol. 7535, pp. 218–230. Springer, Heidelberg (2012)
4. Das, B., Torán, J., Wagner, F.: Restricted space algorithms for isomorphism on bounded treewidth graphs. Information and Computation 217, 71–83 (2012)
5. Datta, S., Limaye, N., Nimbhorkar, P., Thierauf, T., Wagner, F.: Planar graph isomorphism is in log-space. In: Proceedings of 24th Annual IEEE Conference on Computational Complexity, pp. 203–214. IEEE (2009)
6. Datta, S., Nimbhorkar, P., Thierauf, T., Wagner, F.: Graph isomorphism for $K_{3,3}$-free and K_5-free graphs is in log-space. In: LIPIcs-Leibniz International Proceedings in Informatics. vol. 4. Schloss Dagstuhl-Leibniz-Zentrum für Informatik (2009)
7. Diestel, R.: Graph Theory Graduate Texts in Mathematics, vol. 173. Springer, GmbH & Company KG, Berlin and Heidelberg (2000)
8. Elberfeld, M., Jakoby, A., Tantau, T.: Algorithmic meta theorems for circuit classes of constant and logarithmic depth. In: Symposium on Theoretical Aspects of Computer Science, vol. 14, pp. 66–77 (2012)
9. Evdokimov, S., Ponomarenko, I.: Isomorphism of coloured graphs with slowly increasing multiplicity of jordan blocks. Combinatorica 19(3), 321–333 (1999)
10. Kratsch, S., Schweitzer, P.: Isomorphism for graphs of bounded feedback vertex set number. In: Kaplan, H. (ed.) SWAT 2010. LNCS, vol. 6139, pp. 81–92. Springer, Heidelberg (2010)
11. Lindell, S.: A logspace algorithm for tree canonization. In: Proceedings of the 24th Annual ACM Symposium on Theory of Computing, pp. 400–404. ACM (1992)
12. Lokshtanov, D., Pilipczuk, M., Pilipczuk, M., Saurabh, S.: Fixed-parameter tractable canonization and isomorphism test for graphs of bounded treewidth. In: FOCS (2014), http://arxiv.org/abs/1404.0818
13. Nešetřil, J., De Mendez, P.O.: Sparsity: Graphs, Structures, and Algorithms, vol. 28. Springer (2012)
14. Nešetřil, J., de Mendez, P.O.: Tree-depth, subgraph coloring and homomorphism bounds. European Journal of Combinatorics 27(6), 1022–1041 (2006)
15. Reingold, O.: Undirected connectivity in log-space. J. ACM 55(4), 17:1–17:24 (2008)
16. Yamazaki, K., Bodlaender, H.L., de Fluiter, B., Thilikos, D.M.: Isomorphism for graphs of bounded distance width. Algorithmica 24(2), 105–127 (1999)

Author Index